A Text Book Of

UTILIZATION OF ELECTRICAL ENERGY

(Subject Code : ELE 604)

Semester - VI

THIRD YEAR DIPLOMA IN ELECTRICAL ENGINEERING GROUP

As Per SBTE's New Revised Syllabus

SANDEEP M. CHAUDHARI

M.E. (Electrical)

Assistant Professor, Department of Electrical Engg.

AISSMS, IOIT Engg. College

Pune

NIRALI PRAKASHAN

ADVANCEMENT OF KNOWLEDGE

N4742

Utilization of Electrical Energy **ISBN 978-93-89825-94-7**

First Edition : February 2020

Published By :
NIRALI PRAKASHAN
Abhyudaya Pragati, 1312, Shivaji Nagar
Off J.M. Road, PUNE – 411005
Tel - (020) 25512336/37/39, Fax - (020) 25511379
Email : niralipune@pragationline.com

➢ DISTRIBUTION CENTRES

PUNE

Nirali Prakashan (For orders within Pune) : 119, Budhwar Peth, Jogeshwari Mandir Lane, Pune 411002, Maharashtra
Tel : (020) 2445 2044; Mobile : 9657703145
Email : niralilocal@pragationline.com

Nirali Prakashan (For orders outside Pune) : S. No. 28/27, Dhayari, Near Asian College Pune 411041
Tel : (020) 24690204; Mobile : 9657703143
Email : bookorder@pragationline.com

MUMBAI

Nirali Prakashan : 385, S.V.P. Road, Rasdhara Co-op. Hsg. Society Ltd.,
Girgaum, Mumbai 400004, Maharashtra; Mobile : 9320129587
Tel : (022) 2385 6339 / 2386 9976
Email : niralimumbai@pragationline.com

➢ DISTRIBUTION BRANCHES

JALGAON

Nirali Prakashan : 34, V. V. Golani Market, Navi Peth, Jalgaon 425001, Maharashtra,
Tel : (0257) 222 0395, Mob : 94234 91860; Email : niralijalgaon@pragationline.com

KOLHAPUR

Nirali Prakashan : New Mahadvar Road, Kedar Plaza, 1st Floor Opp. IDBI Bank, Kolhapur 416 012
Maharashtra. Mob : 9850046155; Email : niralikolhapur@pragationline.com

NAGPUR

Nirali Prakashan : Above Maratha Mandir, Shop No. 3, First Floor,
Rani Jhanshi Square, Sitabuldi, Nagpur 440012, Maharashtra
Tel : (0712) 254 7129; Email : niralinagpur@pragationline.com

DELHI

Nirali Prakashan : 4593/15, Basement, Agarwal Lane, Ansari Road, Daryaganj
Near Times of India Building, New Delhi 110002 Mob : 08505972553
Email : niralidelhi@pragationline.com

BENGALURU

Nirali Prakashan : Maitri Ground Floor, Jaya Apartments, No. 99, 6th Cross, 6th Main,
Malleswaram, Bengaluru 560003, Karnataka; Mob : 9449043034
Email: niralibangalore@pragationline.com

Other Branches : Hyderabad, Chennai

niralipune@pragationline.com | www.pragationline.com

Also find us on www.facebook.com/niralibooks

Preface ...

Dear students, teachers and practicing engineers, it gives me a great pleasure to present the First Edition of my book on **'Utilization of Electrical Energy'** for Semester VI of Diploma Course in Electrical Engineering Group for the Board of SBTE, Zharkhand. I am thankful to students and teachers as they have highly appreciated and accepted my previous books, which cover cent percent syllabus and gives additional knowledge useful for oral examination also.

In this edition, questions those have been occurred in the previous S.B.T.E. examination question papers have been added for reference and study of students accordingly.

My special appreciation goes to Shri. Dineshbhai Furia, Shri. Jigneshbhai Furia, Shri Kiran Velankar, Ms. Chaitali Takle and Mr. Akbar Shaikh of Nirali Prakashan, Pune for publishing this book in a good and presentable manner within very short time.

I also thanks Mr. M. A. Chaudhari Sir for editing all the Chapters of this book.

Teachers and practicing engineers are requested to suggest for further improvement of this book.

S. M. Chaudhari

■■■

Syllabus ...

Unit I - Illumination

1.1 Definitions of Terms Used in Illumination: Light, Luminous Flux, Luminous Intensity, Lumen, Candle Power, Illumination, Lux or Meter Candle, Mean Horizontal Candle Power (MHCP), Mean Spherical Candle Power (MSCP), Mean Hemi-spherical Candle Power (MHSCP), Reduction Factor, Lamp Efficiency, Specific Consumption, Glare, Space-Height Ratio, Utilization Factor, Maintenance Factor, Depreciation Factor, Waste Light Factor, Absorption Factor, Reflection Factor, Solid Angle.

1.2 Laws of Illumination:
- Law of Inverse Squares
- Lambert's Cosine Law. (No Numericals)

1.3 Sources of Light: Construction, Working and Applications of following Lamps:
- Incandescent Lamps.
- Halogen Lamps.
- Low Pressure Mercury Vapour Lamps (Fluorescent Tube).
- High Pressure Mercury Vapour Lamps.
- Sodium Vapour Lamps.
- Compact Fluorescent Lamps (C.F.L.)
- Metal Halide Lamps
- LED Lamps
- Neon Signs.

1.4 Basic Principles of Light Control.

1.5 Types of Lighting Schemes: Direct, Semi-direct, Semi-indirect, Indirect, General Lighting.

1.6 Design of Lighting Scheme: Objectives of Lighting Scheme.
- Factors to be considered while Designing the Lighting Scheme. (Simple Numericals)

1.7 Factory Lighting:
- General Requirements
- Types of Installations: General Lighting, Local Lighting, Emergency Lighting.

1.8 Lumen or Light Flux Method of Lighting Calculations (Simple Numericals)

1.9 Flood Lighting
- Flood Lighting Purposes.
- Classification of Projectors.
- Location and Mounting of Projectors.

Unit II - Electric Heating and Welding

2.1 Electric Heating

2.1.1 Advantages of Electric Heating.

2.1.2 Modes of Transfer of Heat:
- Conduction, Convection and Radiation.

2.1.3 Classification of Electric Heating Methods:

2.1.4 Resistance Heating: (Construction and Operation)
- Direct Resistance Heating: Salt Bath Furnace.
- Indirect Resistance Heating: Resistance Ovens, Requirements of Heating Element Material, Causes of Failure of Heating Elements, Methods of Temperature Control.
- Applications of Resistance Heating.

2.1.5 Arc Heating: (Construction and Operation)
- Direct Arc Furnace.
- Indirect Arc Furnace.
- Applications of Arc Heating.

2.1.6 Induction Heating: (Construction and Operation)
- Core Type Induction Furnaces: Ajax Wyatt Furnace.
- Coreless Induction Furnace.
- Applications of Induction Heating (Simple Numericals on Melting Furnaces).

2.1.7 Dielectric Heating:
- Principle of Dielectric Heating.
- Advantages of Dielectric Heating.
- Limitations of Dielectric Heating.
- Applications of Dielectric Heating (Simple Numericals on Dielectric Heating).

2.1.8 Eddy Current Heating
- Principle, Advantages and Applications

2.2 Electric Welding

2.2.1 Methods of Electric Welding: Electric Arc Welding, Resistance Welding.

2.2.2 Resistance Welding:
- Principle of Resistance Welding.
- Advantages of Resistance Welding.
- Types of Resistance Welding - (Only List)

2.2.3 Spot Welding Machine.

2.2.4 Electric Arc Welding:
- Formation and Characteristics of Electric Arc.
- Effect of Arc Length.
- Arc Blow.

2.2.5 Polarity in DC Welding

2.2.6 Electrodes for Metal Arc Welding

2.2.7 V-I Characteristics of Arc Welding

2.2.8 Arc Welding Machines: [Only List]
- DC Welding Machines – MG Set, AC Rectified Welding Unit.
- AC Welding Machines – Welding Transformer.

Unit III - Elevators

3.1 Types of Electric Elevators

3.2 Size and Shape of Elevator Car

3.3 Speed of Elevators

3.4 Location of Elevator Machine

3.5 Types of Elevator Machines, Elevator Motors

3.6 Power Transmission Gears Braking

3.7 Safety in Elevators

Unit IV - Electric Drives

4.1 Introduction:
- What is Drive?
- Drives: Mechanical Drive and Electric Drive.

4.2 Advantages and Disadvantages of Electric Drive.

4.3 Factors Governing Selection of Electric Motors.

4.4 Nature of Electric Supply: Three-phase & Single-phase A.C and D.C.

4.5 Type of Drive: Group Drive and Individual Drive.

4.6 Nature of Load: Nature of the Mechanical Load, Matching of the Speed - Torque Characteristics of the Motor with that of the Load, and Starting Conditions of the Load.

4.7 Electrical Characteristics: (Only DC Series, Three-phase and Single-phase Induction Motors are to be dealt)

- **Running Characteristics:** Three Typical Speed - Torque Characteristics: Inverse, Constant Speed and Drooping.
- **Starting Characteristics:** Starting Torque only (No Starters).
- **Speed Control:** Suitability to Economic and Efficient Speed Control Methods (Above and Below Normal Speed).
- **Braking Characteristics:** Plugging, Rheostatic Braking and Regenerative Braking, as applied to DC Series and Three-phase Induction Motor. Only characteristics and applications of following motors: D.C. Series, Single-phase A.C. Series, Single-phase I.M., Three-phase I.M., Universal Motor, Stepper motor. Requirements of Motors used in following applications: Stone Crushing, Textile Industry, Paper Manufacturing Industry, Rolling Mill, Chemical Industry.

4.8 Mechanical Features

- Types of Enclosures as per IS
- Types of Bearings
- Types of Transmission for Drive
- Noise Level.

4.9 Size and Rating of Motor

- Load Cycles for – Continuous Loads, Short Time Loads, Intermittent Loads, Continuous Operation with Short Time Loads and Continuous Operation with Intermittent Loads.
- Duty Cycles.
- Standard Ratings for Motors as per ISS.
- Estimation of Rating of a Motor. (Simple numericals on Estimating Size of Continuously Rated Motor)
- Load Equalization. (No calculations)

4.10 Cost:

- Capital Cost
- Running Cost (Losses, P.F., Maintenance).

Unit V - Electric Traction

5.1 Requirements of an Ideal Traction System

5.2 Traction Systems

- Non-electric Traction Systems
- Electric Traction Systems: Straight Electric Traction, Its advantages and disadvantages. Diesel Electric Traction, Its advantages and disadvantages.

5.3 Systems of Track Electrification: DC System, Composite System – Single-phase to Three-phase System and Single-phase AC to DC System (Kando System). Advantages and Disadvantages of Single-phase 25 kV AC System over DC System.

5.4 Traction Mechanics

- Units Used in Traction Mechanics
- Types of Services
- Speed-Time Curve
- Simplified Speed-Time Curve (No Derivation)
- Average Speed and Schedule Speed
- Factors affecting the Schedule Speed
- Tractive Effort [No Derivation]
- Specific Energy Consumption [No Derivation]
- Factors affecting Specific Energy Consumption
- Coefficient of Adhesion

(Simple Numericals on Simplified Speed - Time Curves and Specific Energy Consumption)

5.5 Traction Motors

- Desirable Characteristics of Traction Motors, Special Features of Traction Motor, Only advantages, disadvantages and applications of following motors:
- Suitability of D.C. Series Motor for Traction
- Suitability of Three Phase Induction Motor for Traction, LIM, Single-phase A.C. Series Motor

5.6 Traction Motor Control:

- Requirements
- Traction Control of D.C. Locomotives and EMUs: Series Parallel Control Combined with Rheostatic Control, Transition from Series to Parallel Combination (Open Circuit Transition, Shunt Transition and Bridge Transition), Energy Efficiency and Limitations of Series Parallel cum Rheostatic Control, Chopper Control of Motors in D.C. Traction Systems
- Traction Control System of A.C. Locomotives: Tap Changer, Stepless Voltage Control through the use of Thyristors, PWM

5.7 Braking:

- Requirements of a Braking System.
- Mechanical Braking: Vacuum Braking, Compressed Air Braking, Hand Brake for Parking.
- Electric Braking: Rheostatic Braking and Regenerative Braking. (No Derivation and No Numericals)
- Sequence of Braking
- Dead Man's Handle

Unit VI - Economic Aspects of Utilizing Electrical Energy

6.1 Economic Aspects of Utilizing Electrical Energy.

6.2 Costing of Electrical Energy: Fixed Charges, Semi-Fixed Charges and Running Charges.

6.3 Formulation of Electrical Tariffs.

6.4 Various Types of Tariffs: Tariffs in force for Domestic, Commercial and Industrial Consumers. Simple, Block rate, Two-part, Three-part, kVA, MD, P.F. Tariffs.

6.5 Power Factor Improvement: Causes of Low Power Factor, Disadvantages of Low Power Factor, Power Factor Improvement by using Static Capacitors, Location of Capacitors for Power Factor Improvement, Most Economical Power Factor. Automatic Power Factor Controller (Derivation and Simple Numericals)

6.6 Energy Conservation: Importance and Need of Energy Conservation, Measures for Energy Conservation in (i) Electric Drives, (ii) Electric Traction, (iii) Electric Heating, (iv) Refrigeration and Air Conditioning, (v) Illumination.

■■■

Contents ...

■■■

Chapter **1...**

Illumination

Syllabus

1.1 Definitions of Terms Used in Illumination: Light, Luminous Flux, Luminous Intensity, Lumen, Candle Power, Illumination, Lux or Meter Candle, Mean Horizontal Candle Power (MHCP), Mean Spherical Candle Power (MSCP), Mean Hemi-spherical Candle Power (MHSCP), Reduction Factor, Lamp Efficiency, Specific Consumption, Glare, Space-Height Ratio, Utilization Factor, Maintenance Factor, Depreciation Factor, Waste Light Factor, Absorption Factor, Reflection Factor, Solid Angle.

1.2 Laws of Illumination:
- Law of Inverse Squares
- Lambert's Cosine Law. (No Numericals)

1.3 Sources of Light: Construction, Working and Applications of following Lamps:
- Incandescent Lamps.
- Halogen Lamps.
- Low Pressure Mercury Vapour Lamps (Fluorescent Tube).
- High Pressure Mercury Vapour Lamps.
- Sodium Vapour Lamps.
- Compact Fluorescent Lamps (C.F.L.)
- Metal Halide Lamps
- LED Lamps
- Neon Signs.

1.4 Basic Principles of Light Control.

1.5 Types of Lighting Schemes: Direct, Semi-direct, Semi-indirect, Indirect, General Lighting.

1.6 Design of Lighting Scheme: Objectives of Lighting Scheme.
- Factors to be considered while Designing the Lighting Scheme. (Simple Numericals)

1.7 Factory Lighting:
- General Requirements
- Types of Installations: General Lighting, Local Lighting, Emergency Lighting.

1.8 Lumen or Light Flux Method of Lighting Calculations (Simple Numericals)

1.9 Flood Lighting
- Flood Lighting Purposes.
- Classification of Projectors.
- Location and Mounting of Projectors.

Objectives

- Explain with sketches the working of the given types of lamp(s).
- Describe with sketches the construction of the given type of lamp fitting.
- Explain with sketches the specified lighting scheme for the given application.
- Select the relevant lamp for the specified application with justification.
- Design simple lighting scheme for the given data.
- Describe the procedure to maintain the given type of lamp and fitting.

INTRODUCTION

- Illumination Engineering is a study of certain concepts. It is in a very general sense, simply an energy to be utilized scientifically for various purposes such as (i) very fine work, (ii) fine work, (iii) medium fine work, (iv) plain work, (v) very plain work.
- **Very fine work:** Such as instrument making, drawing work, dressing, type setting, inspection of high quality products of glass, fibre, leather, ceramics, metals, engraving, polishing, diamond processing, gold or silver smith work, etc.

  This requires the illumination level of 1000 Lux (i.e. 1000 lumens/m^2).
- **Fine work:** Power looms, operation of machine tools in industry, technical drawing (draftsman), fine bench work, tailoring, garment manufacturing.

  This requires the illumination level of 500 to 1000 Lux.
- **Medium fine work:** Such as office work, mechanical processing, plain-bench work, wood-working lathes, spinning work, pattern making, foundry.

  This requires the illumination level of 250 to 500 Lux.
- **Plain work:** Saw mills, paper mills, brick work, casting etc.

  This requires the illumination level of 125 to 250 Lux.
- **Very plain work:** Iron foundry, loading-unloading in industries, textile and glass industries.

  This requires the illumination level of 50 to 125 Lux.

1.1 DEFINITIONS OF TERMS USED IN ILLUMINATION

1. Define each of the following terms of illumination: lux, lumen, luminous flux, illumination.
2. Define the following terms - (i) MHCP, (ii) MSCP, (iii) MHSCP.

- Light unit can be described as "Psychophysical" as they are combination of human response and physical units of power. They stand on their own and cannot be directly related to other physical units.

1. **Light:** Light is a form of radiant energy. Natural source is the sun. Artificial sources are various forms of incandescent bodies depending upon their temperature. Light is defined as radiant energy from a hot body which produces the visual sensation upon the human eye. It is usually denoted by Q and expressed in 'Lume-Hours' which is analogous to watt-hours.
2. **Luminous Flux (ϕ) or (F):** It is defined as the total quantity of light energy emitted (radiated) per second from a luminous body like a lamp. The whole of the electrical power supplied to a lamp is not converted into 'Luminous Flux' but some of the power is lost by heat conduction, heat convection and absorption. Of the remaining radiant flux is a fraction of it lying between the visual range of wavelength. Table 1.1 gives overall efficiency of lamps.

Table 1.1

Tungsten Lamp			Fluorescent Lamp			
Power Input watt	Luminous flux in lumens	Efficiency lumens/watt	Power Input watt	Length in cm	Luminous flux in lumens	Efficiency
10	80	8	4	15.25	75	18.75
40	460	11.5	8	30.50	325	40.53
60	840	14.0	20	61.0	950	47.5
100	1630	16.3	30	91.5	1500	50
200	3660	18.3	40	122.0	2300	57.5
500	9950	19.9	100	152.5	4400	44.0

3. **Color Vision:** Higher the temperature of the body, lower is the wavelength of the radiant energy and higher the radiant efficiency. The sensation of the colour is due to the difference in the wavelengths of the light radiations. The visible light can have the wavelength between 400 A° and 7500 A°. This is shown in Fig. 1.1.

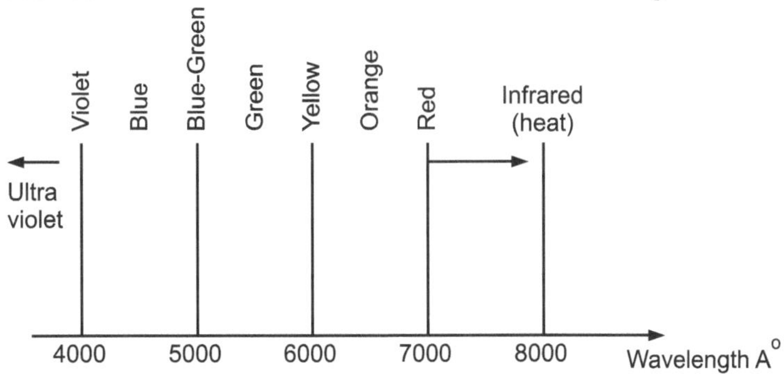

Fig. 1.1: Wavelength/Colour

4. **Luminous Intensity (I):** Consider a point source of light 'O'. Let $d\phi$ be the luminous flux crossing any section of a narrow cone of solid angle $d\omega$ steradians. The apex of the cone so formed is at the source. Then the luminous intensity in the direction of the cone is the ratio of flux $d\phi$ to the solid angle $d\omega$. Or it may be defined as the flux emitted by the source per unit solid angle.

$\therefore$ Luminous intensity $I = \frac{d\phi}{d\omega}$. Its unit is Candela (cd) or Lumens per steradian.

5. **Lumen:** Lumen is the unit of luminous flux. It is defined as the amount of luminous flux given out in a space represented by one unit of 'solid angle' by a source having an intensity of one candle power in all directions.

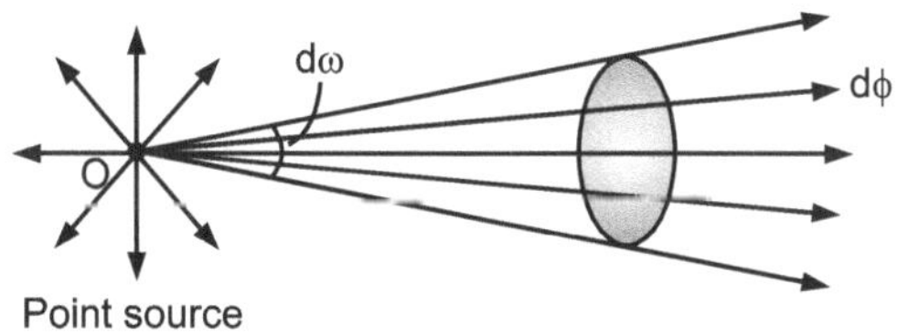

Fig. 1.2: Radiation of Point Source

$\therefore$ Lumen = Candle power × Solid angle

Flux emitted by a source of 1 candle power = 4π lumens.

6. **Candle Power:** Candle power of the source is defined as the number of lumens emitted by that source per unit solid angle in a given direction. **The term 'Candle Power' is used interchangeably with Luminous intensity (I).**

7. **Plane Angle:** It is subtended at a point and is enclosed by two straight lines lying in the same plane. It is expressed in degrees or radians, $\measuredangle\theta$ angle θ in degrees or radians. One radian is the angle subtended by an arc of a circle whose length is equal to the radius of that circle (2π rad = 360°).

Solid Angle: Consider an area 'A' relative to point P. If all points on the boundary of area A joined to P, a cone-like shape is formed at P, at the angle subtended by the area 'A' at point P is known as solid angle.

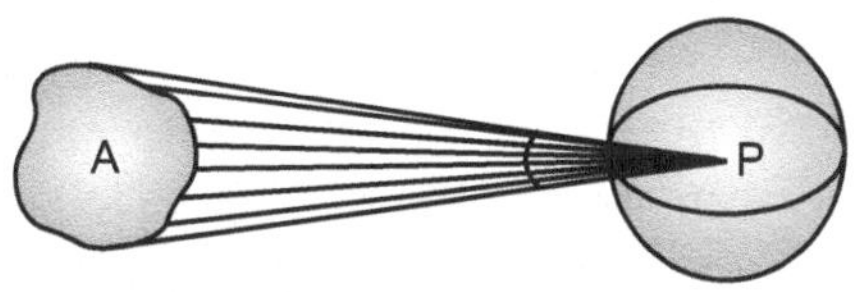

Fig. 1.3: Solid Angle

8. **Illumination [OR Degree of Illumination OR Illuminance] (E):** When light falls upon any surface, the phenomenon is called as illumination, the surface is illuminated. It is defined as the luminous flux (number of lumens) falling on the surface per unit area.

Illumination (E) = $\frac{\text{Flux (Lumens)}}{\text{Area}}$ i.e. Lumens/Area i.e. Lumens per square meter or lux.

If the area is "square feet" then illumination is lumens/sq. ft. or foot candle.

9. **Lux OR Meter Candle:** It is the unit of illumination and it is defined as luminous flux falling per unit area. 1 lux = 1 lumen per square meter.

10. Relation of Lux and Foot Candle:

$$1 \text{ metre} = 3.28 \text{ feet}$$

$$1 \text{ foot candle} = (3.28)^2 \times 1 \text{ lux} = 10.75 \text{ lux}$$

or $$1 \text{ lux} = \frac{1}{10.76} = 0.093 \text{ foot candle.}$$

11. Mean Horizontal Candle Power (MHCP): It is defined as the mean of the candle powers in all directions in the horizontal plane containing the source of light.

12. Mean Spherical Candle Power (MSCP): It is defined as the mean of the candle power of the source of light in all directions and in all planes.

13. Mean Hemi-Spherical Candle Power (MHSCP): It is defined as the mean of the candle powers in all directions within the hemisphere either above or below the horizontal plane.

14. Reduction Factor: Reduction factor of a source of light is the ratio of its mean spherical candle power to its mean horizontal candle power.

$$\therefore \quad \text{Reduction factor} = \frac{\text{MSCP}}{\text{MHCP}}$$

15. Lamp Efficiency (Luminous Efficiency): It is the ratio of luminous flux to the power input to a lamp. It is expressed in lumens per watt. Say 100 watt lamp has luminous flux of 1630 lumens.

$$\therefore \quad \text{Luminous efficiency} = \frac{\text{lumens}}{\text{watt}} = \frac{1630}{100} = 16.3 \text{ lumens/watt}$$

16. Specific Consumption: It is defined as the ratio of the power input to the average candle power. It is expressed in watts per candela.

17. Brightness or Luminance (L): It is defined as the luminous intensity per unit projected area of either a surface source of light or a reflected surface. It is in candela/m^2. Other unit is foot-lambert and it is defined as the brightness of an area emitting or reflecting light at the rate of 1 lumen per square foot.

18. Glare: It can be defined as the brightness within the field of vision of such a character as to cause annoyance, discomfort, interference with vision or eye fatigue.

19. Space to Height Ratio: It is defined as the horizontal distance between adjacent lamps to the height of their mountings.

$$\therefore \text{Space to height ratio} = \frac{\text{Horizontal distance between adjacent lamps}}{\text{Mounting height of lamps from working surface}} = \frac{\text{Distance (a)}}{\text{Distance (b)}}$$

When reflectors are used, value of this ratio is in between 1 to 2. The height of lamp should be between 2.2 to 2.45 m from the ground level.

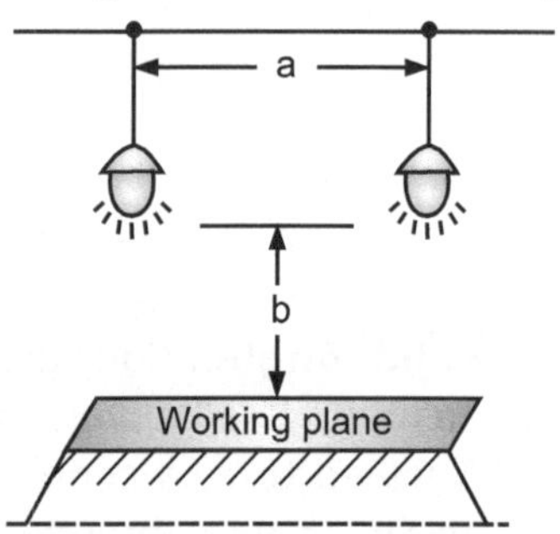

Fig. 1.4: Space to Height Ratio

20. Utilization Factor (also called as Co-efficient of Utilization): It is the ratio of total lumens reaching the working plane to the total lumens given out by the lamp. This ratio is less than unity (between 0.25 to 0.5 for direct lighting and between 0.1 to 0.3 for indirect lighting).

$$\therefore \text{Utilization factor} = \frac{\text{Lumens received on working plane}}{\text{Lumens produced by lamp}}$$

This factor depends on:

(a) Area to be illuminated.
(b) Height of lamp fittings.
(c) Surrounding colours.
(d) Type of lighting-direct/indirect.

21. Maintenance Factor: Due to dirt, dust, smoke deposition on the surfaces of the lamps, walls and ceilings, full produced lumens are not utilized on the working plane. Maintenance factor is defined as the ratio of illumination under normal working condition to the illumination when everything is perfectly clean. This ratio is less than one. It is about 0.8.

$$\text{Maintenance factor} = \frac{\text{Illumination in normal working condition}}{\text{Illumination when everything is perfectly clean}}$$

22. Depreciation Factor: This factor is inverse of maintenance factor and it is more than unity.

$$\text{Depreciation factor} = \frac{\text{Illumination when everything is perfectly clean}}{\text{Illumination in normal working condition}}$$

Following Table 1.2 shows light reflected in % from the coloured surfaces.

Table 1.2: Light reflected in % from the coloured surfaces

Colour of Surface	% of light reflected
Light White	81%
Light Cream	69 to 75%
Light Green	65%
Light Grey	58%
Medium Grey	55%
Dark Tan	46%
Dray Grey	25%
Dark Olive Green	15%
Dark Red	12%
Natural	23%

23. Waste Light Factor: When so many light sources leave overlapping lights, there ought to be wastage of light. This fact is taken into account by multiplying the theoretical value of lumens required by 1.2 (for rectangular shaped working area) and by 1.5 (for irregular shaped areas).

24. Absorption Factor: Sometimes the particular areas are suffered by fumes and smokes such as bakery, smithy shops, foundries etc. where full lumens are not utilized and we have to provide more lumens for proper illumination. In such cases, 'absorption factor' has to be considered:

$$\text{Absorption factor} = \frac{\text{Total lumens available after absorption}}{\text{Total lumens emitted by light source}}$$

This factor may be 0.5 to 0.6 and for clean atmosphere it is unity.

25. Beam Factor: This is to be considered in case of projectors where some lumens are absorbed by reflector and front glass of projector lamp. This factor varies between 0.3 to 0.6.

26. Reflection Factor: As shown in Fig. 1.5, a light is incident on the surface and it is reflected from the surface. In such a case some light is absorbed by the surface. Reflection factor is the ratio of reflected light to the incident light. This ratio is less than unity.

$$\text{Reflection factor} = \frac{\text{Reflected light}}{\text{Incident light}}$$

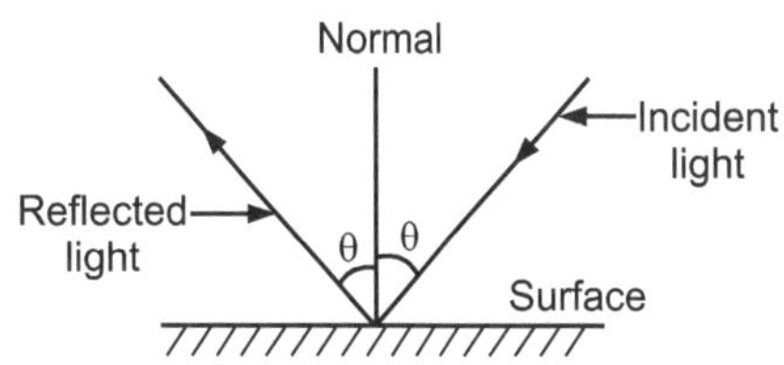

Fig. 1.5: To find Reflection Factor

SOLVED EXAMPLE

Example 1.1: *A lamp working on an A.C. voltage of 250 volts takes a current of 0.8 ampere. It radiates a flux of 3000 lumens. Calculate its luminous efficiency. Find MSCP and express it in per watt.*

Solution: Data:

$$\text{Flux } \phi = 3000 \text{ lumens}$$

$$\text{Power of lamp} = W = VI = 250 \times 0.8 = 200 \text{ watts}$$

$$\text{MSCP} = \frac{\phi}{4\pi} = \frac{3000}{4\pi} = \mathbf{240}$$

$$\text{MSCP per watt} = \frac{240}{200} = \mathbf{1.2}$$

$$\text{Luminous efficiency} = \frac{\text{lumens}}{\text{watt}} = \frac{3000}{200} = \mathbf{15\ lumens/watt}$$

1.2 LAWS OF ILLUMINATION

There are two laws of illumination: 1. Law of inverse squares, 2. Lambert's cosine law.

1. Law of Inverse Squares:

- **Statement:** *"The illumination of a surface is inversely proportional to the square of the distance between the surface and the light source provided that the distance between the surface and the source is sufficiently large so that the source can be regarded as a point source".*
- **Explanation:** If a source of light which emits light equally in all directions be placed at the centre of a hollow sphere, the light will fall uniformly on the inner surface of the sphere, i.e. each square mm of the surface will receive the same amount of light. If the sphere be replaced by one of the larger radius, the same total amount of light is spread over a larger area proportional to the square of the radius. The amount which falls upon any square mm of such a surface will, therefore, diminish as the radius increases, and will be inversely proportional to the square of the distance.

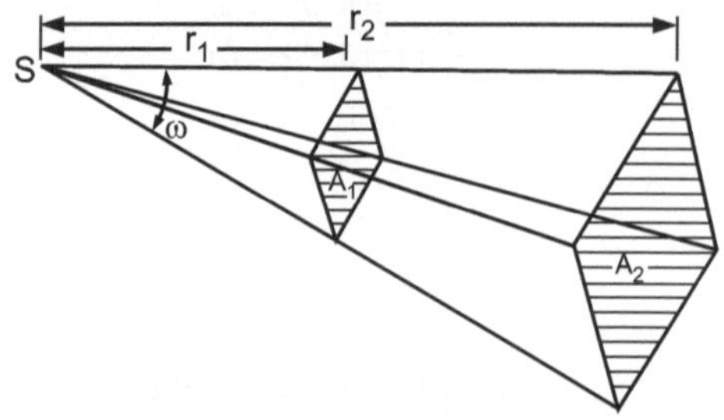

Fig. 1.6: Inverse Square Law

- A similar relation holds if we have to deal with a beam of light in the form of a cone or pyramid, as shown in Fig. 1.6. If we consider parallel surfaces which cut the pyramid at different distances from the source, the areas of these surfaces are proportional to the square of these distances and, therefore, the amount of light which falls on one unit of the area of these surfaces, is inversely proportional to the square of the distance from the source. This relationship is referred to as the law of *inverse squares*.
- Mathematically, it can be proved as follows:

  Let us consider surface area A_1 and surface area A_2 at distances r_1 and r_2 respectively from the point source S of luminous intensity I and normal to the rays, as shown in Fig. 1.6.

  Let the solid angle subtended be ω steradians.

  Luminous flux radiated per steradian = I.

  Total luminous flux radiated = Iω lumens

  Illumination on the surface of area $A_1 = \frac{I\omega}{A_1}$ lumens/unit area

  and area $A_1 = \omega r_1^2$

  ∴ Illumination on the surface of area A_1,

$$E_1 = \frac{I\omega}{\omega r_1^2} = \frac{I}{r_1^2} \text{ lumens per unit area}$$

  Similarly, illumination on the surface of area A_2,

$$E_2 = \frac{I\omega}{A_2} = \frac{I\omega}{\omega r_2^2} = \frac{I}{r_2^2} \text{ lumens per unit area}$$

2. Lambert's Cosine Law:

- **Statement of Law:** According to this law the illumination at any point on a surface is proportional to the cosine of the angle between the normal at that point and the direction of luminous flux.

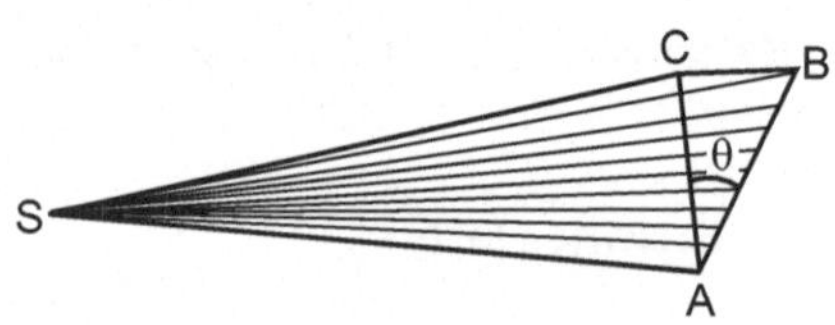

Fig. 1.7: Lambert's Cosine Law

- Many times the illuminated surface is not normal to the direction of the light as shown by AC in Fig. 1.7 and it makes an inclination as shown by line AB. The area over which the light is spread is then increases in the ratio of $\frac{AB}{AC} = \frac{1}{\cos\theta}$ or illumination decreases in the ratio of $\frac{\cos\theta}{1}$.

$$\therefore \quad \text{Illumination } E = \frac{I\cos\theta}{r^2}$$

SOLVED EXAMPLES

Example 1.2: *(a) A lamp emits 2000 lumens. Find its MSCP.*

(b) Find the average intensity of illumination on the surface which is inclined at an angle of 60° to the rays 5 m away from a lamp of 100 C.P.

Solution: (i) $\text{MSCP} = \frac{\text{flux}}{4\pi} = \frac{2000}{4\pi} = \mathbf{160}$

(ii) $\text{Average intensity} = \frac{I \cdot \cos\theta}{r^2} \; I = CP = \frac{100 \times \cos(90° - 60°)}{(5)^2} = \mathbf{3.464\ Lux}$

Example 1.3: *Find (i) MSCP, (ii) Luminous intensity in lumens per watt, (iii) MSCP per watt of 250 volts lamps which takes a current of 0.4 ampere and has a total flux of 1500 lumens.*

Solution: Wattage of the lamp = V × I = 250 × 0.4 = **100 watts**

(i) $\text{MSCP} = \frac{\text{Flux}}{4\pi} = \frac{1500}{4\pi} = \mathbf{120}$

(ii) $\text{Lumens/watt} = \frac{1500}{1000} = \mathbf{1.5\ luminous\ intensity}$

(iii) $\text{MSCP/watt} = \frac{120}{100} = \mathbf{1.2}$

Example 1.4: *A light source of 1000 watts having MSCP = 2500 is suspended 3 m above the working plane. Find the following:*

(a) Illumination in Lux directly below the lamp on the working plane.

(b) Lamp efficiency in Lumens/watt.

(c) Illumination 3 m away on the horizontal plane from vertically below the lamp.

Solution: As per the given distances, the position of the lamp and a point 'P' is shown in Fig. 1.8 as follows.

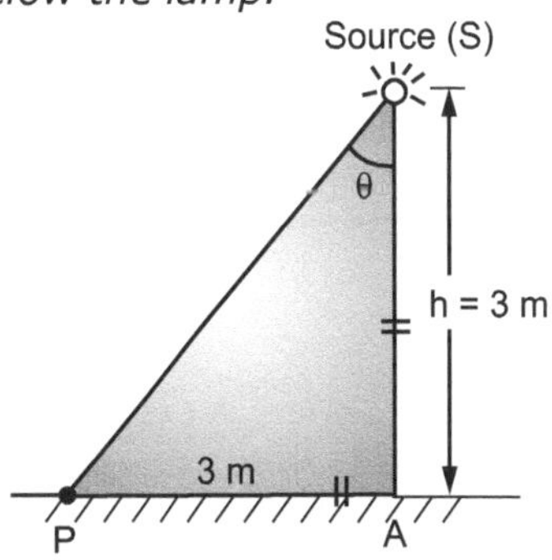

Fig. 1.8

Given I = 2500, h = 3 m.

(a) Illumination directly below the lamp on plane at point 'A' is given by

$$E_A = \frac{I}{h^2} = \frac{2500}{(3)^2} = \mathbf{277.77\ Lux}$$

(b) $\text{Lamp } \eta = \frac{\text{Luminous flux}}{\text{Wattage}} = \frac{4\pi \times \text{MSCP}}{\text{Wattage}} = \frac{4\pi \times 2500}{1000}$

$= \mathbf{31.416\ lumens/watt}$

(c) Illumination at point 'P', $E_P = \frac{I \cdot \cos^3\theta}{h^2} = \frac{I \cdot (\cos 45°)^3}{(3)^2} = \frac{2500 \times (0.707)^3}{9} = \mathbf{98.165\ Lux}$

Example 1.5: *Prove the following relations:*

(a) Illumination $E = \frac{I \cdot \cos^3\theta}{h^2}$ *at point 'P'.*

(b) Illumination $E = \frac{I}{h^2}$ *at point 'O'.*

for the position of lamp shown in Fig. 1.9.

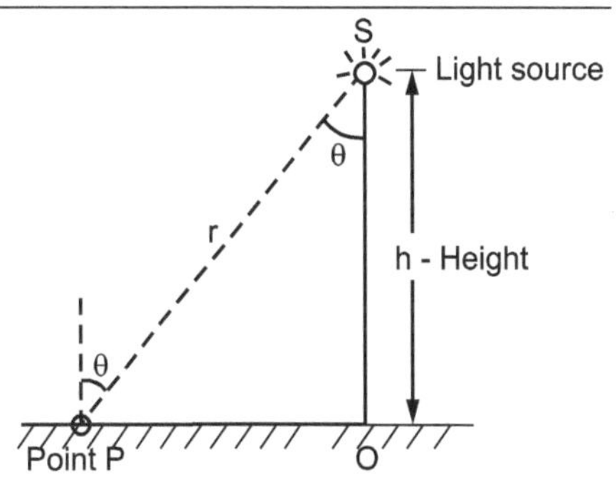

Fig. 1.9

Solution: (a) Illumination at point 'P' $= \frac{I \cdot \cos\theta}{r^2}$ ... general relation

But, $\cos\theta = \frac{h}{r}$

∴ $r = \frac{h}{\cos\theta}$

∴ $E = \frac{I \cdot \cos\theta}{\left[\frac{h}{\cos\theta}\right]^2} = \frac{I \cdot \cos^3\theta}{h^2}$

(b) If point is vertically below the light source then at point 'O', $\theta = 0$

$\therefore \cos\theta = 1 \qquad \therefore \cos^3\theta = 1$

$$\therefore \qquad E = \frac{1}{h^2} \text{ i.e. } \frac{CP}{h^2}$$

Example 1.6: *Two light sources S_1 and S_2 each of 500 CP and spaced on the horizontal plane surfaces are shown in Fig. 1.10.*

Find the maximum spacing between them, so that illumination on the ground midway between the lamp shall be half of directly under the lamps.

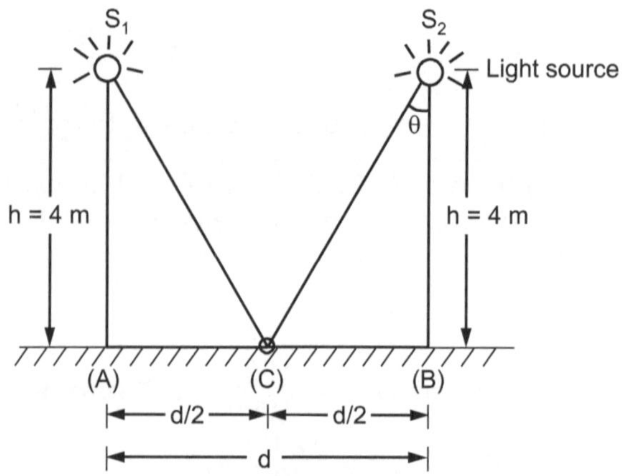

Fig. 1.10

Solution: Let, AB be the distance between poles = 'd' metres, h = 4 m given.

Illumination midway between the lamp, S_1 and S_2 at point 'C'.

$$E_C = 2 \times \text{Illumination due to either lamp}$$

$$= 2 \times \frac{500}{(4)^2} \times \frac{(4)^3}{\left(4^2 + \frac{d^2}{4}\right)^{3/2}} = \frac{4,000}{\left(4^2 + \frac{d^2}{4}\right)^{3/2}}$$

Illumination under either of the lamps, say under lamp S_2.

$$E_B = \text{Illumination due to lamp } S_1 + \text{Illumination due to lamp } S_2$$

$$= \frac{500}{(4)^2} \times \frac{(4)^3}{(4^2 + d^2)^{3/2}} + \frac{500}{(4)^2} = \frac{2,000}{(4^2 + d^2)^{3/2}} + 31.25$$

Since,

$$E_C = \frac{1}{2} E_B$$

$$\therefore \qquad \frac{4,000}{\left(4^2 + \frac{d^2}{4}\right)^{3/2}} = \frac{1}{2}\left[31.25 + \frac{2,000}{(4^2 + d^2)^{3/2}}\right]$$

or

$$\frac{4,000}{\left(4^2 + \frac{d^2}{4}\right)^{3/2}} = 15.625 + \frac{1,000}{(4^2 + d^2)^{3/2}}$$

or

$$d = 9.56$$

$\therefore$ Maximum spacing between two sources, d = **9.56 m**

Example 1.7: *A small light source with intensity of light uniform in all directions is mounted at a height of 8 m above the horizontal plane surface. The two points 'A' and 'B' both lie on this surface with point 'A' directly beneath the source. How far is point 'B' from point 'A' if illumination of 'B' is only one half as great as at 'A'?*

Solution: Let the distance between points 'A' and 'B' be s.

The following sketch shows this arrangement.

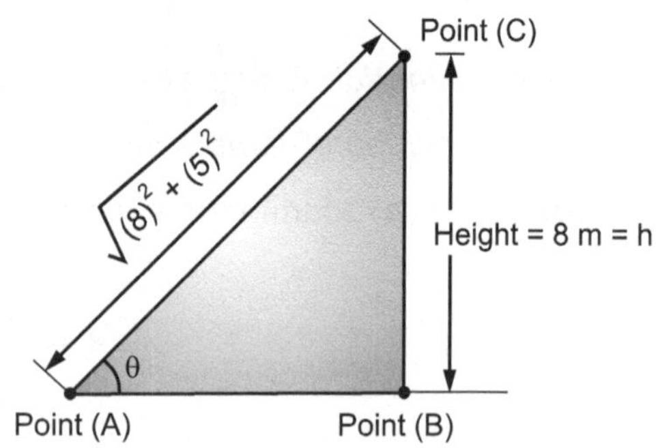

Fig. 1.11

Let intensity of the source = I.

By law of inverse squares, we get,

$$\text{Illumination at 'A'} = E_A = \frac{I}{(h^2)} = \frac{I}{(8^2)} = \frac{I}{64}$$

Now, to find illumination at point 'B', we use $\cos^3\theta$ law.

$$\therefore \qquad E_B = \frac{I}{h^2}\cos^3\theta = \frac{I}{(8)^2} \times \left(\frac{8}{\sqrt{(8)^2 + s^2}}\right)^3$$

$$= \frac{8 \times I}{(8^2 + s^2)^{3/2}}$$

But given that, $E_B = \frac{E_A}{2}$

$$\therefore \quad \frac{8 \times I}{(8^2 + s^2)^{3/2}} = \frac{I}{64 \times 2}$$

Solving this for distance (s) we get,

s = **6.15 m** i.e. distance of B from A.

Example 1.8: *A lamp 'L' is at the height of 10 m from the horizontal plane. The point 'B' is at the vertically down the lamp and B is away from the lamp on the same plane. Find the distance AB if illumination at B = 0.1 of that of at A.*

Solution: Let the intensity of the lamp L be I candela and the distance between points A and B be x metres.

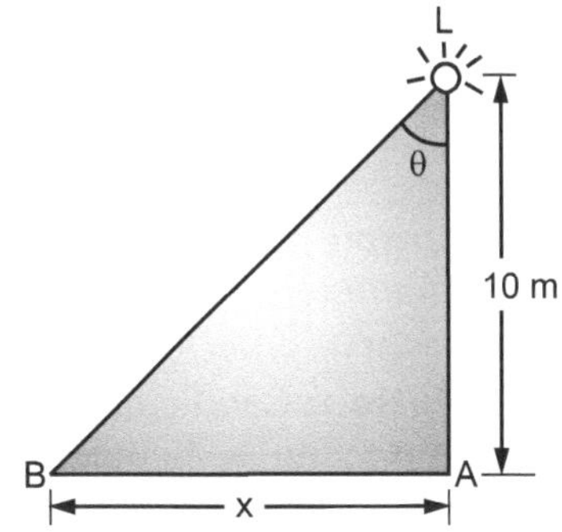

Fig. 1.12

Illumination at point A, $E_A = \frac{I}{(10)^2} = \frac{I}{100}$

Illumination at point B, $E_B = \frac{I}{(10)^2} \times \left(\frac{10}{\sqrt{(10)^2 + x^2}}\right)^3$

$$= \frac{10I}{(100 + x^2)^{3/2}}$$

Since, $E_B = \frac{1}{10} E_A$

$$\therefore \quad \frac{10\,I}{(100 + x^2)^{3/2}} = \frac{1}{10} \times \frac{I}{100}$$

or $(100 + x^2)^{3/2} = 10{,}000$

or $(100 + x^2) = (10{,}000)^{2/3} = 464$

or $x = \sqrt{464 - 100} = 19.1$ m

$\therefore$ Distance AB = x = **19.1 m**

Example 1.9: *A lamp of 500 W having a MSCP of 1,250 is suspended 2.7 m above the working plane. Calculate:*

(i) Illumination directly below the lamp at the working plane.

(ii) Lamp efficiency.

(iii) Illumination at a point 3 m away on the horizontal plane from vertically below the lamp.

Solution: Luminous flux intensity of lamp, I = 1,250.

Height of lamp, h = 2.7 m

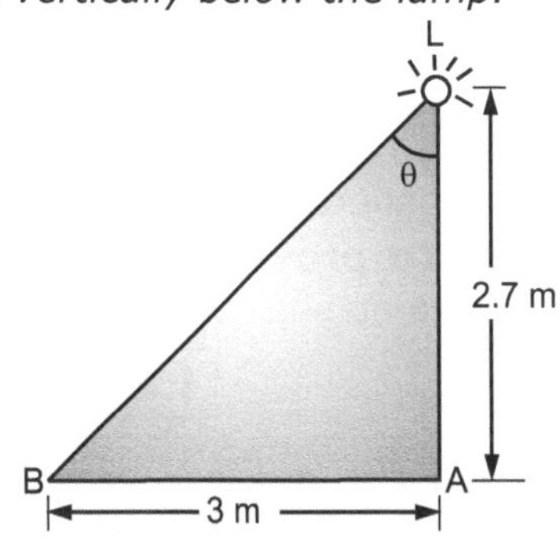

Fig. 1.13

(i) Illumination directly below the lamp at the working plane,

$$E_A = \frac{1.250}{(2.7)^2} = \mathbf{171.47\ lux}$$

(ii) Lamp efficiency $= \frac{\text{Luminous flux}}{\text{Wattage of lamp}}$

$$= \frac{4\pi \times \text{MSCP}}{W} = \frac{4\pi \times 1{,}250}{500}$$

$$= \mathbf{31.42\ lumens/watt}$$

(iii) Illumination at a point 3 m away on the horizontal plane from vertically below the lamp,

$$E_B = \frac{I}{h^2} \cos^3 \theta = \frac{1{,}250}{(2.7)^2} \times \frac{2.7^3}{[3^2 + 2.7^2]^{3/2}} = \mathbf{51.33\ lux}$$

Example 1.10: *The illumination at a point on a working plane directly below the lamp is to be 100 lumens/m^2. The lamp gives 256 CP, uniformly below the horizontal plane. Determine the height at which the lamp is suspended.*

Also find illumination at a point on the working table 1.2 m away from the vertical axis of the lamp.

Solution: Luminous intensity of the lamp, I = 256 cd

Illumination at a point on a working plane directly below the lamp,

$$E = 100 \text{ lumens/m}^2$$

Let the height at which the lamp is suspended be h metres.

Since, $E = I/h^2$

$\therefore$ $100 = 256/h^2$

or $h = \sqrt{\frac{256}{100}} = \mathbf{1.6\ m}$

Illumination on the working table 1.2 m away from the vertical axis of the lamp,

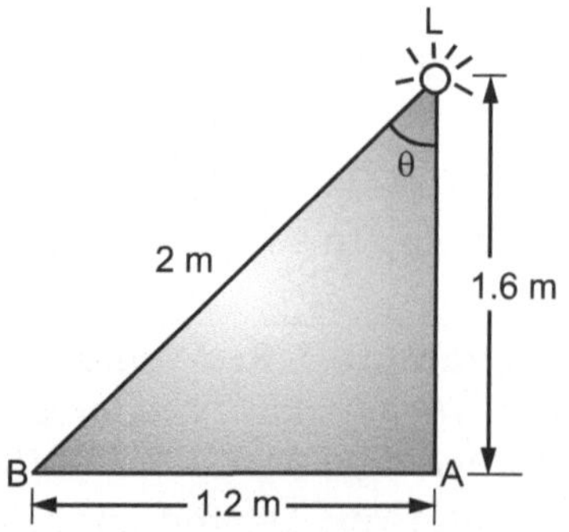

Fig. 1.14

$$E_B = \frac{I}{h^2}\cos^3\theta = \frac{256}{(1.6)^2} \times (0.8)^3 = \mathbf{51.2\ lux}$$

Example 1.11: *A lamp with a reflector is mounted 12 m above the centre of a circular area of 24 m diameter. If the combination of the lamp and reflector gives a uniform CP of 1,000 over the circular area, determine the maximum and minimum illumination produced on the area.*

Solution: Candle power of the lamp, CP = 1,000.

The maximum illumination will occur directly below the lamp, i.e. at point A and $= \frac{CP}{h^2} = \frac{1,000}{(12)^2} = 6.94$ lux

Minimum illumination will occur at the periphery of the circular area, i.e. at point B and $= \frac{CP}{h^2}\cos^3\theta = \frac{1,000}{(12)^2} \times \left(\frac{12}{\sqrt{12^2 + 12^2}}\right)^3$

$$= \frac{1,000}{(12)^2} \times \frac{1}{\left(\sqrt{2}\right)^3} = \mathbf{2.46\ lux}$$

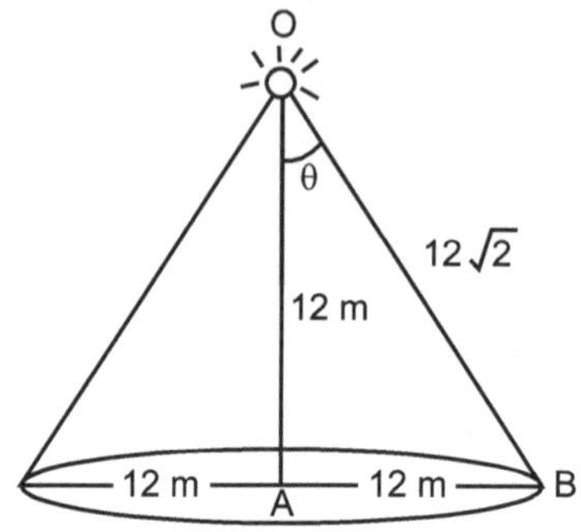

Fig. 1.15

1.2.1 Features of Good Illumination Scheme

- Following are the features of good illumination scheme:
  1. **To provide adequate illumination:** For different purposes, different levels of illumination are required, that requirement shall be fulfilled.
  2. **To avoid glare and hard shadows:** Disposition of lamps and the method such as direct or indirect shall be adopted, such that hard shadows and glare on the eyes shall be avoided.
  3. **To provide light of suitable colour:** Different lamps emit light of different colours, generally whitish day-light colour is desirable.
  4. **To provide light distribution uniformly over the working plane:** Disposition of the lamp shall be proper to achieve this objective.

1.2.2 Advantages of Good Illumination Scheme

1. Enlist any four features of good illumination scheme.

1. Different types of work require different level of illumination.
2. Scientific illumination scheme allows the persons to see easily and accurately to work at an efficient speed.
3. Good illumination scheme avoids fatigue and eye stains.
4. Good illumination scheme avoids excessive brightness, contrast, glare and disturbing shades.
5. Good illumination scheme helps to create a pleasant atmosphere.
6. Good illumination scheme provides the light of correct colour appearance and renders to enable the workers to judge details of their work such as colour, shape, etc. quickly and accurately.

7. Good illumination schemes saves work hours of the job which is beneficial.
8. Good illumination scheme increases productivity.
9. Good illumination scheme improves morale and motivation.
10. Good illumination scheme reduces product rejects.
11. Good illumination scheme avoids accidents.
12. Good illumination scheme provides safety and security.
13. Good illumination scheme provides light in economical manner.
14. Good illumination scheme utilizes day-light and blend it properly with electric light.
15. Good illumination scheme provides good lighting maintenance.
16. Good illumination scheme provides a perfect Lux value not less, not more for the respective purposes.

1.3 SOURCES OF LIGHT

- Light is a very very important factor in human life on which all the activities of the human depend.
- God gifted natural light from the sun is very useful and helpful which give the sensation of objects to our eye in the hours of the day. But night is dark.
- Where there is no natural light, artificially produced light is helpful.
- Light produced electrically is superior than other ways of producing like fire etc.
- Electrically produced light has various advantages like:
  - Cleanliness
  - Ease in controlling
  - Reliable
  - Steady output
  - Low cost
- Apart from its aesthetic and decorative aspects, good lighting has a strictly utilitarian value in increasing production, reducing worker's fatigue, protecting their health, eyes and nervous system and reducing accidents.
- Hence, science of "Illumination Engineering" has became of major importance today.

Nature of Light (Visible, Ultraviolet and Infrared)

- Light is a form of "Radiant Energy".
- Various incandescent bodies are sources of light.
- The light emitted from such bodies depends on the temperature of bodies.
- Radiated energy goes into the surrounding medium which is cooler than the heated bodies.
- Heat by the bodies may be (i) Red hot or (ii) White hot.
- Radiated energy by the red hot has a sufficiently large **wavelength** and energy is in the form of **heat.**
- But when the temperature is further increased the body changes from red hot to **white hot.** The wavelength of energy radiated becomes smaller and enters into the range of wavelength of light.
- Actually hot bodies emit heat as well as light energy. The velocity of these is 3×10^8 m/s.
- The light waves have wavelengths varying from 0.00075 mm to 0.00045 mm.
- The unit of expression of light wavelength is "Angstrom".

$$1 \text{ Angstrom} = 10^{-10} \text{ m}$$

- $$\text{Radiant efficiency} = \frac{\text{Energy emitted in the form of light}}{\text{Total energy emitted by a body}}$$

  This depends on temperature.

  Higher temperature → Lower wavelength of radiant energy → Higher the radiant efficiency

- Radiant efficiency will be maximum when the temperature of a body will be such that the wavelength of shortest wave radiated by a body is 0.0004 mm (4000 A°).
- Maximum radiant efficiency is at the temperature of 6250°C.
- Radiant efficiency of artificial light sources is lower.
- The sensation of colour is due to the difference in wavelengths of the light radiations.

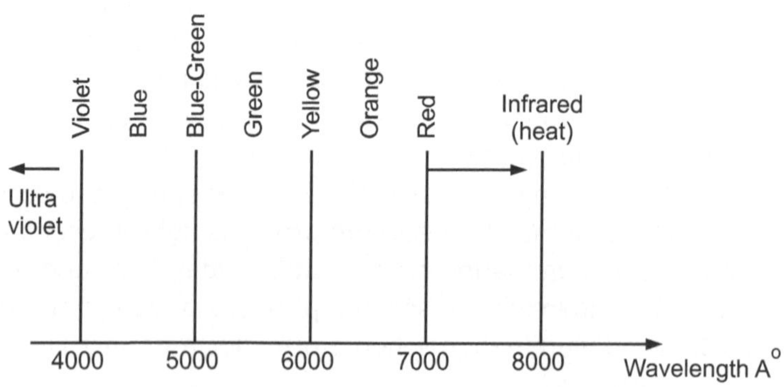

Fig. 1.16: Wavelength/Colour

- The visible light can have the wavelength between 4000 A° to 7500 A°.
- Refer Fig. 1.16.

  For ultraviolet light wavelength < 4000 A°

  Between 5000 A° to 4000° – Blue

  5000 A° – Blue green

  Between 6000 – 5000 A° – Green

  6000 A° – Yellow

  Between 7000 A° to 6000 A° – Orange

  7000 A° – Red

  Beyond 7000 A° and 8000 A° – Infrared

  So less than 4000 A° and more than 7000 A°, no sensation of colour.

  Ultraviolet and infrared are beyond our sensation.

Types of Lamps

- There are different types of lamps working on the normal supply voltage and produce light for illumination. We will study in this topic the construction, working and application of these lamps.

1.3.1 Incandescent Lamp

1. State any four disadvantages of incandescent lamp and advantages of LED lamps.

- There are two types of incandescent lamps. They are:

  (A) Vacuum lamps (B) Gas filled lamps.

(A) Vacuum Lamps:

- These lamps can operate upto the temperature of 2000°C. The construction is shown in Fig. 1.17. The filaments used for this type of lamp are carbon, osmium, tantalum or tungsten. Tungsten is best of all. The filament materials have the following properties:
  - (i) High melting point.
  - (ii) Low vapour pressure.
  - (iii) High resistivity.
  - (iv) Low temperature coefficient.
  - (v) Ductile.
  - (vi) Mechanical strength.
- The bulb is evacuated so that filament is not oxidized even at higher temperatures. Melting point of tungsten is 3400°C. The filament heats up immediately and its resistance increases immediately. Its luminous efficiency is 10 lumens/watt.
- As the time passes, aging of tungsten starts and its efficiency decreases. The total depreciation of light output is roughly 15% over the useful life range.

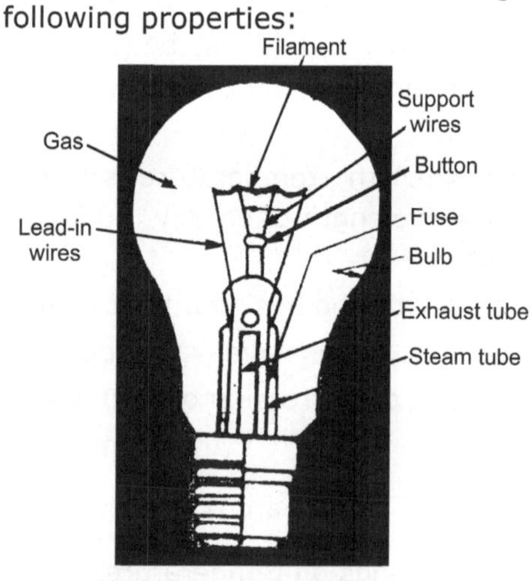

Fig. 1.17: Vacuum/Gas Filled Incandescent Lamp

- There is effect on the performance of the lamp due to voltage variations. 5% increase in voltage increases lumens output by 20% but shortens the life by 50%.
- 5% decrease in voltage reduces 20% lumens output but increases the life of lamp by two times.
- The normal life of filament lamp is 1000 working hours.

Advantages:

1. Operating power factor is unity.
2. Operates directly on normal voltages.
3. Good radiation characteristic in luminous range.
4. Available in various shapes and sizes.
5. Rating 15 watts, 25 watts, 40 watts, 60 watts, 100 watts, 200 watts and 500 watts.
6. Very cheap in cost.
7. Handy and easily fixed up.

(B) Gas Filled Incandescent Lamps (Construction is same as type A)

- A metal filament can work in an evacuated bulb upto 2,000°C without oxidation and if it is worked beyond this temperature it vaporizes quickly and blackens the lamp. For higher efficiency it is necessary to use working temperature more than 2,000°C keeping down the evaporation, which is possible by filling the bulb with an inert gas-argon with a small percentage of nitrogen.
- Nitrogen is added to reduce the possibility of arcing. Krypton is the best gas for this purpose but it is so expensive that it is used only in special purpose lamps, such as miner lamps. The tungsten filament can safely be burn at a temperature of 2,400°C to 2,750°C according to the size of the lamp. However, due to presence of gas there is heat loss due to convection currents. This loss depends upon the surface area of filament. As such, coiled coil filaments, which take much less space compared with coiled filaments are used with such lamps.
- A coiled coil filament is made by winding tungsten wire on a fine iron wire to form a spiral which is again wound on to a thick iron wire to form a coiled coil (iron is later on dissolved out by acid). Helical filaments also have a slower rate of tungsten evaporation.
- This evaporation eventually causes bulb blackening because the tungsten vapour condenses as a black film on the inner surface of the bulb. In a gas-filled lamp, the hot gas carries the tungsten vapour upward. Therefore, a black spot forms at the top of the bulb instead of spreading over the entire inner surface, as in a high-vacuum bulb. Chemicals called 'getters' are often placed inside the bulb to capture tungsten vapour and thereby reduce the rate of blackening.
- A piece of wire mesh called a collector grid may also be attached to each lead in wire to attract the particles of tungsten vapour. Efficiency of coiled coil lamp is about 12 lumens/watt.
- For low wattage lamps, however, the heat loss due to introduction of gas is more than in medium wattage lamps, so for low wattage (upto 40 watts), vacuum type lamps are used.

(a) Coiled filament **(b) Coiled coil filament**

Fig. 1.18

1.3.2 Arc Lamps

- **The basic principle:** Two electrodes touching to each other are connected to the source of electricity as shown below. Both are carrying currents. Refer Fig. 1.19 (a).
- If now electrodes (1) and (2) are separated a bit, even then they carry an electric current and in the gap separation an 'arc' is formed.

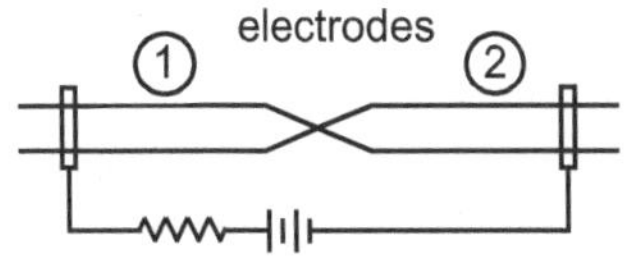

Fig. 1.19 (a)

- In the gap, the arc has tremendous heat and temperature and light is produced as shown in the following Fig. 1.19 (b).
- On this principle different arc lamps are designed.
- The arc lamps are (i) carbon arc, (ii) flame arc or (iii) magnetic arc lamps.

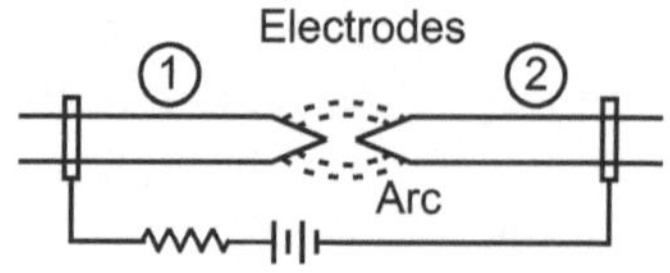

Fig. 1.19 (b)

1.3.2.1 D.C. Carbon Arc Lamps

1. Explain with neat sketch the working of carbon arc lamp with its applications.

- In this type, carbon electrodes are used.
- This is the old type lamp.

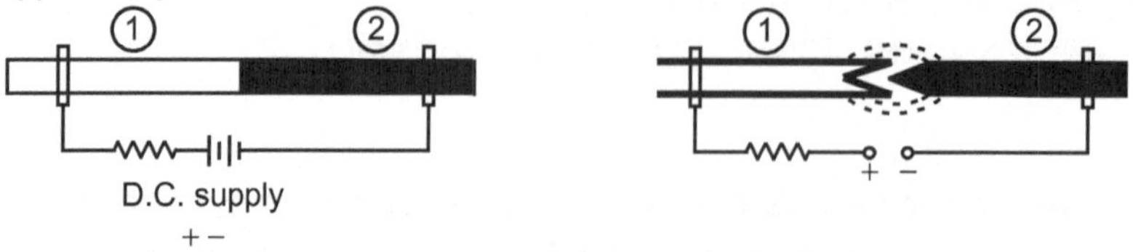

(a) (b)

Fig. 1.20: D.C. carbon arc lamp

- Two electrodes (1) and (2) placed touching to each other face to face are as shown in Fig. 1.20 (a).
- D.C. supply is about 40 to 60 volts.
- When current flows, rods soon become incandescence due to high resistance of carbon.
- Now if they are slightly pulled, so that a gap of about 2 to 3 mm is formed, an arc is formed between them in the gap and a white light is produced.
- Transfer of carbon particles from one rod to the other rods maintain the arc in the gap.
- Due to transfer of particles of carbon the positive rod after some period becomes hollow and negative rod becomes as a pointed pencil as shown in Fig. 1.20 (b).
- Both rods burn out after some period. The rate of burning of positive rod is twice the negative rod. So cross-section of positive rod is made twice of the cross section of negative rod.

1.3.2.2 A.C. Carbon Arc Lamps

- In its case, burning of both positive and negative rod is same and hence their cross-sectional area is same in A.C. operation.
- Since the carbon is consumed during operation, frequent adjustment for the gap between the rods is required.
- An automatic device is provided for feeding the carbon into the arc for big lamps.
- **Requirement of voltage:** $V = (39 + 2.8\, l)$ where l: length of the arc in mm.
- A series resistance as shown in Fig. 1.18 is used for stabilizing the arc.
- Overall luminous efficiency is very low which is 12 Lumens/watt.
- In A.C. arc lamp instead of a series resistance, inductor is used for stability.
- **% Production of light:**

  85% of light by positive carbon.

  10% of light by negative carbon.

  5% of light by the air.
- Temperature of positive carbon is about 3500°C to 4000°C and that of negative carbon is about 2500°C.

Uses of Arc Lamps:

1. Cinema projectors.
2. Search lights.

1.3.2.3 Flame Arc Lamp

- The principle of operation is similar to that of carbon arc lamp.
- The electrode of such a lamp has 5 to 15% fluoride (called the flame material), 85 to 95% carbon.

- Generally core type carbon electrodes are used, the cavities are filled with fluoride.
- The fluoride has a characteristic which radiates light energy efficiently from a very high heated arc stream.
- Fluoride turns into vapour along with carbon.
- Vapours cause a very high luminous intensities.
- In addition to fluoride, there are some other flame materials.
- Different flame materials which produce different colours.
- Most of these colours do not appeal to eyes at the same time they produce strain on them.
- The arc can be drawn out to one side with the help of a magnet.

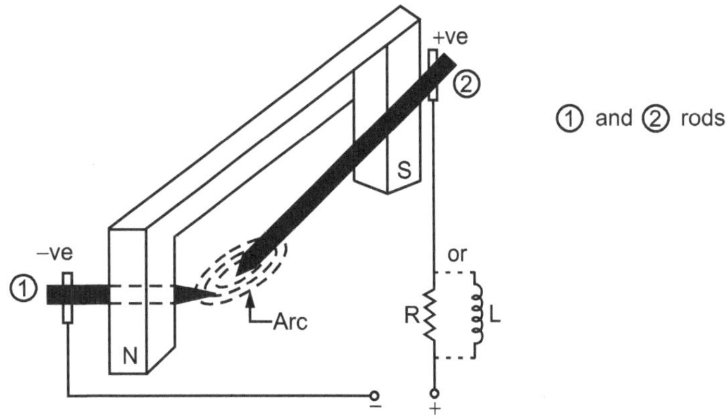

Fig. 1.21: Flame arc lamp

- Though arc is very efficient but owing to objection to its colours, it has now been superseded by the electric discharge lamps.
- The luminous efficiency of such a lamp is 8 lumens per watt.

Magnetic Arc Lamp

- In such a lamp, positive electrode made of copper and negative electrode made of magnetic oxide of iron are used.
- The arc is struck in the similar way as in case of carbon arc lamp, such lamps are rarely used.

1.3.3 Fluorescent Lamp

1. Explain with neat diagram, construction and working of fluorescent tube light.

1. **Principle:** "When fluorescent materials are subjected to electromagnetic radiations of particular wavelength, they get excited and in turn gives out radiations at some other length, and persist in giving out radiations even if exciting radiations are removed".
2. **Basic operation:** Fluorescent lamp, in principle is low pressure mercury discharge lamp with internal surface coated with suitable fluorescent material.
   Tube contains argon or krypton gas to ease starting with small quantity of mercury under low pressure. At start, current is passed through filaments which get heated up and emit electrons which strike mercury to give out radiations which fall on fluorescent material thus emitting light.

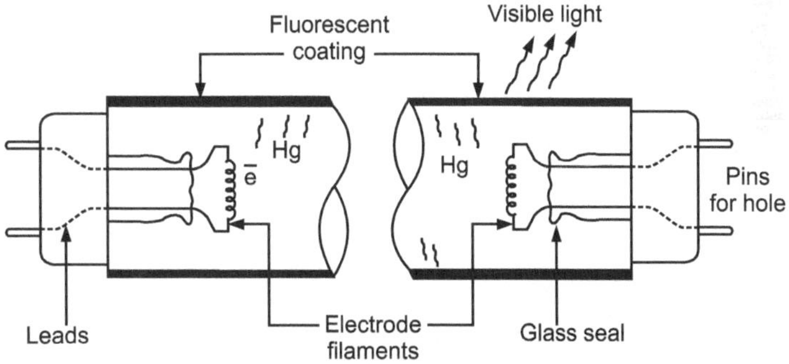

Fig. 1.22: Fluorescent tube

3. Tube contains argon gas in small amount in vapour form and mercury in globular form i.e. not in pure vapour form. Argon gas is used to initiate the arc.
4. Circuit diagram for fluorescent lamps with:
   (a) Glow type starter
   (b) Thermal type starter.

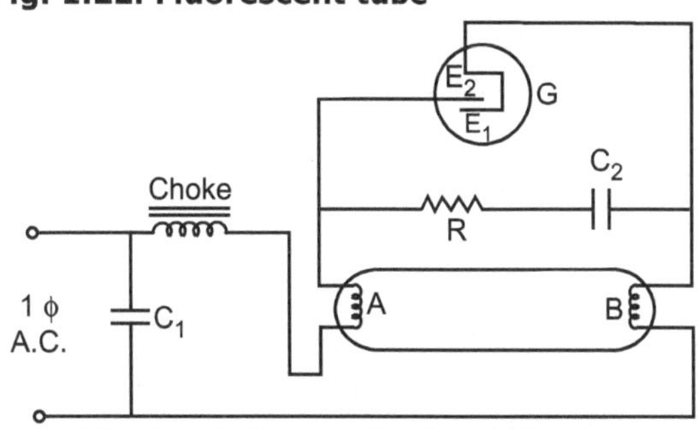

Fig. 1.23: Tube with Glow Starter

(a) Glow type starter:

- In circuit diagram, R and C are used for protection purpose:
- R - is to avoid welding of starter contacts and C - is to reduce radio interference.
- Starter is made of two electrodes E_1 and E_2 out of which E_1 is fixed and E_2 is of bimetallic strip. They are fitted into a glass tube having helium and hydrogen at low pressure.
- Normally, the contacts E_1 and E_2 are open and when supply is applied, full voltage is received by glow switch, this voltage is sufficient to start glow discharge between bimetallic strips E_2 and E_1 and heat generated bends E_2 to make contact with E_1 thus completing main circuit through choke and lamp electrodes A and B.

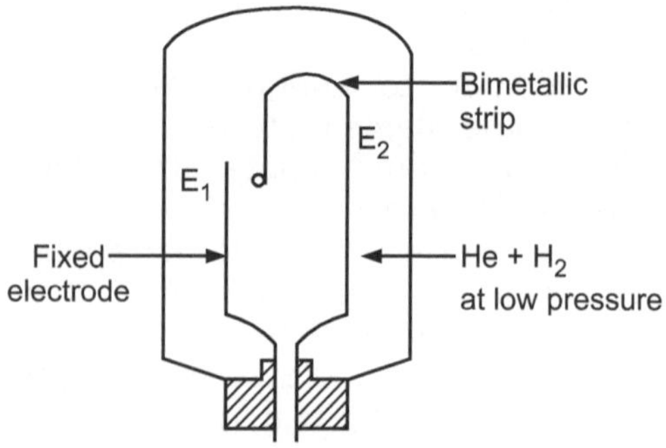

Fig. 1.24

- Current starts flowing through A and B and are heated in incandescent and argon gas which is in immediate vicinity is ionized, by this time glow is shorted out and so E_1 and E_2 cools and contacts open again.
- Sudden rate of change of current induces a voltage surge of about 1000 V across choke and initiates a discharge of argon gas and heat is produced. This heat is sufficient to vaporize mercury and potential difference falls to 100 to 110 V.

(b) Thermal type starter:

- Two bimetallic strips are initially closed so that current flows through filaments, heater coil, thus the electrodes A and R are heated to incandescence and argon gas is ionized.
- Meanwhile, one of the bimetallic strip near to heater coil gets heated and bends thus interrupting filament current producing inductive voltage surge across electrode; establishing discharge through mercury vapour.

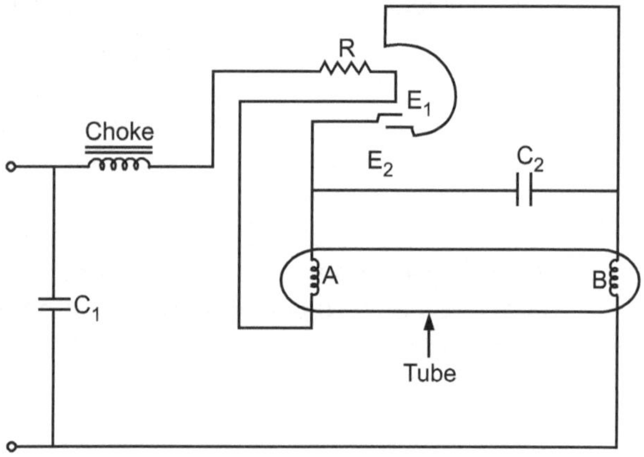

Fig. 1.25: Tube with Thermal Starter

- Starting time is very much less than the flow type starter.

5. Choke exclusively performs two functions:
   - Supply large potential for starting arc.
   - Limit arc current to a safe value.

Advantages:

1. If condenser is connected across surplus, efficiency is about 90.5%.
2. Its life is three times more than filament lamp.
3. It has low surface brightness – so less glare and it casts soft shadow.

Disadvantages:

1. High initial cost due to use of choke and starter.
2. Available in small wattage rating than incandescent lamp.

Rating of fluorescent lamps – 20 watts, 40 watts, 65 watts, 80 watts.

Note: Low ambient temperature increases starting time of fluorescent lamp.

1.3.3.1 Fault Finding in Case of Fluorescent Lamp

Table 1.3

Type of fault	Reasons	Actions of repairs
1. Tube does not start.	1. Defective and loose holders. 2. Defective starter. 3. Defective choke. 4. Defective tube.	1. Turn tube through 90°. Try another starter and check connections. 2. Replace starter. 3. Replace choke. 4. Replace tube
2. Sometimes end of a tube remain lighted.	Short-circuited starter.	Check starter and replace.
3. Blinking ON and OFF.	1. Wrong connections. 2. Loose contact. 3. Voltage is low. 4. Defective starter. 5. Defective tube.	1. Check all connections. 2. Check all connections. 3. Change ballast tapping if required. 4. Replace starter. 5. Replace tube.
4. Blackening of the end portion.	Too low or too high voltage.	Adjust ballast tapping if provided.
5. Dark streaks along lamp.	Mercury globules.	Rotate tube through 180°.
6. Burn out electrodes.	Control unit in circuit or choke short-circuited.	Test with a new choke.
7. Snaking spiraling (with glow type of starter)	Insufficient heating due to either starter opening quickly or wrong tappings made on the ballast.	Replace starter or ballast. In case of new lamp, it cures after few days.

1.3.3.2 Comparison between Tungsten Filament Lamps and Fluorescent Tubes

Table 1.4

Tungsten Filament Lamps	Fluorescent Tubes
1. Luminous efficiency increases with the increase in voltage of the lamp.	1. Luminous efficiency increases with the increase in wattage and increase in length of tube.
2. It gives light close to natural light, therefore, objects are properly seen.	2. It does not give light close to natural light, therefore, colour rendering is defective.
3. Voltage fluctuation has comparatively more effect on the light output.	3. Voltage fluctuation has comparatively low effect on light output as the variations in voltage are absorbed in the choke.
4. Luminous efficiency of coloured filament lamps is poor because coloured glasses are used for this purpose.	4. Different colour lights can be obtained by using different composition of fluorescent powder. Hence, efficiency is high and better colours are obtained.
5. Due to comparatively high working temperature, heat radiations are also present.	5. Due to low working temperature, heat radiation is low.

contd. ...

6. Due to aging, light output is reduced.	6. Due to aging, light output is gradually reduced.
7. Brightness is more.	7. Brightness is less.
8. Though the life of filament lamps varies with the working voltage, however, its normal life is 1,000 working hours.	8. Life of fluorescent tubes is not affected so much by variations in voltage but it depends on the frequency of starting. The life of the tube is about 7,500 working hours.
9. No stroboscopic effect.	9. It has objectionable stroboscopic effect.
10. Cheaper in cost.	10. Its cost per tube is more.
11. For same lumens output, more lamps are required and wiring cost is more. Life of the lamp is also less. Hence, overall cost of maintenance is more.	11. For same lumens output, lesser number of tubes is required, the wiring cost is low. Since, its life is comparatively more, replacement cost is low. Hence, overall maintenance cost is low.

1.3.4 Types of Other Lamps

1.3.4.1 High Pressure Mercury Vapour Lamp

1. **Principle:** "When mercury discharge under low pressure gives mainly ultraviolet radiations. If pressure is increased to one or two atmospheres, its proportion of radiations in visible spectrum is increased and we get light having bluish tinge".
2. **Construction and Working:**
   - It consists of two tubes. Inner tube contains neon or argon gas under low pressure, two main electrodes in the form of oxide coated coils and starting or auxiliary electrode near the cap end side of main electrode.
   - Inner tube is made of hard glass and in an evacuated outer tube, which maintains high operating temperature of the inner arc tube.
   - When supply is given, electric field is setup between the starting electrode and the adjacent main electrode. This causes discharge first to take place between them through limiting carbon resistance.
   - This discharge through argon gas vaporizes all mercury and for the lamp to reach full brilliance requires 4 to 8 minutes depending upon the design.

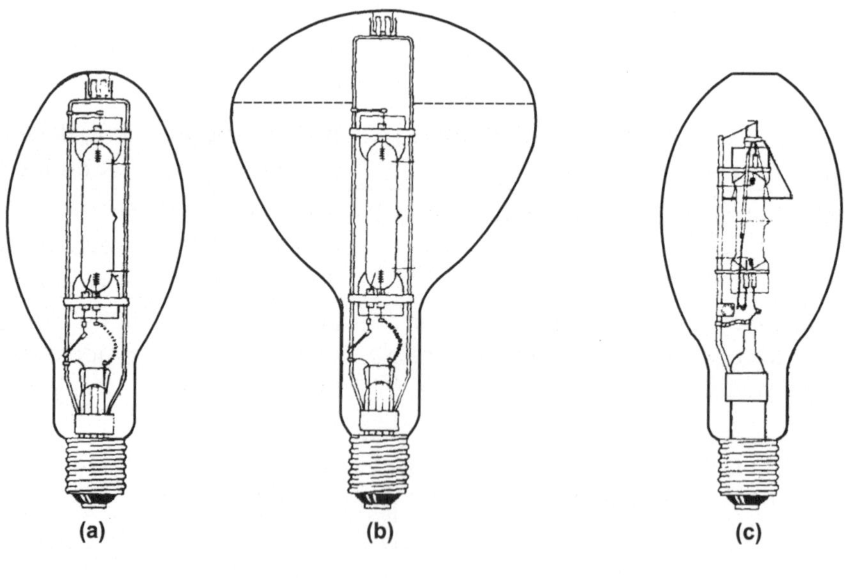

(a) (b) (c)

contd. ...

- Once arc tube is filled with mercury vapour a low resistance path is created for current to flow between the main electrodes. The arc then shifts to main electrodes. Path of starting electrode automatically becomes inactive.
- Electrodes are kept in electron emitting condition by the bombardment of heavy mercury ions.
- Should the lamp go out while in services, it will not restart till mercury vapour pressure has fallen low enough to allow for restarting of discharge between main and auxiliary electrodes.

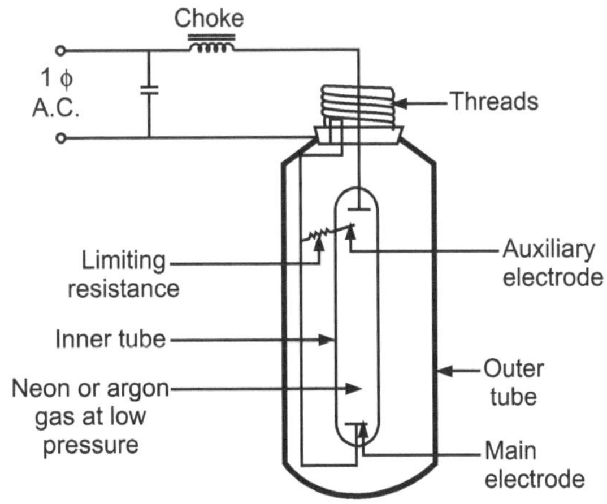

Fig. 1.26: Construction of High Pressure Mercury Vapour Lamp

3. Mercury disacharge lamps are made in five sizes viz. 80, 125, 250, 400 and 1000 W.
4. In case of mercury vapour lamps, magnitude of the striking voltage is very much affected by ambient temperature.
5. Efficiency of these lamps is of 30 to 40 lumens/watt and life is about 5000 hr.
6. **Application:** These are used for high ceiling manufacturing bays street lighting, stage lighting, sports grounds, garages and petrol station lighting, railway marshalling yards and park lighting.

Mercury Iodide Lamps:

- Construction is same as that of the HPMV lamp.
- In HPMV lamp only mercury is added whereas in mercury iodide lamp mercury and also number of iodides are added for filling the gap in light spectrums to improve the colour characteristic of light so produced.
- As compared to HPMV lamp, its efficiency is higher (70 to 90 lumens/watt).
- In HPMV lamp for ignition only choke is used whereas in mercury iodide lamp, choke as well as a separate ignition device is used.

Advantages:

1. Good light quality.
2. Higher efficiency (70 to 90 lumens/watt).
3. Better performance.
4. Ignition is effective.

Applications:

1. In the fields of flood lighting.
2. Industrial lighting.
3. Public lighting.

1.3.4.2 Comparison of HPMV Lamp and Mercury Iodide Lamp

Table 1.5

HPMV Lamp	Mercury Iodide Lamp
1. Construction is same.	1. Construction is same.
2. Filling gas is Hg.	2. Hg + iodides.
3. Luminous efficiency is 60 to 65 lumens/watt.	3. Luminous efficiency is 70 to 90 lumens/watt.
4. Ignition facility choke only.	4. Choke and additional ignition device.
5. Cost moderate.	5. More than HPMV lamp.
6. Used in street lighting, stage lighting, sports ground garages and petrol station lighting, railway marshalling yards, park lighting.	6. Used for flood lighting, industrial lighting, public lighting.

1.3.4.3 Sodium Vapour Discharge Lamp

1. **Principle:** "Electrons are emitted from cathode by subjecting the surface of the cathode to very high electric stress so that electrons are forcibly ejected from orbits of atom against nuclear centripetal force".
2. **Construction and Working:**
   - It consists of inner tube of U shape fitted with neon gas and globules of sodium. For good performance, 300°C temperature is to be maintained.
   - To protect discharge tube it is contained in evacuated double walled glass jacket. This prevents heat loss.
   - At starting, electric discharge is to be established through neon gas till sodium gas vapourized, so at start for 10-15 min. discharge through lamp will be pink and then as sodium vaporizes, it will turn to yellow.
   - For this voltage of about 410 is required white normal voltage is about 165. Supply is given through a leak transformer which has high leakage reactance.
   - As soon as discharge through the tube starts, heavy voltage drop takes place in the transformer so that required operating voltage is automatically impressed across the cathodes.
   - P.F. of these lamps is 0.3 lagging and therefore condenser can be used for P.F. connection.

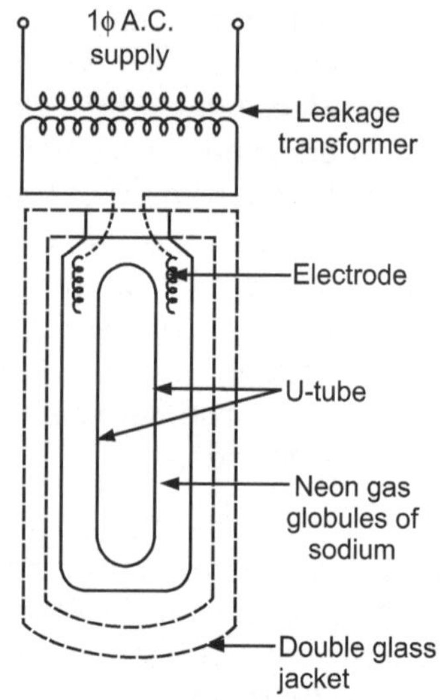

Fig. 1.27: Sodium Vapour Lamp

3. **Precaution:** Care should be taken in handling these lamps, for example, if inner tube is broken and sodium comes in contact with moisture, fire will result.
4. **Life:** Normal average life of sodium vapour lamps is 2500 hrs.
5. **Applications:** Due to yellow light its use is limited to street lighting, flood lighting of docks, quarries etc.
6. **Sizes available:** 45, 60, 85, 140 W.

1.3.4.4 LED Lamp

- **Introduction:** Electroluminescence is the emission of light due to the interaction of an electric-field with a suitable solid. These solids may be panels, Light Emitting Diodes (LEDs), light emitting films. The panels employ microcrystalline powder phosphors based on the compounds like zinc sulphide, whereas in case of LEDs, compounds like "gallium phosphide" is used. Panels require high voltage to active where as LEDs require a small voltage about 2 volts D.C. and their current densities are of higher order.
- In this chapter our study is limited to LEDs only.
- **LED:** It was noted in 1962 that a gallium-arsenide junction biased in the forward direction was found to be an efficient emitter of radiation. Electroluminescent p-n junctions are similar to conventional silicon or germanium diodes so far as electrical characteristics are concerned. However, when biased in the forward direction, some of the energy dissipated is converted into light rather than into resistive heating and hence called as "semiconductor lamps".
- A single crystal material is produced in the form of a rod by slowly withdrawing a seed crystal of gallium phosphide or gallium arsenide from a saturated melt.
- LEDs have similar electrical characteristics as the normal rectifying diodes have.
- LEDs are virtually point sources of high intensity radiation life of LED device is very long (i.e. upto 100,000 hours).
- These are mechanically robust, low current consuming.

Fig. 1.28: LED Lamp

Applications:

(i) Indicating lamps.
(ii) Bi-colour indicators.
(iii) Fault-indicators.
(iv) Display-boards.
(v) Decorative lighting.
(vi) Wrist watches.
(vii) In automobile industries as electro-mechanical control.
(viii) In instrumentation.

1.3.4.5 CFL Lamp

1. These are lamps of choice for those looking for an energy efficient alternative to incandescent lamps; they are paired for consuming a little as $1/5^{th}$ of the power and lasting upto 13 times longer than incandescent lamps.
2. Increasing variety in shape and colour and small size of CFL have made them more versatile and acceptable than traditional long tube fluorescent lamp.
3. **Construction and Working:**
   - A typical CFL consists of a gas filled gas tube with two electrodes mounted in an end cap. It contains a low pressure mix of argon gas, mercury vapour and liquid mercury and is coated on the inside with three different phosphors.

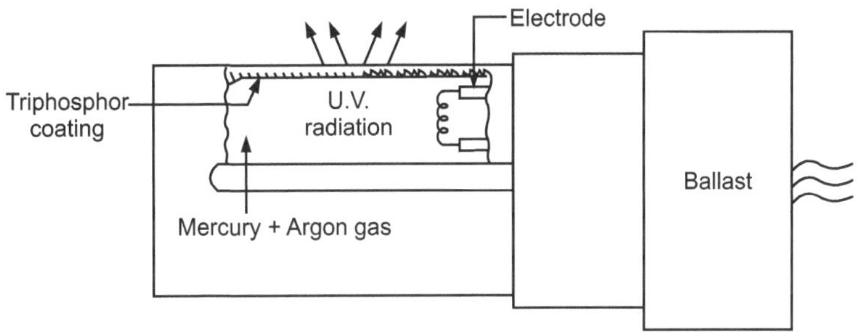

Fig. 1.29: CFL Construction

   - They give off light when exposed to ultraviolet radiation released by mercury atoms as they are bombarded by electrons. The flow of electrons is produced by an arc between two electrodes at the end of the tube.
   - Ballast provides the high initial voltage required to create the starting arc and then limit current to prevent the lamp from self destruction.
   - Ballast themselves can be "standard" electromagnetic, "energy efficient", electro-magnetic or electronic and may have power factor correction, radio interference, suppression or other features unheard of in the world of incandescent light.
4. **Colour:**
   - With development of more efficient "tri-phosphor" coatings came smaller 'CF' with light in red, blue and green giving white light when blended together.
   - By changing relative balance of these phosphors, manufacturers can produce range of apparent colour temperature from a cool 4100°C to a warm 2700°K.
   - Colour rendering index of lamp reflect how accurately colour of an object can be determined under given light source. CFL have CRI of 82 out of 100 while incandescent have 97 which is considered excellent for fluorescent sources.
5. **Temperature Effects on Performance:**

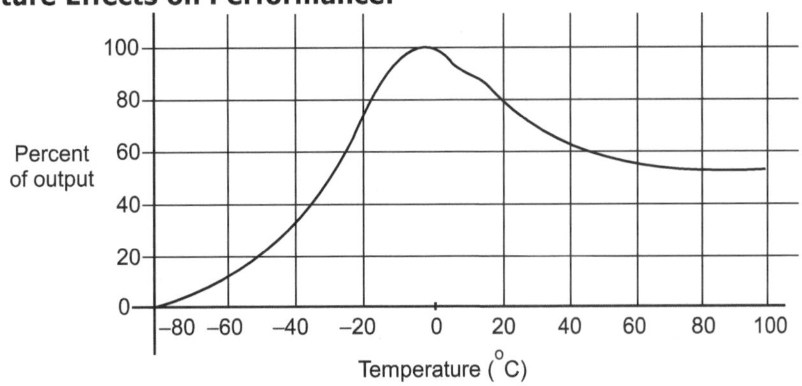

Fig. 1.30: Output versus Temperature Graph (Characteristics)

- The ambient temperature around a 'CFL' can have significant effect on light output and lamp efficiency. The optimum lamp wall temperature for CFL is generally 100°F (38°C). At very low temperature (below 32°F or 0°C) lamp output can decline to one third of the rated value or less. Because of temperature below optimum, mercury vapour will condense at cold spot, reducing number of mercury atoms available to emit u.v. radiations, thus light output drops.

6. Thus compact fluorescent lighting system offers potential for significant economic and environmental savings. Therefore, versatile range of different lamp-ballast configurations is available that can provide a comfortable, productive and well-illuminated space if properly used.

1.3.4.6 Halogen Lamp

1. Halogen lamps are developed filament lamps and are free from drawbacks of filament lamps.
2. **Principle:** "Amount of radiations in visible spectrum increases as temperature of incandescent material increases upto 6000°C absolute".

   Therefore, any material will be suitable and will have high radiating efficiency which has high melting point. Carbon and tungsten are better choices for filament lamp, having melting points of 3500°C and 3400°C respectively.
3. **Drawbacks for filament lamps:** Carbon vaporizes very fast beyond its working temperature of 1800°C, so large electrical input is required to produce given c.p. (Candle Power). Thus efficiency is low about 3 lumens/watt.

   Vaporisation of carbon results into blackening of the inside of the bulb. This phenomenon is more pronounced with carbon that in tungsten employed bulbs.
4. In 1959, along with the argon gas filling, iodine was added and the drawback was eliminated.
5. The evaporated tungsten of filament on reaching the relatively low temperature near the wall of the envelope forms a compound of tungsten iodine and is represented by the chemical reaction:

$$W + nX \rightarrow WX_n$$

   where, W → Tungsten

   X → Halogen

   n → Number of atoms.
6. Tungsten iodine which is very volatile suffers thermal diffusion in the direction of filament and on reaching the filament at high temperature it decomposes into the tungsten and halogen as follows:

$$WX_n \rightarrow W + nX$$

   The tungsten so released is deposited back on the filament. Iodine, therefore, works as a transport gas. In this way, evaporated tungsten is restored to the filament by iodine regenerative cycle.
7. In order that regenerative cycle should function properly, it is necessary to maintain relatively high bulb wall temperatures of 250°C. For this reason, quartz is chosen as the envelope material and its size is kept very small, and quartz also facilitates high gas filling pressure, which is capable of giving higher light output.
8. Iodine being corrosive in nature created problem during filling. So CH_3Br was used which was easy to fill; non-corrosive in nature and has less atomic weight than that of iodine.

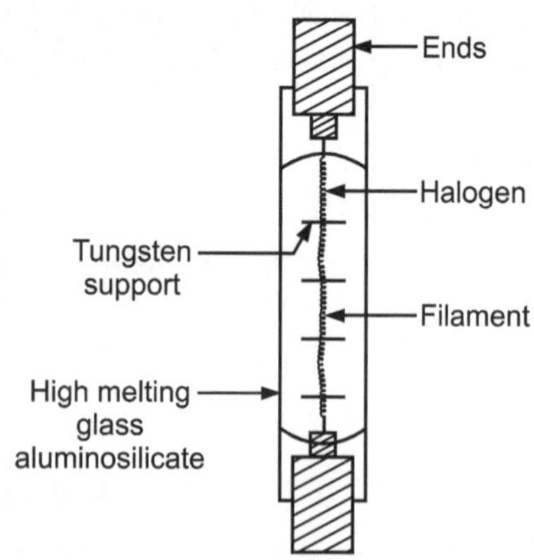

Fig. 1.31: Double Ended Linear Lamp

9. **Advantages of halogen lamp:**
   - It is available in miniature size, single ended or double ended lamps.
   - No blackening of lamps.
   - It has higher lumen output 22-33 lumens/watt and high operating temperature.
   - It has longer life about 2000 hrs and approximately double that of filament lamp.
   - It has better colour radiation.
10. **Applications:**
    These are used in public halls, factories, photofilm OHP, car lighting, large gardens, airport runways, sport halls, photo films, T.V., studios etc.

1.3.4.7 Ultraviolet Lamps

- Ultraviolet (UV) designates a band of the electromagnetic spectrum with wavelength from 10 nm to 400 nm, shorter than that of visible light but longer than X-rays. UV radiation is present in sunlight, and contributes about 10% of the total light output of the Sun. It is also produced by electric arcs and specialized lights, such as mercury-vapor lamps, tanning lamps, and black lights. Although long-wavelength ultraviolet is not considered an ionizing radiation because its photons lack the energy to ionize atoms, it can cause chemical reactions and cause many substances to glow or fluoresce. Consequently, the chemical and biological effects of UV are greater than simple heating effects, and many practical applications of UV radiation derive from its interactions with organic molecules.

Short-wave Ultraviolet Lamps

- Shortwave UV lamps are made using a fluorescent lamp tube with no phosphor coating, composed of fused quartz, since ordinary glass absorbs UVC. These lamps emit ultraviolet light with two peaks in the UVC band at 253.7 nm and 185 nm due to the mercury within the lamp, as well as some visible light. From 85% to 90% of the UV produced by these lamps is at 253.7 nm, whereas only 5-10% is at 185 nm. The fused quartz tube passes the 253.7 nm radiation but blocks the 185 nm wavelength. Such tubes have two or three times the UVC power of a regular fluorescent lamp tube. These low-pressure lamps have a typical efficiency of approximately 30-40%, meaning that for every too watts of electricity consumed by the lamp, they will produce approximately 30-40 watts of total UV output. They also emit bluish-white visible light, due to mercury's other spectral lines. These "germicidal" lamps are used extensively for disinfection of surfaces in laboratories and food-processing industries, and for disinfecting water supplies.

1.3.4.8 Neon Lamp

1. **Principle:** "Electrons are emitted from cathode by subjecting the surface of the cathode to very high electric stress so that electrons are forcibly ejected from orbits of atoms against nuclear centripetal force".
2. **Construction and Working:** It has two closely spaced electrodes sealed within a soft glass envelope containing neon, at pressure of 8 torr (1 torr = 133.322 Pascal), (1 torr = 1 mm of Hg).
   - In these, cathodes are heated for emission of electrons and so they are said to have cold cathode but requires very high voltage depending upon distance between the electrodes.

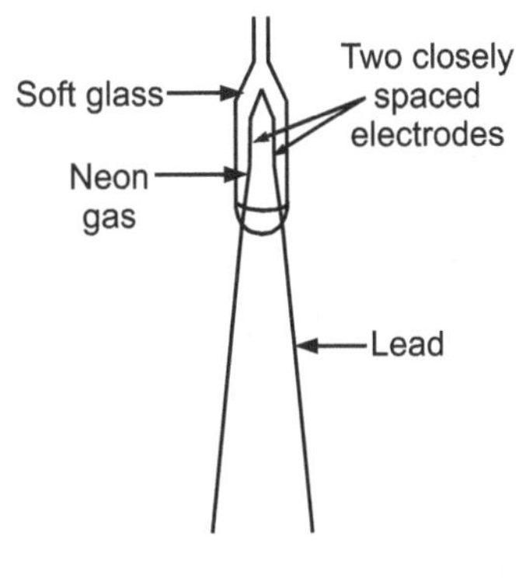

(a) Miniature Neon Lamp

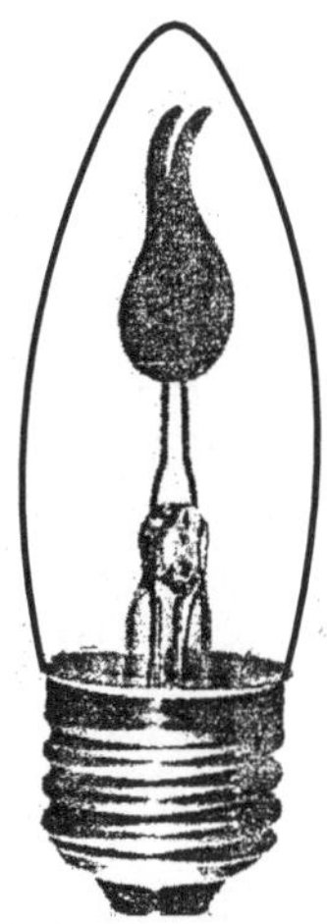

(b) Flicking Neon Lamp

Fig. 1.32: Miniature Size Neon Lamp

- To built up electric stress i.e. to discharge neon gas starting voltage of about 1.5 to 2 times of the normal running voltage.
- Electric field strength near the cathode is very high and ions under the influence of stray field will gain lot of momentum which will cause cathode sputtering or disintegration.
- In a closely spaced electrode system, where the negative glow emits light, a series resistance limits the lamp current, the discharge source having a negative flow characteristic.
- Too low a current will give flickering of the discharge and a high current will give a bright stable discharge, but life will be shorter and blackening of the envelope will occur. Lamp current is therefore specified to identify a standard or high brightness lamp being less than 1 mA or upto about 2.5 mA respectively.
- If A.C. voltage is applied, each electrode will turn +ve and –ve alternatively continuous glow is achieved.
- Thus such cold cathode lamps are suitable for high operating voltage and low current.

Advantages:

- It has instant glow.
- Available in miniature configuration.

Applications:

- It is used as indicator source as unit cost is low.
- Display applications, letters, numericals.
- Simulation of candle flame in flickering lamp.
- Red light for hazardous warning beacons on tall structures.

1.3.4.9 Neon Sign Tubes

Construction:

- In the form of tube even upto 9 m length and can be bend in manufacturing process to any shape/shapes.
- It is a glass tubing consisting of two electrodes normally cylindrical in shape of iron, steel or copper.
- It contains neon.
- Different colours are obtained by varying the composition of glass and adding different substances to neon gas different colours such as orange yellow, red, green etc. are obtained.
- Diameters of tubing – 10, 15, 20, 30 mm
- Carrying currents – 25, 35, 60, 150 mAmp. respectively.
- Operating voltage – 300 to 1000 volts/m of tube length.
- Starting voltage – 1.5 times the applied voltage.
- Voltage source – Step-up transformer of high reactance to get drooping characteristic.
- Operating voltage – 6000 volts.
- Supporting structure – Mounted on wooden frame or on metallic base.
  Matched with step-up transformer by connecting suitable tappings for the rated current.
- Connections between letters – Made by nickel wires and glass tubings being slipped over them.
- Power factor – Very low, but improved by condensers to be connected on low voltage side of transformer.

Flickering Drawbacks in Neon Sign Tubes and their Rectification:

Table 1.6

Drawback due to	Its rectification
1. Transformer secondary voltage is too low.	1. Can be rectified by adjustment of tap-changer.
2. Reduction in gas pressure in tube (due to absorption of gas in electrodes).	2. Remove tube, refilling by manufacturer.

Neon Tube Circuit

Precautions to be taken:

1. If installed in open - frequent cleaning is required, say 4 times in a year.
2. Keep the key of open lock system with operator while work is being done for safety. (Refer Fig. 1.33)

Applications of Neon Sign Tubes:

1. Advertising.
2. For signs.
3. Decorative treatment of buildings.

1.3.4.10 Metal Halide Lamp

- This lamp is known as mercury halide or metal halide lamp. It has a short arc tube. Inside it there are series of lamps combined within a single arc tube with output characteristics of different mixtures of elements. When such metals (like thallium, indium, sodium, dysprosium) are introduced into mercury arc tube, the mercury spectrum is suppressed reducing the ultraviolet radiation but added metal atoms partly compensate by causing a wide range of spectral bands to be emitted.

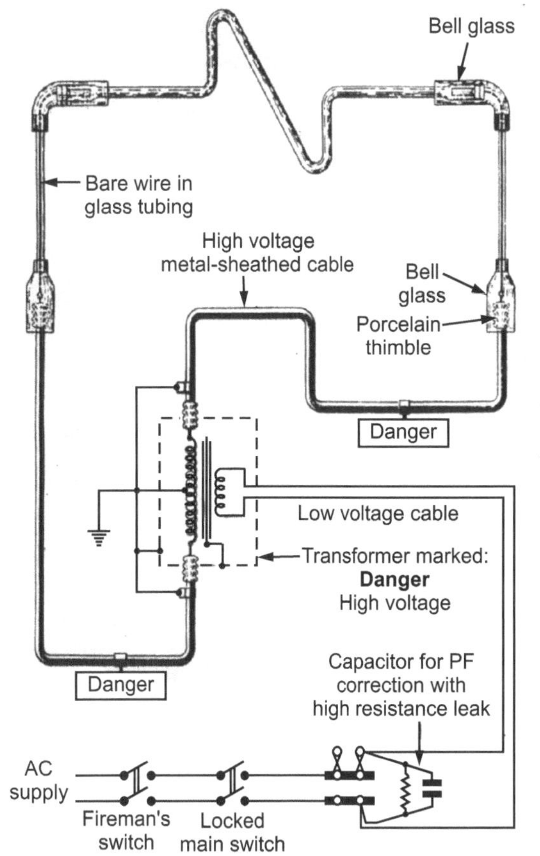

Fig. 1.33: Neon tube circuit with accessories

- It is possible to produce white or green light. If the metals are used in the element form they attack the discharge tube. To avoid this instead of metals in element form its halide is used like iodide i.e. sodium iodide. The lamps require high starting voltage. This is provided by ignitor as shown connector in Fig. 1.34. Ignitor is connected across the lamp terminals. This type of starter (ignitor) enables the lamp to be restruck within a minute of being switched off.

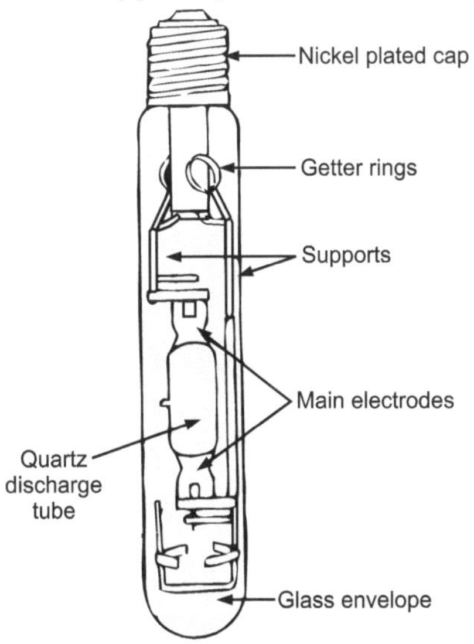

(a) Metal Halide Lamp

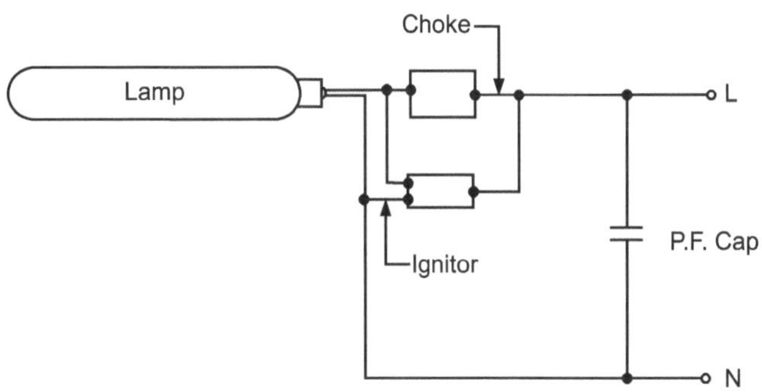

(b) Control Circuit with Ignitor for a Metal Halide Lamp

Fig. 1.34

Uses:

1. Used for high bay interior and high mast exterior installations.
2. Sports stadium where coloured T.V. transmission is expected.
3. Flood lighting projectors.

1.3.4.11 Laser

- **Long form of LASER is Light Amplification by Simulated Emission of Radiation.**
- When external energy is applied on an atom, electron jumps from lower energy level to higher energy level state.
    - It emits electromagnetic radiations and light.
    - Its colour depends on frequency and wavelength.
    - Phonon energy is released.
    - Afterwards electron goes to its previous lower energy level emitting photon in the process.
    - The laser beam intensity depends on size of laser rod, optical cavity, location reflectivity of mirrors.

Types of Lasers:

(i) Liquid lasers

(ii) Gas lasers

(iii) Chemical lasers

(iv) X-ray lasers

(v) Semiconductor lasers

(vi) Optically pumped solid state lasers.

Advantages:

(i) Laser travels very long distances.

(ii) Laser can be used for beautification.

(iii) Used in medical equipments.

(iv) Useful in metal processing industries.

Disadvantages:

(i) Much costly.

(ii) Safety precautions needed.

(iii) Can damage to organs.

(iv) Other equipments like operating equipments and control gears are needed.

Applications:

(i) For special light effects in media-work.

(ii) In medical field – for surgeries.

(iii) In defence applications.

(iv) Metal cutting, welding work.

(v) In computers, CD drive.

1.3.4.12 Comparison of Different Lamps

Sr. No.	Point of Comparison	Incandescent lamp	Fluorescent lamp	Sodium vapour lamp	Mercury vapour lamp	Halogen lamp	Compact fluorescent lamp	Metal halide lamp	Neon lamp	LED	LASER
1.	Working principle	Heating effect of electric current	Gas discharge	Gas discharge	Gas discharge	Heating effect of electric current	Gas discharge	Electric arc through gaseous mixture	Electron emission	Electro luminescence	Light amplification by simulated emission of radiation.
2.	Construction	Fig. 1.17	Fig. 1.22	Fig. 1.27	Fig. 1.26	Fig. 1.30	Fig. 1.31	Fig. 1.33	Fig. 1.28	Fig. 1.34	-
3.	Requirement of starting gear	No	Yes	Yes	Yes	No	No	No	Yes	No	Yes
4.	Starting time	Can start immediately	Requires some time (5 max. second)	Requires some time (10 mins.)	Requires sometime	Can start immediately	Can start immediately	Requires small time 3-5 seconds	Requires small time for starting	Can start immediately	Can start immediately
5.	Environ-mental friendly	Yes	No (due to mercury)	Yes	No (due to mercury)	Yes	No	Yes	No	Yes	Yes
6.	Life in hours	2000	12000	20000 to 30000	6000 to 1000	2000	800	10,000 to 20,000	25000	50000	Depends on type and application
7.	Lumens per watt	15 to 20	50 to 60	80 to 150	28 to 50	22 to 33	50 to 75	60 to 115	50 to 65	50 to 100	Depends on type

... *(Contd.)*

Sr. No.	Point of Comparison	Incandescent lamp	Fluorescent lamp	Sodium vapour lamp	Mercury vapour lamp	Halogen lamp	Compact fluorescent lamp	Metal halide lamp	Neon lamp	LED	LASER
8.	Colour of light	Yellow	White	Yellow (pale)	White (diffused)	Yellow	White	Yellow	Yellow	Available in many shades	Available in many shades
9.	Cost	Less	More	More	More	More	More	More	Less	High	High
10.	Heat produced during operation	More	Less	Moderate	Moderate	High	Moderate	Moderate	Moderate	Least	More
11.	Use	Residential indoor lighting	Residential commercial lighting	Street lighting	Street lighting, Garden lighting	Flood lighting, outdoor lighting	Residential, commercial lighting	High bay interior, sports stadium, flood lighting	Display indicators	Decorative lighting, residential, commercial, street lighting	Used for media lighting, metal process industries, medical instruments.

1.3.5 Selection Criteria of Different Lamps

Table 1.7

Illumination Areas	Suitable Lamp Selection
1. General house illumination.	1. (i) Incandescent lamps. (ii) Fluorescent lamps. (iii) Compact Fluorescent Lamps (CFLs) (iv) LED

contd. ...

2. Cinema projectors.	2. Arc lamp.
3. Search light.	3. Arc lamps.
4. Outdoor illumination of building.	4. Halogen lamps.
5. Large gardens.	5. Mercury vapour lamps, Halogen lamps, Sodium vapour lamps.
6. Car parks.	6. Halogen lamps.
7. Air port run ways.	7. Halogen lamps.
8. Street lighting.	8. Fluorescent tubes, Sodium vapour lamps, Mercury vapour lamps, LED.
9. Rail yards/Storage yards.	9. Sodium vapour lamps, HPMV lamps.
10. Industrial lighting.	10. HPMV lamps, Mercury iodide lamps.
11. Medial instrument	11. LASER
12. Shopping centre commercial places.	12. HPMV lamps, Green lighting, Fluorescent tubes, LED
13. Spots/Stadiums.	13. Metal halide lamps.
14. Indicator, night lamp.	14. Neon lamps.
15. Metal processing industry.	15. LASER
16. Advertising	16. Neon tubes
17. Decorative treatment of buildings.	17. Neon tubes, LED.
18. Flood lighting.	18. Halogen lamps, Metal halide lamps.

Thus, different lamps are selected for different purposes of illuminating the areas.

1.4 BASIC PRINCIPLES OF LIGHT CONTROL

- For different areas to be illuminated, different lighting systems are employed and different lamps are used (lighting sources are selected).
- For different areas like buildings, workshops, commercial places, streets, yards, sport grounds, proper light source is selected as listed in the table in chapter two.
- Refer table in chapter two regarding selection of light sources.
- Though proper lighting source is selected yet for various reasons, the light controlling methods are needed for the control of brightness and colours for the purposes.
- Basic principle of light control, mirror, reflection, diffused reflection is introduced in the following section.
- In controlling the sources, different dimmers like:
  (i) Resistance type dimmer
  (ii) Salt water dimmer
  (iii) Electronic dimmer

  are used. These are explained with circuit arrangements in the preceding sub-sections of this chapter.
- Also lamp connections and their controls by switches are explained with suitable circuit diagrams.
- Lighting control is an integral part of illumination engineering. The lighting control is used as per the aesthetic requirements and can change moods of human beings. The lighting control has to perform these functions in an efficient way.

- The lighting control is broadly classified into two categories.

Lighting control

Dimming control | ON/OFF control

- **Dimming control:** Dimming means smooth control of the light. The intensity of light is gradually increased or decreased as per the requirements by using different types of devices.
- **ON/OFF control:** The simplest form of ON/OFF control is one lamp controlled by one switch. The lamp is turned ON when required and it gives 100 % output and when it is turned OFF it gives 0% output. The lamp can be controlled from one switch or two switches or more than that. These circuits are discussed in detail in last part of this topic.
- The ON/OFF control is also possible by using sensors and relays. For example, occupancy sensors are used to detect the presence of a person in a room. If there is occupancy of a person the lights are turned ON by using relay and if person occupancy is not there, the lights are turned OFF.
- In the similar way LDR (light dependant resistance) sensors are used for street light control. The LDR senses light or illumination level. If the natural light or sunlight is below the prescribed value of illumination, then LDR circuit operate the relay and street lights are turned ON. And if the natural light or sunlight is above the prescribed value of illumination, the LDR circuit is operated the relay and street lights are turned OFF.

Basic Principles of Light Control:

- We have already defined the term reflection factor which is the ratio of reflected light energy to the incident light energy. When the light falls on the surface due to the nature of the surface, some light is reflected, some is transmitted through medium or surface and the rest is absorbed.
- The reflections are of two types:
  1. Mirror or Specular reflection
  2. Diffuse reflection.

1. Mirror or Specular reflection

- In this type as shown in Fig. 1.35 the light beam from the lamp is reflected but it is not scattered.
- The observer only sees the light when eye is in the path of reflected beam, otherwise he is not able to see it. He sees the light source. This type of surface is silvered mirrors or highly polished surfaces of metals.

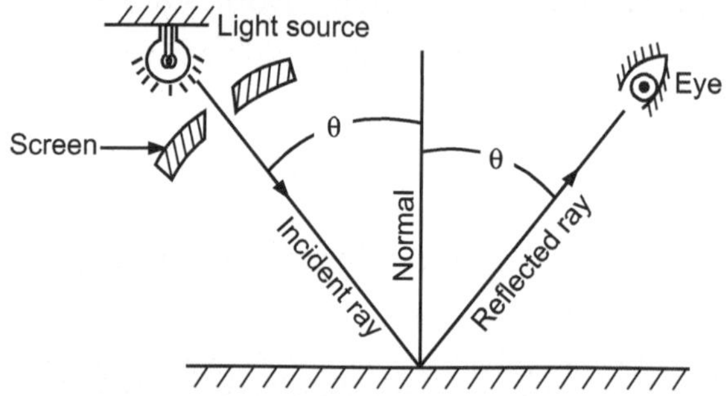

Fig. 1.35: Specular/Mirror Reflection

2. Diffuse reflection

- In this case, the reflected light is scattered in all directions. The viewer sees the illuminated surface but not the light source. Such surfaces are of the paper, frosted glass, dry earth, chalk, plaster etc.
- If the surface that is uniformly illuminated by a beam of light appears to be equally bright when viewed from possible angles, the deflections are said to be perfectly diffused.
- Perfectly diffused or perfect mirror surfaces are not possible in nature, but actual surface is partly mirror and partly diffused. The surface free from mirror is a mat surface.

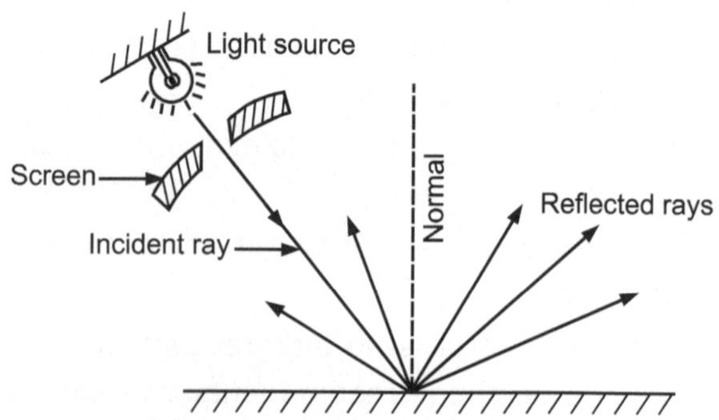

Fig. 1.36: Diffuse reflection

Generally following Illumination Levels are suggested for Different Locations:

- Efficiency of an electrical installation depends on the following factors:
  1. Type of lamp.
  2. Accessories.
  3. Cleanliness of bulbs.
  4. Reflection factors of walls, ceiling, furniture.
  5. Lighting system (Direct, Indirect, Semi).
  6. Space to height ratio.
- The light control is necessary for the following purposes:

1. Uniformity of illumination:

- If not uniform, then use the controlling circuit and adjust more or less luminaries.
- This can be adjusted by general lighting in addition to localized lighting.
- The modern trend is towards "localized lighting plus general lighting" and towards the adoption of "general lighting oriented towards the working surface" especially in mass production factory, office, drawing office, shops etc.

2. Colour of light:

- The appearance of the body colour entirely depends upon the colour of the incident light.
- Generally, the composition of the light should be such that the colour appears 'natural'.
- In other words, its appearance by artificial light is not appreciably different from the day light.
- Use of day-light fluorescent tubes is now a days increasing which economically illuminates large areas.
- For street light, discharge lamps are suitably used.

3. Shadows and glare are the hurdles in lighting:

- These can be avoided (Refer Chapter 2) for details.

4. Mounting height:

- This plays an important part.
- It is different for different types of lighting schemes.
- For large floor areas of rooms the luminaries are mounted close to ceilings. Lower heights will not give uniform illumination on the floor area, moreover a trouble of shadows and glare will be invited.
- In case of indirect/semi-indirect lighting systems, the luminaries shall be suspended far enough down from the ceiling to give uniform illumination.

5. Spacings of luminaries:

- Correct spacing has a great importance for uniform illumination. Space to height ratio is to be correctly selected.

6. Walls/Fencing walls/Ceilings:

- The colours of these are equally important factors in illumination as reflections from these play important part.

  Dull colour - inferior reflection

  White colour - good reflection
- The above different points are to be considered which are also the parts of illumination/light control in addition to the different methods of light control like dimmers.

Working Principle and Operation of Dimmer:

- **Different Types:** (1) Resistance type, (2) Salt water type, (3) Transformer dimmer, (4) Auto transformer dimmer, (5) Two winding transformer dimmer (6) Thyristor operated dimmer, (7) TRIAC operated dimmer.
- It is a device used to control the brightness of luminary by controlling electric power fed to the luminary. It is generally used in theatres, photo studios where special light effects are required.

Resistance Type Dimmer:

- The resistance dimmers were most popular and were extensively used till 1980. The main advantage is simplicity of the circuit. The resistance dimmer consists of a rheostat connected in the lamp or luminary circuit. By varying the resistance, the current is controlled and thus brightness of the lamp can be controlled.
- Fig. 1.37 shows the details of resistance dimmer. A variable resistance is arranged on a former as shown. The sliding contact is made by means of carbon brushes. The lamp is connected in series with this resistance.

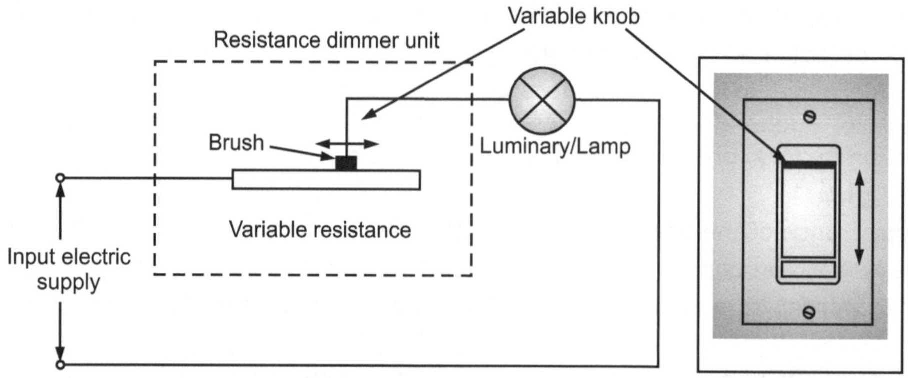

(a) Resistance dimmer circuit **(b) Photograph of resistance dimmer**

Fig. 1.37

Advantages:

- Simple operation.
- Maintenance required is less.

Disadvantages:

- Considerable amount of power is wasted in the resistance.
- Proper heat dissipation arrangement is necessary for the resistance.

Salt Water Dimmer:

- It makes use of salt water for controlling the brightness of luminary/lamp. Fig. 1.38 shows the arrangement of salt water dimmer.

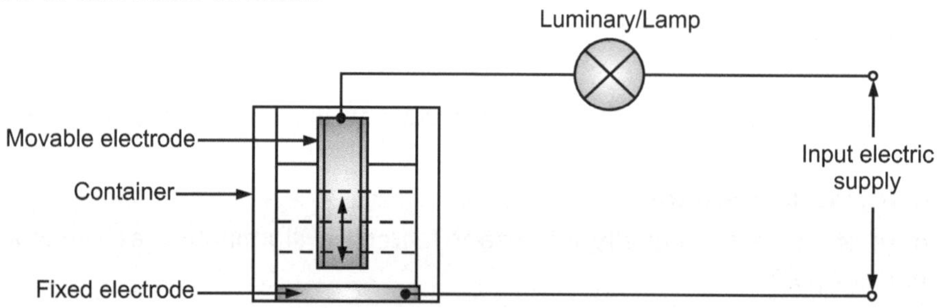

Fig. 1.38: Salt water dimmer

- Salt water dimmers rely on salinity and distance between metal contacts to control the brightness of the luminary/lamp. It consists of a container in which salt water is poured. There is one electrode fixed at the bottom and the other electrode can move up and down. The electric supply is connected across these electrodes. Movable electrode is connected in series with the lamp.
- The electric power fed to the lamp can be controlled by the movement of this electrode i.e. by increasing and decreasing its distance from the fixed electrode. As the movable electrode moves upwards, the distance between the two electrodes increases. The resistance between these electrodes increases, therefore voltage applied to the luminary increases, hence the brightness of the lamp decreases.

- As the movable electrode moves upwards, the distance between the two electrodes increases. The resistance between these electrodes increases, therefore voltage applied to the luminary decreases, hence the brightness of the lamp decreases. The closer the contacts are to each other, the brighter the luminary. The effectiveness of the control depends on the smoothness of the movement of raising and lowering the movable electrode.
- Fig. 1.39 shows salt water dimmer used in old days. It consists of number of wheels arranged on a common shaft.
- The wheels can be rotated by moving the control rod. As the control rod is moved, the motion of wheel is transferred to the belt. The moving electrode is connected to the belt, thus moving rod also moves. The motion of the moving rod can be controlled with control rod.

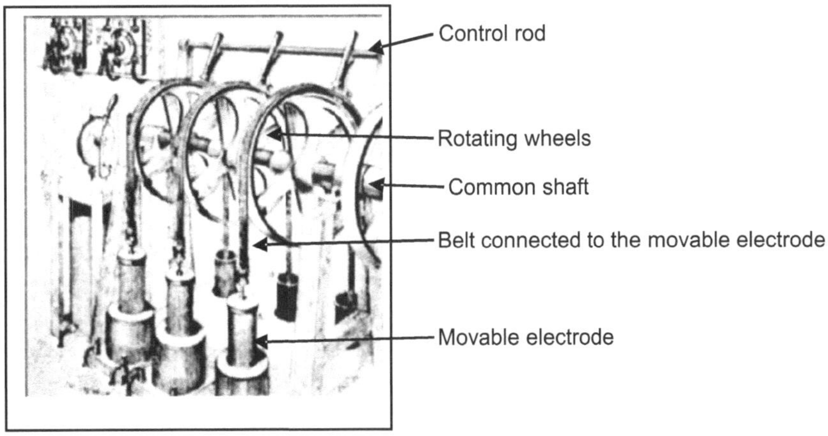

Fig. 1.39: Photograph of salt water dimmer

Advantages:

- Smooth control of brightness.
- Better control.

Disadvantages:

- Operation is not safe. Leakage current can be present.
- The current can produce toxic chlorine gas or flammable hydrogen and oxygen due to chemical reaction in salt water.
- Considerable attention is required to keep the water level constant.
- Operating mechanism has to be designed for precise movement to get better control.

Transformers and their Types

Dimmer Transformer:

- The illumination control or dimming can be obtained by varying the voltage applied to the luminary. The variable voltage can be obtained from transformers which are known as dimmer transformers. These are used in theatre lighting. The dimmer transformers are of two types:
  - Autotransformer dimmer
  - Two winding transformer dimmer

Autotransformer Dimmer:

- Autotransformer is a single winding trans-former. The luminaries/lamps (L_1, L_2, L_3, L_4) are connected to the secondary side of the auto-transformer. Fig. 1.40 shows the arrangement. If the illumination level is to be increased, the voltage applied to the luminaries should be more, the variable knob of the autotransformer is moved upwards and the voltage applied to the luminaries increases, thus brightness will be more.

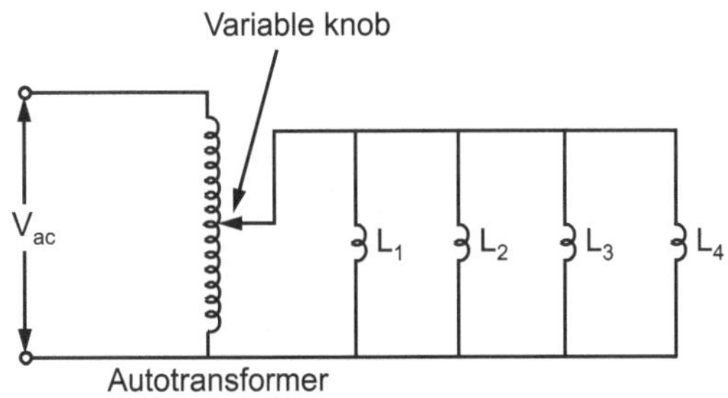

Fig. 1.40: Autotransformer dimmer

- These are widely used in theatres for light control. The autotransformer rating should be carefully selected.
- The advantage is that single device can control number of luminaries.

Two Winding Transformer Dimmer:

1. List different dimmers used in illumination control. Explain the operation of any one type.

- The two winding transformer dimmer is of two types:
  - Normal two winding transformer dimmer.
  - Coil-rotation transformer dimmer.

Normal Two Winding Transformer Dimmer

- Normal transformer is used to control the illumination. The primary winding is given AC voltage, the secondary consists of number of tapings. So a variable voltage is available across the secondary depending upon the selection of tappings. The luminaries are connected across the secondary, the dimming can be controlled by applying variable voltage.

Coil-Rotation Transformer Dimmer

- The details of coil-rotation transformer dimmer are shown in Fig. 1.41. It consists of two coils or windings which are kept close to each other. The AC supply is connected to the primary winding.
- The primary coil is connected to motorized system. The primary coil can be rotated. Due to rotation, the flux linkage between primary and secondary changes.
- When the coil axis are parallel to each other, the flux linkage is maximum. Thus maximum e.m.f. is induced in the secondary coil. The luminaries receive maximum voltage and glow bright. If the primary coil is rotated such that its axis is 90° to the secondary coil, the flux linkage is minimum. Thus minimum e.m.f. is induced in the secondary coil. The luminaries receive minimum voltage and glow dim. For the angular positions between 0° to 90°, the brightness increases gradually. The motorized set can be operated through remote control.

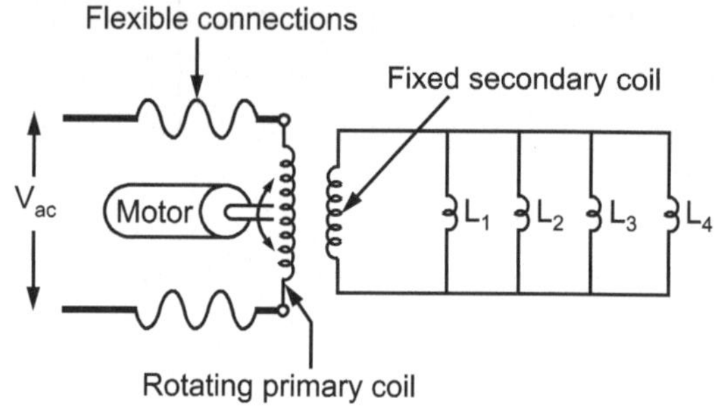

Fig. 1.41: Coil-rotation transformer dimmer

Electronic Dimmer:

- These make use of solid state devices such as SCR (silicon controlled rectifier), TRIAC, DIAC to control the intensity of lamp. The power loss is reduced to great extent. These are also compact. Some circuits using solid state devices are discussed below.

Thyristor Operated Dimmer:

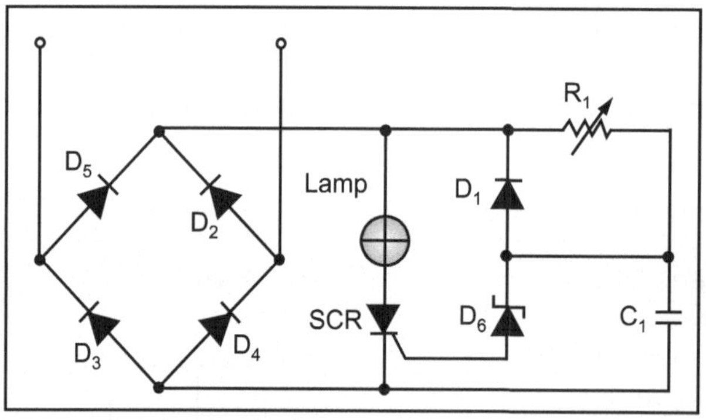

(a) Thyristor operated dimmer circuit

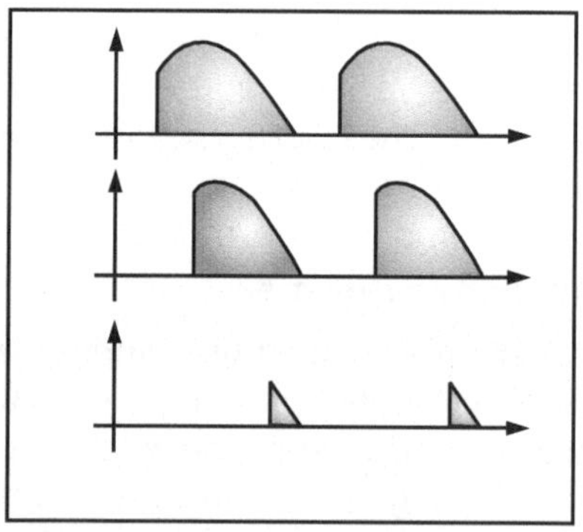

(b) Waveforms of controlled voltage applied to the dimmer

Fig. 1.42

- These make use of thyristor or SCR. The thyristor controls phase angle i.e. voltage applied to the luminary. In case of resistance dimmers, considerable power loss takes place but in this type of dimmer the power loss is minimized.
- The circuit consists of diodes D_2, D_3, D_4, D_5 forming bridge rectifier. When rectified output voltage of bridge rectifier increases, capacitor C_1 is charged. The voltage across the capacitor increases till it causes zener diode D_6 to conduct.

- When zener diode conducts, the SCR is turned ON. The capacitor C_1 now discharges through SCR. The SCR is turned OFF when this current comes to zero. The combination of R_1 and C_1 decides the conduction current of SCR. Thus voltage applied to the lamp is varied, hence the lamp intensity can be controlled.
- In Fig. 1.42 (b) the sample waveforms of voltage applied to the lamp are shown. For first case the brightness will be maximum and for third case it will be minimum since voltage applied to the lamp is very less.

TRIAC Operated Dimmer:

1. Explain TRIAC operated dimmer for light control.

- TRIAC operated dimmer circuit is shown in Fig. 1.43. The lamp intensity is controlled using TRIAC and DIAC.
- The resistance R_1 and capacitor C_1 are connected in series. The capacitor is made to charge through the R_1 resistance. When the voltage across reaches above the breakover voltage of DIAC, it starts conducting. The voltage across the capacitor now gets applied to TRIAC gate and the TRIAC is turned ON. The charging of the capacitor can be controlled by the variable resistor R_1, which indirectly controls the firing delay of TRIAC. Thus the intensity of the lamp can be controlled.

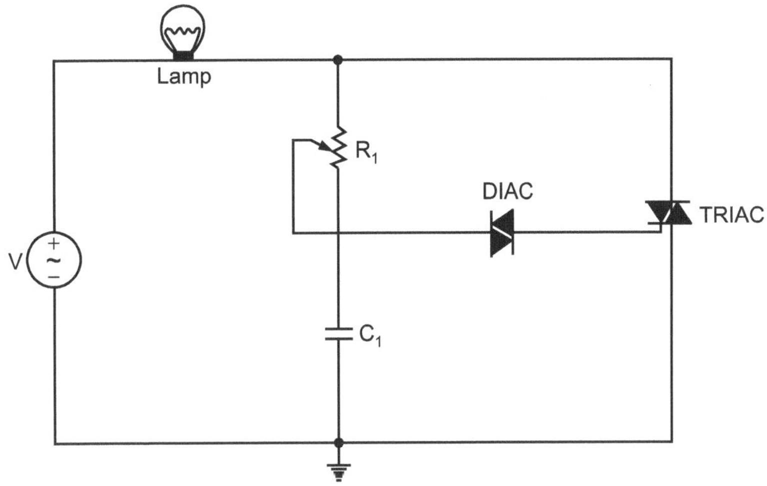

Fig. 1.43: TRIAC operated dimmer

Control of Enhance Lighting:

Refer previous sections.

Methods Used for Light Control:

- These methods are required for different types of discharge lamps. The methods are used for (i) Ballast, (ii) Starter, (iii) Igniter, (iv) Holding devices, (v) Starter for low pressure lamps, (vi) Capacitors for power factor improvement.

Control Circuits for Lamps:

1. Explain with neat sketch the following lighting control circuits. (i) Single lamp controlled by single switch.

(i) One lamp controlled by one switch:

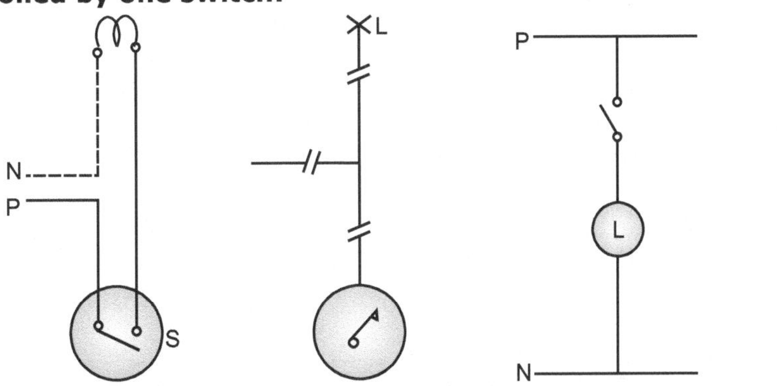

Fig. 1.44: (a)

(ii) Two lamps controlled by two switches (one lamp controlled by one switch separately):

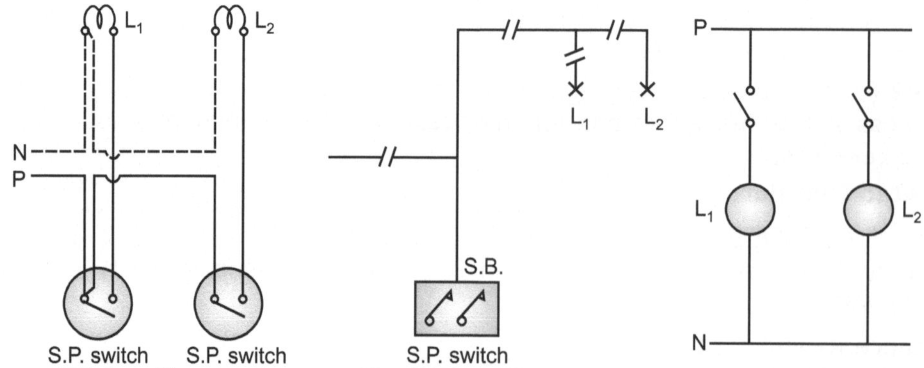

Wiring diagram **Single Line diagram** **Schematic diagram**

Fig. 1.44: (b)

(iii) Control of two lights and one fan by their individual switches, show control on one board:

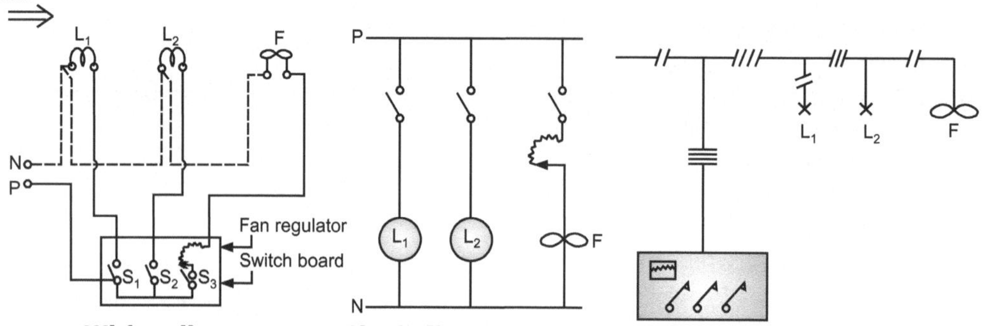

Wiring diagram **Single line diagram** **Schematic diagram**

Fig. 1.44: (c)

(iv) Control of one lamp and one plug point (three pin socket) on one board and one bell from another board, each point controlled separately by its switch:

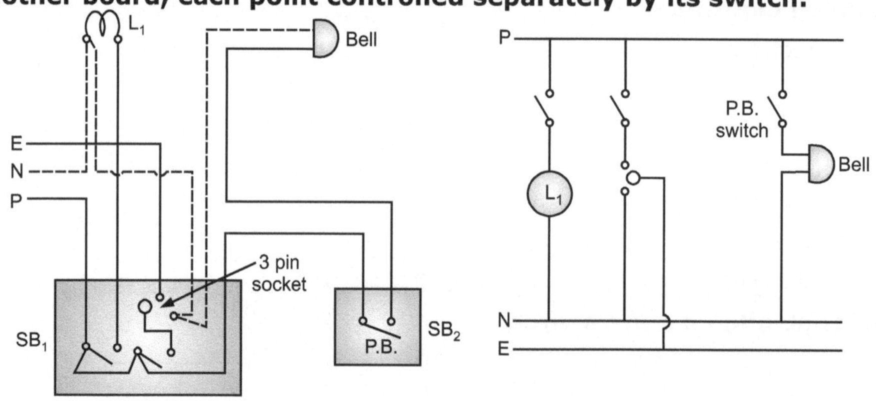

Wiring diagram **Schematic diagram**

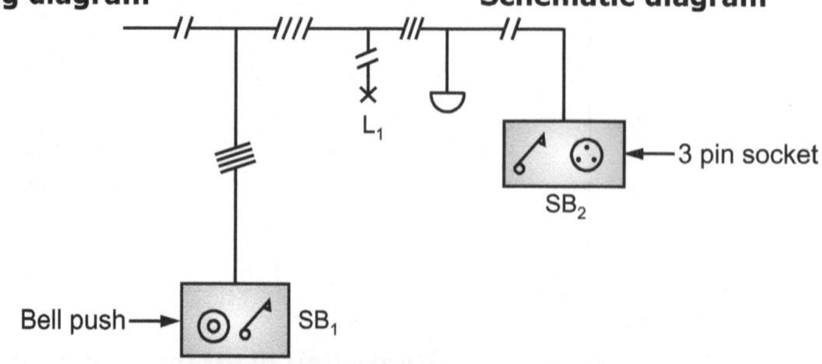

Single line diagram

Fig. 1.44: (d)

Single Lamp Controlled by Two Point Method, Three Point Method and Four Point Method:

1. Explain with neat sketch the following lighting control circuits:
   (i) Single lamp controlled by single switch.
   (ii) Single lamp controlled by two point method.
   (iii) Single lamp controlled by three point method.
   (iv) Auto transformer dimmer.

Series parallel circuit: Such type of connections are obtained by the following ways.

(i) With two-way and off switch: In this type of connection in one position of switch, one lamp will get full voltage, hence it is fully bright. In the second position, the switch is in off position, hence, no lamp is lighted. While in the third position both lamps are connected in series across the supply, hence voltage is divided across them and they are dim.

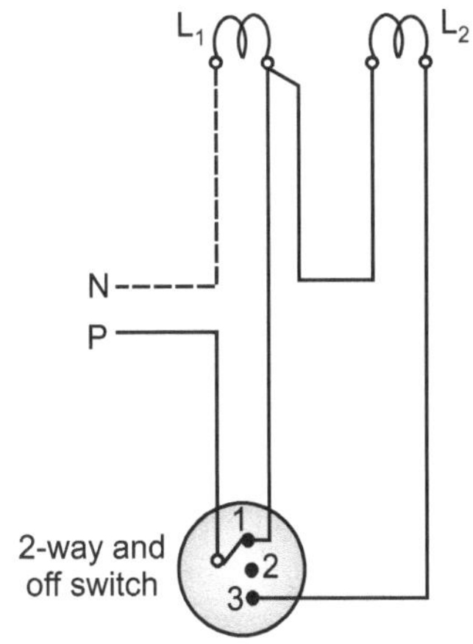

Fig. 1.44: (e-i)

(ii) With one S.P. switch and one two way and off switch: In this case, in position one lamp L_1 is full bright, in position 3 of switch contact lamp L_2 is full bright, while in position 2 both lamps are connected in series, and are dim. In all cases, S.P. switch is assumed to be in on position.

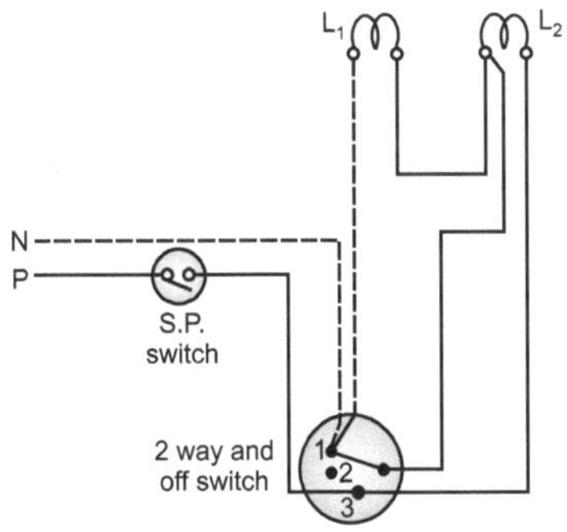

Fig. 1.44: (e-ii)

(iii) One lamp controlled by two-way switches or stair-case wiring: Lamp can be made 'on' or 'off' from two different places where two-way switches are fitted i.e. on ground floor and on first floor.

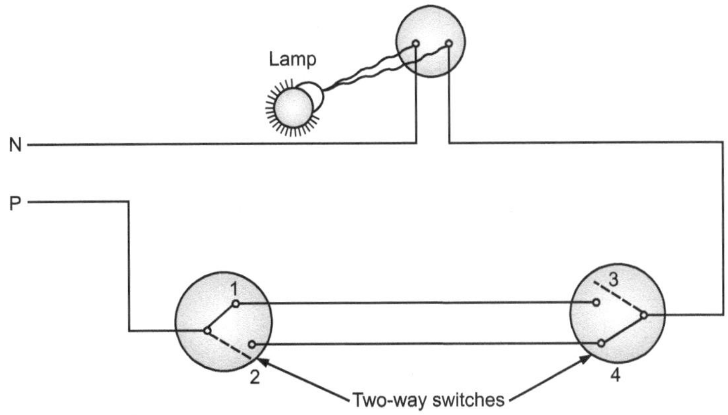

Fig. 1.44: (f)

(iv) Corridor lighting: In this system, every lamp is having two controlling switches, one will be in corridor and the other inside the bungalow.

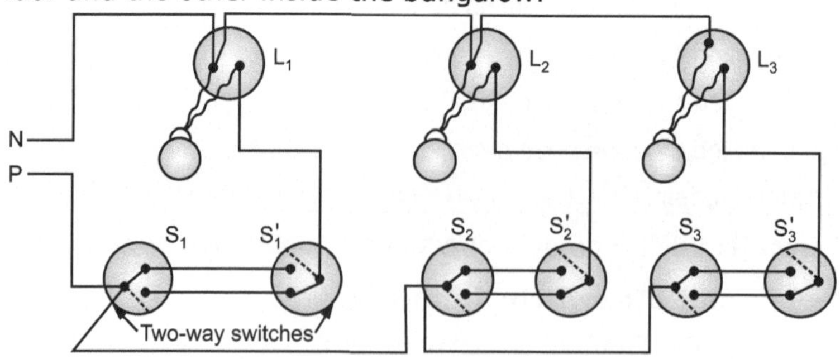

Fig. 1.44: (g)

(v) Lamps in series and parallel: In upper position of switch, the lamps L_1 L_2 are in series and in down position, L_1 and L_2 are in parallel.

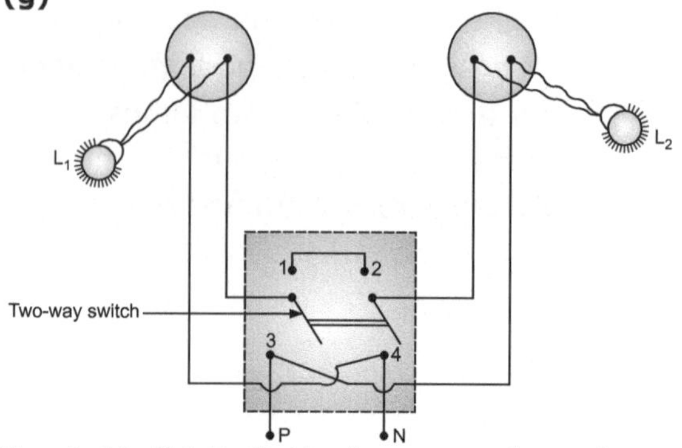

Fig. 1.44: (h) Both the lamps are in series or both the lamps are in parallel

Control Circuits for Lamps:

(i) Godown wiring: S_1 is 'on' position, L_3 is lighted. Then S_3 is changed, L_2 is lighted, L_3 becomes 'off'. Then S_2 is operated, L_2 is off and L_1 is lighted. Then S_1 is 'off' and all lamps 'off'.

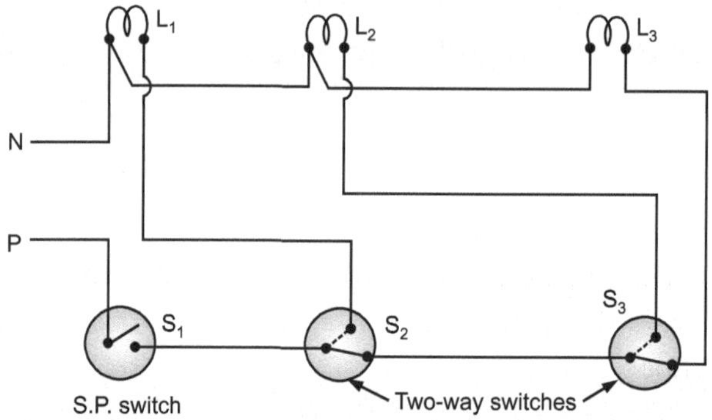

Fig. 1.44: (i)

(ii) Master on control: By master 'on' switch, all lamps are lighted and No 'off' by other switches S_1, S_2, S_3.

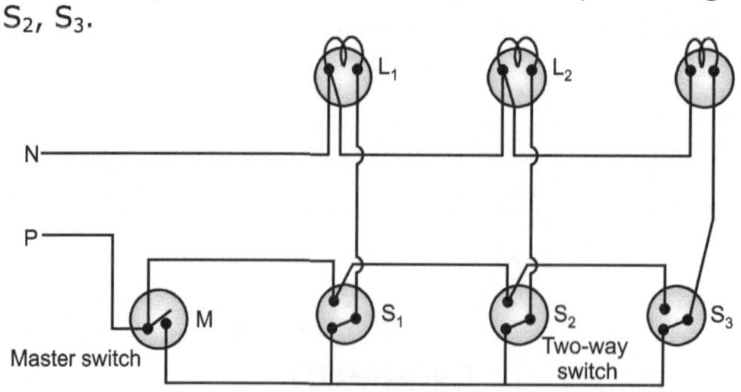

Fig. 1.44: (j)

(iii) Master off circuit:

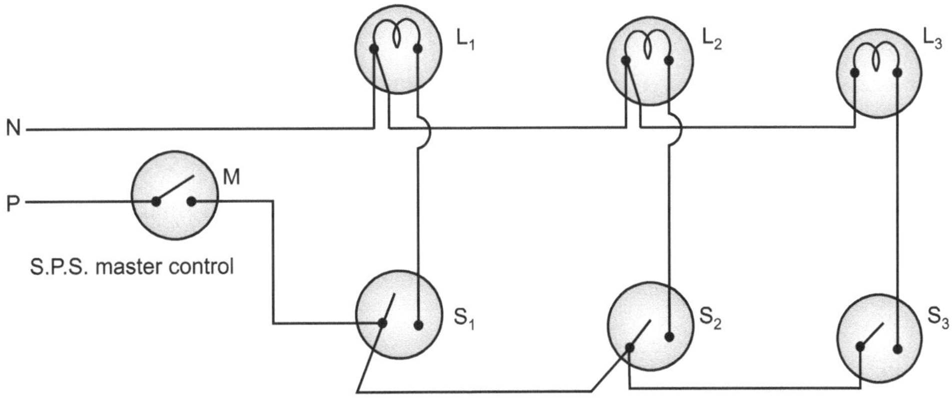

Fig. 1.44: (k) By master 'off' switch No - On by S_1, S_2 or S_3

(iv) Arrangement of single-phase supply at the premises of consumer: Shown below EM – Energy meter, F – Fuse, ICDP = Iron Clad Double Pole Switch.

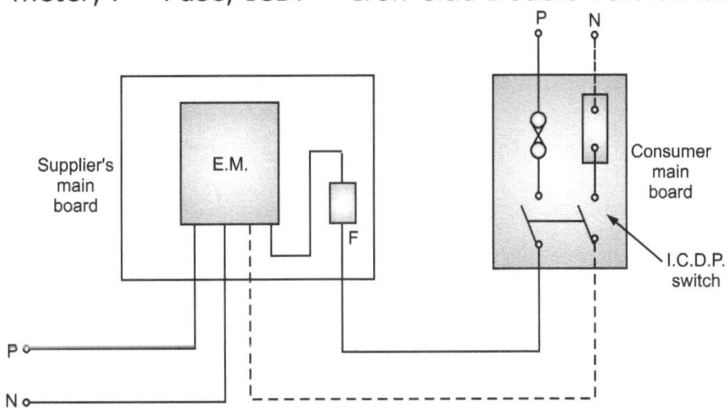

Fig. 1.44: (*l*) Supply at consumer's premises

(v) Wiring diagram showing connection of supplier's main board, consumer's main board with D.P. MCB and single pole MCB showing sub-circuits.

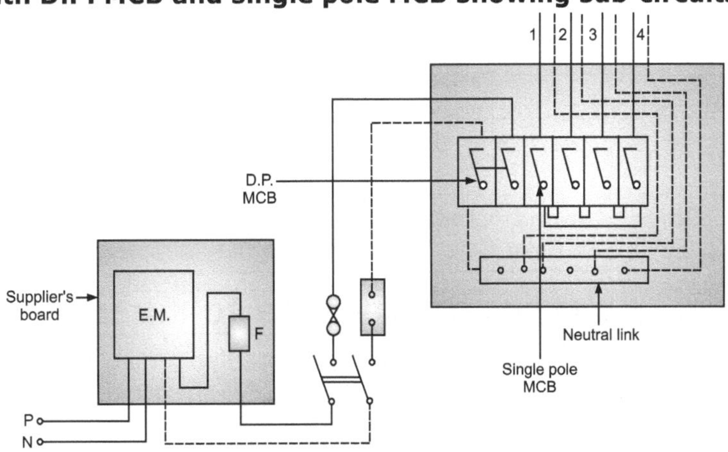

Fig. 1.44: (m)

(vi) Use of two-way switches and intermediate switch to control lamp from three points (places):

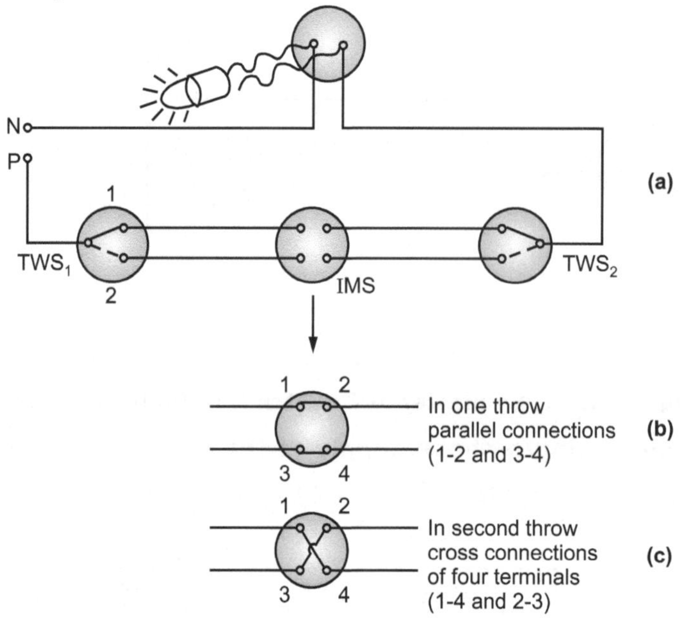

TWS_1, TWS_2 =Two-way switches, IMS = Intermediate switch, L = Lamp

Fig. 1.45

- Fig. 1.45 (a) shows that lamp 'L' is 'ON'.
- This can be made OFF by changing the throw of any of the three switches SW_1, SW_2 or IMS.
- Similarly, it can be made 'ON' by any of the three switches.
- **Control from 4 points (places):** This is also possible by using 2 numbers of two-way switches and two numbers of intermediate switches (Students can develop above diagram for this operation).

Differences between Dimmer Control and ON/OFF Control:

1. Explain differences between dimmer control and ON/OFF control in lighting control.

Dimming control	ON/OFF control
1. Light control very smooth and uniform.	1. Control is by steps, so not so uniform.
2. Cost is very high.	2. Chear in cost.
3. Dimmer transformer and electronic circuit needed.	3. Only switching ON and OFF of series and parallel combination of lamps.
4. Applications - Stage, drama, theatre, decorative places, exhibition halls, etc.	4. Its use is very limited.

1.5 TYPES OF LIGHTING SCHEMES

- For various purposes different areas are to be illuminated providing different lighting schemes.
- The different types of lamps studied so far are selected for different illumination lighting schemes.

Types of Lighting Schemes:

- As studied in the previous topics, if the light falls on the polished surface the light is reflected back. This is good for illuminated but it produces glare on the eyes which is harmful.
- If light falls on the coarse surfaces like painted ceilings, frosted glass or paper the light is diffused in all directions so no effect of glare on the eyes. So this method is used for internal or external lighting.

Types of Lighting are:

1. Direct Lighting
2. Semi-direct Lighting
3. Semi-indirect Lighting
4. Indirect Lighting
5. General Lighting.

1. Direct Lighting:

- By this method, most of the light source is made available on the working surface and very few percent is wasted. The light energy is lighting sources, are hung and light is diverted by proper shaped reflectors or globs. So number of light units needed are less and overall cost of the illumination scheme is reduced. Though this system is most efficient, the drawback of the system is that dark shadows fall on the working plate, moreover, there is a glare on the eyes.

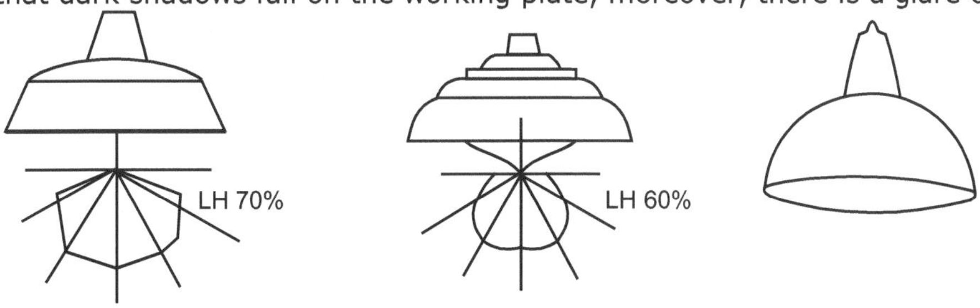

(a) Dispersive Type Reflector **(b) Diffused Fitting** **(c) Concentrated Reflector**

HL – Lower hemisphere **UL – Upper hemisphere**

Fig. 1.46: Direct Lighting

2. Semi-Direct Method:

- In this scheme, about 60 to 90% of the total light flux is made to fall on the working surface and 10 to 40% is allowed to fall on the ceiling and walls. This is achieved by providing semi-direct reflectors. Such a scheme is best suited to rooms having bigger heights. Glare is avoided and diffused bulbs are used.

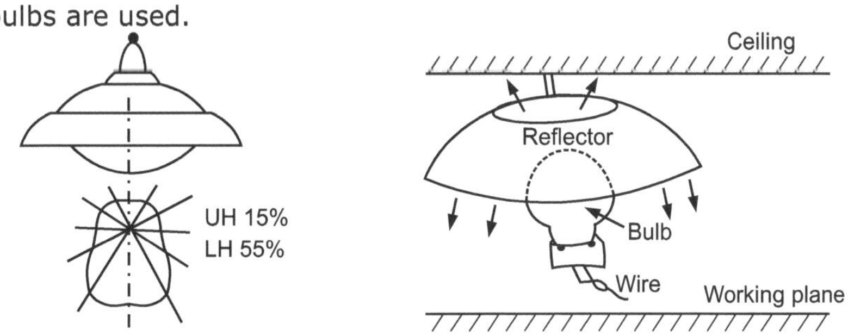

(a) Made of Opal Glass **(b)**

Fig. 1.47

3. Semi-Indirect Lighting:

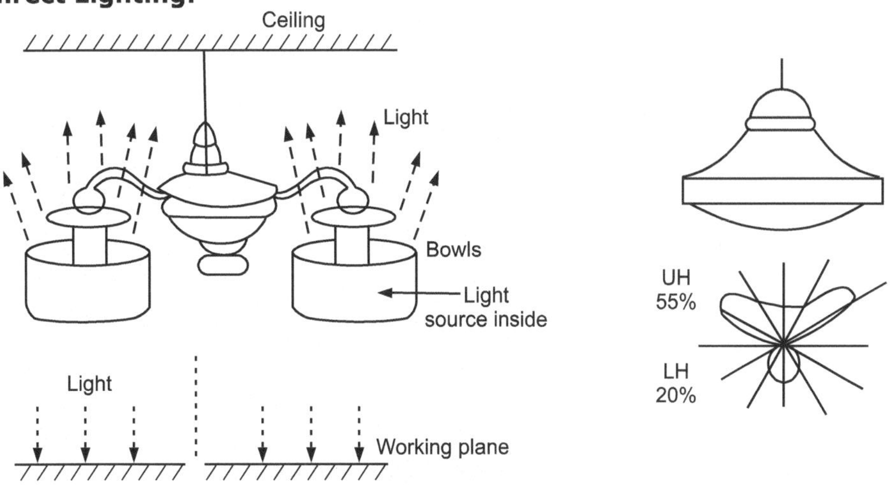

(a) (b)

Fig. 1.48: Semi-Indirect Lighting

- In this system, 60 to 90% of the total light flux is diverted to fall on the ceilings from where the light is directed on the working surface by diffused reflection. Only 30 to 40% flux reaches the working plane. Some is absorbed by the bowls. This provides pleasant lighting, free from glare and shadows are very very soft. Scheme is costly. But it is used for indoor decoration purposes. Bowls are semi-transluscent type.

4. Indirect Lighting:

- The light does not reach the working surface directly. The maximum light (about 90%) is thrown towards the ceilings for diffused reflection by using inverted reflectors or bowl reflectors. As if the ceiling acts as a light source therefore the glare is reduced to minimum value. Illumination is pleasant, diffused and very soft and shadows are illuminated. Scheme is decorative. This scheme is suggested for function halls, cinema theatres, three star hotels. Also suggested for big workshops in the industries to avoid accidents due to shadows.

  The scheme requires more light sources and hence costly.

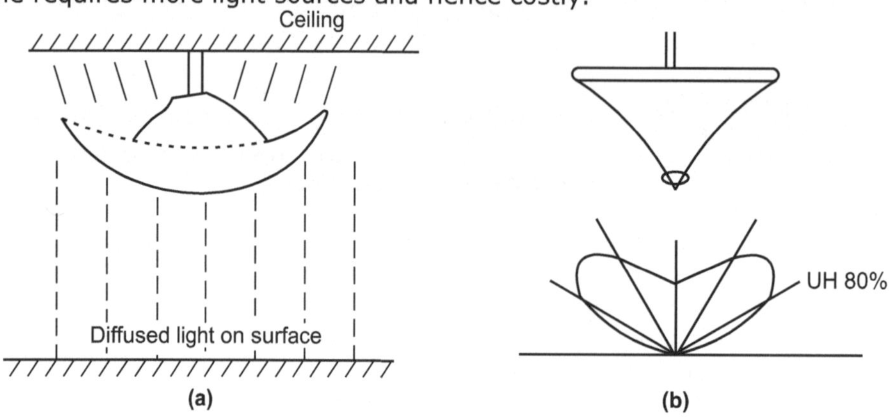

Fig. 1.49: Indirect Lighting

5. General Lighting:

- Special diffusing glasses are used for lamps which sprade the light uniformly on the working plane. The light illumination is equal in all directions.
- Following Fig. 1.50 gives the idea of all above five types in a schematic diagram with percentage division of light illumination.

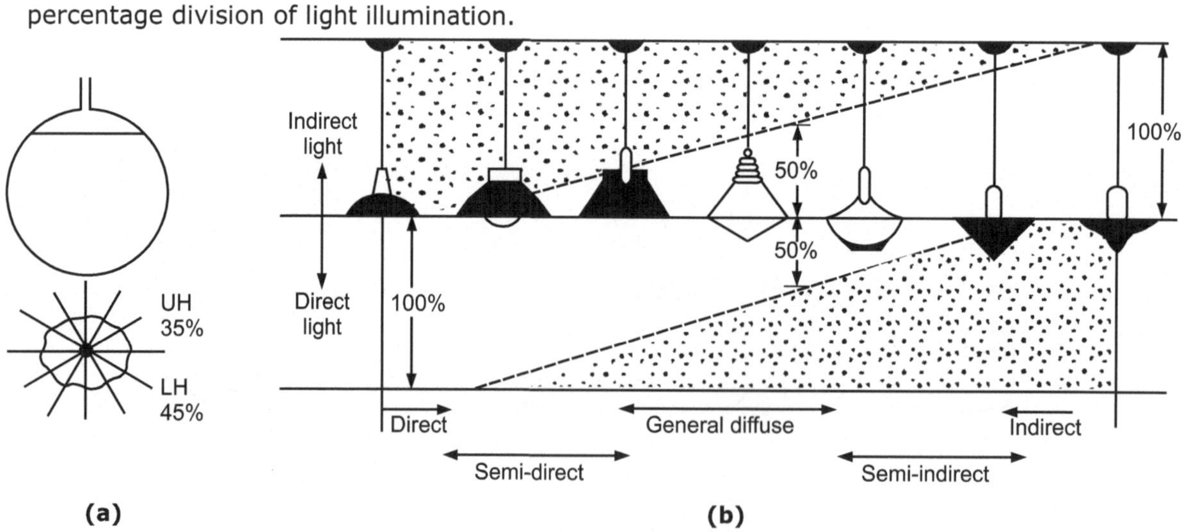

Fig. 1.50: General classification of Luminaries

1.6 DESIGN OF LIGHTING SCHEMES

1.6.1 Objectives of Lighting Scheme

1. **To provide adequate illumination:** For different purposes, different levels of illumination are required that requirement shall be fulfilled.
2. **To avoid glare and hard shadows:** Disposition of lamps and the method such as direct or indirect shall be adopted such that hard shadows and glare on the eyes shall be avoided.
3. **To provide light of suitable colour:** Different lamps emit light of different colours, generally whitish day-light colour is desirable.
4. **To provide light distribution uniformly over the working plane:** Disposition of the lamp shall be proper to achieve this objective.

1.6.2 Factors to be Considered while Designing the Lighting Scheme

1. **Illumination level:**

- For different types of works different ranges of brightness are required so that the eye can visualize the objects in shape, size, appearance, colour etc. The degree of illumination to give necessary brightness to the objects depends on:
  1. size of the object to be seen
  2. distance of the object from the observer.
- For smaller size of objects greater illumination shall be provided. If the distance is much more then illumination level required is much more. Similarly, greater the contrast between objects and its background, greater will be the illumination required. For causal work less illumination is sufficient but for objects to be seen for a length of fine then sufficient illumination shall be provided like industry working, drawing office etc. Moving objects require greater illumination to visualize than the stationary objects. So illumination level shall be provided properly. This factor is most important in illumination design.
- Following levels of illuminations are suggested:

Place	Illumination required
1. Living room, bed room, table games, kitchen sink, mirror, study room	= 300 Lux
2. Dressing table, kitchen, laundry, workshop	= 200 Lux
3. Dining room	= 150 Lux
4. Entrance, recreation room, bathroom, staircase	= 100 Lux
5. Garage	= 70 Lux
6. Sewing, precision work, stitching	= 700 Lux

2. **Uniformity of Illumination:**

The human eye adjusts itself automatically to the brightness within the field of vision. If there is a lack of uniformity the pupil of the eye has to adjust frequently. This creates fatigue to the eye. This reduces productivity. General lighting is provided irradiation to localized lighting to get good results. If lighting is not uniform then it may create psychological feelings, of loneliness, unfriendliness and gloom. In mass production units, drawing offices, general offices therefore need localized lighting plus general lighting scheme to be provided to avoid drawbacks.

3. **Colour of Light:**

The appearance of the object colour – solely depends upon the colour of the light incident on the objects. Therefore, care shall be taken to provide the light colour which will create the natural appearance of the object bodies. Now-a-days newly invented tube lights which give the natural day lights are used for illumination of large spaces. For street lights, highly efficient lamps are sufficient.

4. **Shadows:**
Illumination level and type should be such that long and hard shadows be avoided which disturb the working and accident may take place in production of engineering factory. But it is also not desirable the scheme shall provide shadowless illumination. Because, soft shadows are necessary which help to give shapes to the solid objects and make them easily recognized. In shadowless lights, objects appear to be flat and uninteresting, colours are lost and it becomes difficult for the eyes to form a correct judgement of the shape of the objects. For the work of drawing, shadowless light is expected as the drawing paper is flat.
Long and hard shadows can be avoided by using:
(i) Large number of small luminaries mounted at the height nearly 2.5 metres.
(ii) By using indirect lighting systems or use of globes over filament lamps.
5. **Glare:**
It may come directly from the light sources or indirectly by reflection from desk of shining surfaces or from calendered papers. Direct glare causes uncomfort and reflected also causes some less effect. Metal reflectors in the industries shall be provided with a skirt around the rim of the reflector.
6. **Mounting Height:**
Height is different from different shape and size of the buildings. If the direct light system is used to provide illumination to a room of large floor area, the luminaires shall be mounted as near as possible to ceilings. If in such cases the luminaires are discended more than they may come to the level of eyes and create glare problems more over the light may not be uniform on the working surface. In case of higher ceilings, the use of filament lamps is suggested lowering them to some extent and fitted with focussing reflectors.
In case of indirect and semi-indirect systems, the luminaires are suspended sufficiently to result in getting uniform light on the working plane reflected from ceilings. Space to height rooms, the light on the ceiling is unavoidably ununiform which can be ignored.
7. **Spacing in Luminaires:**
This is the important point to be taken into consideration for getting a desirable uniform lighting on the working plane, otherwise mis-spacing may result darkness at some portion of the area. Space to height ratio shall be properly selected, such as unity to 0.6 for fluorescent tubes and for incandescent lamps it should be approximately 0.6 to 0.65. This ratio is for the system of direct and semidirect type. But in case of indirect lighting system, in no case horizontal spacing shall exceed 1.5 times the height of lamps from the working surfaces.
8. **Colours of Surrounding walls:**
As shown in the table of extent of reflection as per the colours, the walls and ceilings shall be white painted or faint green painted to have proper reflection of light.

1.6.3 Domestic and Industrial Lamp Fittings

- Light sources used in such systems:
  (i) Fluorescent tubes with electronic choke.
  (ii) Fluorescent tubes with chokeless.
  (iii) Compact fluorescent tubes (CFL).
  (iv) Triphosphere Fluorescent lamps.
  (v) LED Lamps/Tubes/Newly invented system.
- Various fittings and accessories with such systems are:
  (i) Pendants
  (ii) Floor mounted
  (iii) Wall brackets
  (iv) Ceiling mounted
  (v) Corner attached
  (vi) Table mounted
  (vii) Using fiber optics.
- As explained in Section 1.6 various lighting systems are used in domestic and industrial places.

- However newly designed luminaries and their location and fittings (sketches) are shown below:

Sr. No.	Sketches	Description	Uses
1.	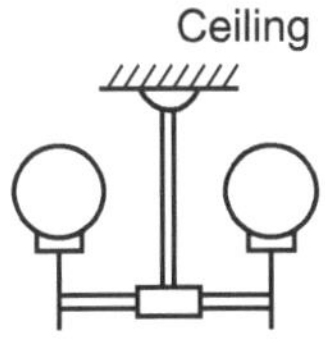 **Fig. 1.51 (a)**	**General Diffused distribution:** Dimmer can be provided to change illuminance	(i) For functions. (ii) Decoration purposes.
2.	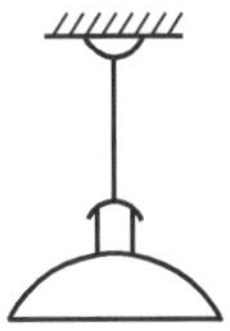 **Fig. 1.51 (b)**	**Direct-Indirect type:** Opaque or Luminous shades or shields provided. Provide upward light. Height from working space/table adjustable.	(i) Room lighting. (ii) Functional lighting. (iii) For dining tables. (iv) Game tables.
3.	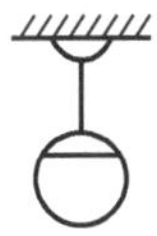 **Fig. 1.51 (c)**	**Direct-Down lighting:** These are located at or below eye level. Used to provide ascending illumination.	(i) In stair-ways. (ii) At entrance. (iii) In Halls along with general lighting.
4.	 **Fig. 1.51 (d)**	**Exposed Lamps:** Various attractive and decorative designs of various shapes.	(i) To provide decorating high lights, sparkle, emitting sparks, stars. It also requires general room lighting.

1.6.4 Light Sources for Room Lighting

1. Fluorescent tubes.
2. CFLs.
3. Triphosphor fluorescent lamps.
4. Accessories/Fittings/Luminaries:
   - (i) Wall brackets
   - (ii) Ceiling mounted
   - (iii) Floor mounted
   - (iv) Table mounted - glass type or metal based finish
   - (v) Corner attached
   - (vi) Pendants
   - (vii) Use of fibre optics for high lighting show pieces, bathrooms etc.

1.6.5 Illumination Levels for Domestic Lighting

1. Estimate the number and wattage of lamps which is required to illuminate a workshop space 70 m × 20 m by means of lamps mounted 5.5 m above the working plane. The average illumination required is 90 lux, coefficient of utilization is 0.45, luminous efficiency is 18 lumens per watt. Assume a space height ratio of unity and maintenance factor 0.85

Table 1.8

Location	Illumination level, lux
Entrances, hallways	100
Living room	300
Dining room	150
Bed room	–
General	300
Bed room, Dressing tables, bed heads	200
Games or recreation room	100
Table games	300
Kitchen	200
Kitchen sink	200
Laundry	100
Bathroom	300
Bathroom mirror	700
Sewing	200
Workshop	200
Stairs	100
Garage	70
Study	300

1.6.6 Methods to Plan and Execute Illumination for the Residential/House/Home Lighting

- The methods of lighting calculations are elaborated in chapter 1, under sub-topics 1.6.1, 1.6.2 and 1.6.3 under the head "Lighting Calculation Method".
- The readers are advised to go through this theory explained for 3 types of methods.
  (i) Watt/m^2 method.
  (ii) Lumens or light lux method.
  (iii) Point to point methods.
- The numerical examples in chapter 1, Example 1.1 to Example 1.22 can also be useful for this chapter 4.

Example 1.12: *A room of size 15 × 6 metres is to be illuminated by twenty 200 W lamps. The mscp of each lamp is 250. Assume a depreciation factor 1.2 and utilization factor 0.6. Find the average illumination produced on the floor.*

Total area to be illuminated, A = 15 × 6 = 90 m^2.

Solution:

$$\text{Total lumens given out by lamps} = \text{Number of lamps} \times \text{mscp of each lamp} \times 4\pi$$

$$= 20 \times 250 \times 4\pi = \mathbf{62.832\ lumens}$$

$$\text{Lumens reaching on the working plane} = \phi \times \frac{\text{Utilization factor}}{\text{Depreciation factor}} = \frac{62832 \times 0.6}{1.2} = \mathbf{31{,}416\ lumens}$$

$$\text{Illumination on the working plane} = \frac{\text{Lumens reaching on the working plane}}{\text{Total area to be illuminated}} = \frac{31{,}416}{90} \simeq 350\ \text{Lux}$$

∴ Average lumens on the floor = **350 Lux**

Example 1.13: *An illumination on the working plane of 75 lux is required in a room 72 m × 15 m in size. The lamps are required to be hung 4 m above the work bench. Assuming a suitable space - height ratio, a utilization factor of 0.5, a lamp efficiency of 14 lumens per watt and a candle power depreciation of 20%, estimate the number, rating and disposition of lamps.*

Solution: Area to be illuminated, $A = 72 \times 15 = 1080 \text{ m}^2$

$$\text{Illumination required, } E = 75 \text{ lux}$$

$$\therefore \quad \text{Lumens required} = A \times E = 1080 \times 75 = 81000$$

$$\text{Utilization factor} = 0.5$$

$$\text{Maintenance factor} = 1 - \text{Candle power depreciation} = 1 - 0.2 = 0.8$$

$$\therefore \quad \text{Gross lumens required} = \frac{A \times E}{\text{Utilization factor} \times \text{Maintenance factor}}$$

$$= \frac{81000}{0.5 \times 0.8} = \boxed{202500}$$

$$\text{Total wattage required} = \frac{\text{Gross lumens required}}{\text{Lamp efficiency}} = \frac{202500}{14} = \mathbf{14{,}464}$$

Rating and disposition of lamps:

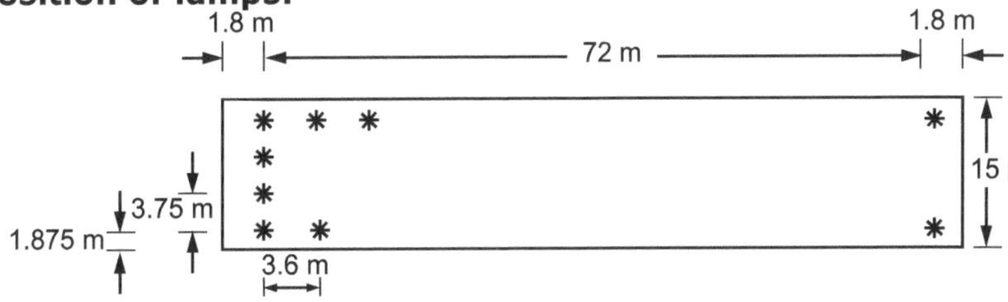

Fig. 1.52

80 lamps in 4 rows, each row having 20 lamps can be used giving spacings of 3.6 m in length and 3.75 m in width and space-height ratios of 0.9 and 0.9375 respectively.

$$\text{Horizontal space-height ratio} = \frac{\text{Distance between two lamps}}{\text{Height of lamp above the ground}} = \frac{3.6}{4} = 0.9$$

$$\text{Vertical space-height ratio} = \frac{3.75}{4} = 0.9375$$

This ratio should be approximately equal to 1.

$$\text{Wattage of each lamp} = \frac{14464}{80} = 180.8 \approx \mathbf{200\ W}$$

Example 1.14: *Estimate the number and wattage of lamps which would be required to illuminate a workshop spaced 60 × 15 m by means of lamps mounted 6 m above the working plane. The average illumination required is about 100 lux, coefficient of utilization is 0.4, luminous efficiency is 16 lumens per watt. Assume a space - height ratio of unity and a candle power depreciation of 20%.*

Solution: Gross lumens required,

$$\phi = \frac{\text{Area to be illuminated} \times \text{Average illumination required}}{\text{Coefficient of utilization} \times \text{Maintenance factor}}$$

$$= \frac{60 \times 15 \times 100}{0.4 \times (1 - 0.2)} = \mathbf{2{,}81{,}250}$$

$$\text{Total wattage required,} \quad W = \frac{\phi}{\text{Luminous efficiency in lumens/watt}}$$

$$= 2{,}81{,}250 = \mathbf{17{,}578\ watts}$$

Disposition

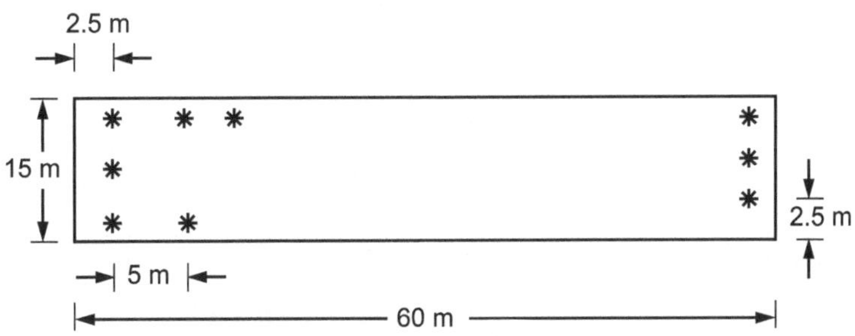

Fig. 1.53

For space - height ratio unity, three lamps can be mounted along the width of the rooms. 12 lamps can be arranged along the length of the room.

$\therefore$ Total number of lamps = 12×3 = **36**

Wattage of each lamp = $\frac{17578}{36} = 488.3$

$\therefore$ Standard lamp of 500 watts can be located.

$\therefore$ **Wattage of lamp = 500 W**

and **Number of such lamps = 36**

Example 1.15: *A hall 30 m by 15 m with a ceiling height of 5 m is to be provided with a general illumination of 120 lumens/m*^{2}*. Taking a coefficient of utilization of 0.5 and depreciation factor of 1.4, determine the number of fluorescent tubes required, their spacing, mounting height and total wattage. Take luminous efficiency of fluorescent tube as 40 lumens per watt and for 80 watt tubes.*

Solution: Total lumens required = $A \times E = 30 \times 15 \times 120$

$$\therefore \quad \text{Gross lumens} = \frac{A \times E}{UF \times MF} = \frac{30 \times 15 \times 120}{0.5 \times 0.714} = 151260$$

$$MF = \frac{1}{\text{Depreciation factor}} = \frac{1}{1.4} = 0.714$$

$$\text{Total wattage required} = \frac{151260}{40} = \mathbf{3781.5\ W}$$

$$\therefore \quad \text{Number of tubes required} = \frac{3781.5}{\text{Wattage of each tube}} = \frac{3781.5}{80} = \mathbf{47.26 \approx 48}$$

Say tubes required 48 Nos.

Disposition of tubes total 48 in numbers.

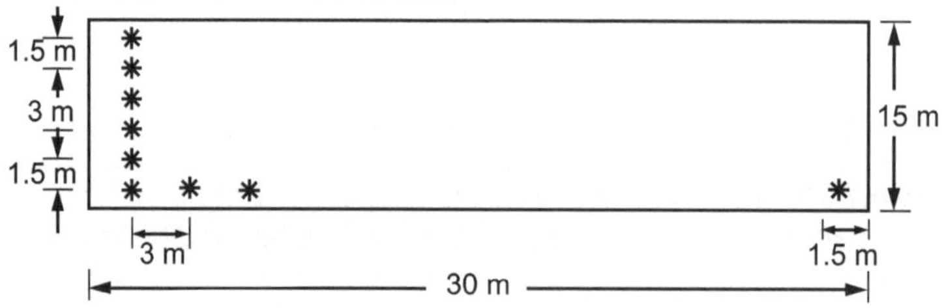

Fig. 1.54

To maintain space to height ratio equal to unity, the tubes must be mounted at height of 3 m above the ground. Space to height ratio = $\frac{3}{3}$ = **1**.

Example 1.16: *A hall 30 m long and 12 m wide is to be illuminated and illumination required is 50 cp. Five types of lamps having lumens output as given below are available.*

Watts	100	200	300	500	1000
Lumens	1615	3650	4700	9950	21500

Taking depreciation factor of 1.3 and utilization coefficient of 0.5, calculate the number of lamps needed in each case to produce required illumination. Out of above five types of lamps, select most suitable type and design a suitable scheme and make a sketch showing location of lamps. Assume a suitable mounting height and calculate space - height ratio of lamps.

Solution: Gross lumens required = $\frac{A \times E}{UF \times MF} = \frac{A \times E \times DF}{UF}$

$$DF = \frac{1}{MF} = \frac{30 \times 12 \times 50 \times 1.3}{0.5} = \mathbf{46800\ lumens}$$

If 100 W lamps are used, then number of lamps required

$$= \frac{46800}{1615} = 28.97 = \mathbf{29\ Nos.}$$

If 200 W lamps are used, then number of lamps required

$$= \frac{46800}{3650} = \textbf{13 Nos.}$$

If 300 W lamps are used, then number of lamps required

$$= \frac{46800}{4700} = \textbf{10 Nos.}$$

If 500 W lamps are used, then number of lamps required

$$= \frac{46800}{9950} = \textbf{5 Nos.}$$

If 1000 W lamps are used, then number of lamps required

$$= \frac{46800}{21500} = \textbf{2 Nos.}$$

Since mounting height/ceiling height is not given, we will assume it to be equal to 5 m.

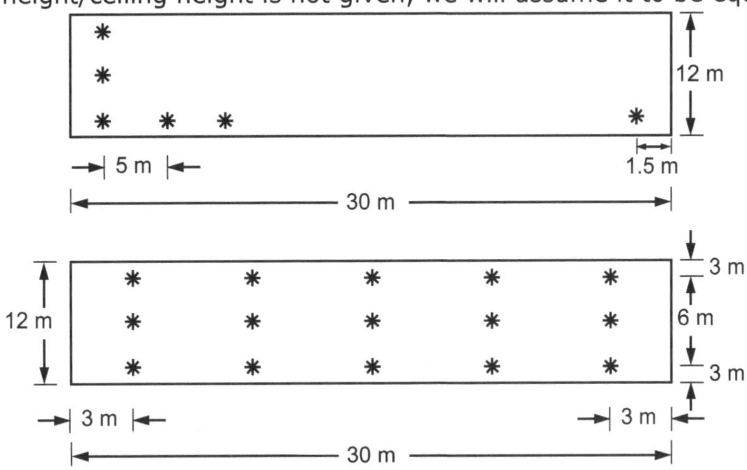

Fig. 1.55

Most suitable type of lamps will be **300 W lamps requiring 10 lamps in two rows, each having 5 lamps** giving space of 6 m in length as well as width and space to height ratio of $\frac{6}{3}$ i.e. 1.2.

Lamps of lower wattage will require more number of fittings and therefore will increase the cost whereas lamps of higher wattage will be few in requirement giving space-height ratio much more than desirable.

1.6.7 Commercial Lighting Stepwise Procedure

1. State and explain any six factors while considering designing the illumination for interior location of commercial lighting.
2. State the design considerations for interior locations of residential unit. (Any four)

- From plan the number of light points, fan points, plug points i.e. general lights and fan load is determined in addition the requirement of power loads for motors and other equipments is determined.
- From this load requirement the number and size of sub-circuits is determined.
- Generally, every sub-circuit should have a load of 800 watts or 8-10 points (maximum), for light and fan, here we are required to make provision of spare sub-circuits for future expansions, otherwise a load of 500 W or so per circuit with nearly 6-8 points per circuit will not need any spare sub-circuits for future expansion.
- On each power sub-circuit a load of 2000 watts or 2 outlets should be provided.
- **Load Calculations and Selection of Size of Service Connection and Nature of Supply:** All buildings are supplied power by the authorized agency such as Electricity Boards etc.

- They have their distribution network from which a 230 volts single phase or 400 volts three-phase supply is provided by means of underground cable or overhead service.
- For very big complex or high buildings having demand more than 50 kVA, the supply agency makes the consumer to adopt H.T. supply at 11 kV, 6.6 kV or 3.3 kV. Depending upon the feeder supply. They have to erect their own sub-station to feed the desired voltage.

Example 1.17: *A uniform illumination of 50 lux is to be obtained on the floor of a room measuring 10 m × 10 m by arranging electric light suitably. Calculate the number of lamps and watt rating of each lamp if the efficiency of lamp is 15 lumens/watt. Assume and write suitable values required in this calculation.*

Solution: Area to be illuminated $= 10 \times 10 = 100 \text{ m}^2$

Assume, Utilization factor $= 0.6$

Assume, Depreciation factor $= 1.2$

$\therefore$ Total lumens on the working plane $=$ Illumination in lux × Area

$= 50 \times 100 = 5000$ lumens

Lumens from lamp is more than this $= \dfrac{5000 \times \text{Depreciation factor}}{\text{Utilization factor}}$

$= \dfrac{5000 \times 1.2}{0.6} = 10{,}000$ lumens

Given that lamp $\eta = 15$ lumens/watt

$\therefore$ Total wattage of lamps $= \dfrac{10{,}000}{15} = 666$ watts

As the area is in square shape 10 × 10, 4 lamps each of 200 watts are suitable.

4 × 200 = 800 watts which matches with 666 watts.

Sea the disposition of lamps for uniform illumination on the area.

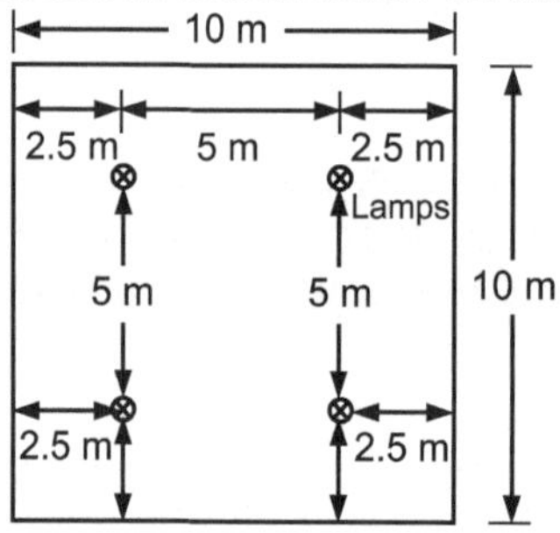

Fig. 1.56

Along the border 2.5 m and distance between the lamps = 5 m. This will give uniform illumination.

Example 1.18: *Estimate the number and wattage of lamps which is required to illuminate a workshop space 70 m × 20 m by means of lamps mounted 5.5 m above the working plane. The average illumination required is 90 lux, coefficient of utilization is 0.45, luminous efficiency 18 lumens per watt. Assume a space height ratio of unity and maintenance factor 0.85.*

Solution: Total lumens required $= \dfrac{\text{Area} \times \text{Average illumination}}{\text{Utilization factor} \times \text{Maintenance factor}}$

$= \dfrac{(20 \times 70) \times (90)}{0.45 \times 0.85} = 329412$ lumens

$\therefore$ Total wattage required $= \dfrac{329412}{18} = 18{,}300$ watts

For uniform illumination to suit for the room dimensions 12 lamps @ length and 3 lamps @ width be taken so that 12 × 3 = 36 lamps can be well accommodated.

To find wattage of lamp $= \dfrac{\text{Total watts}}{\text{Number of lamps}} = \dfrac{18300}{36} = 508 \simeq 500$ W

So 36 number of 500 watt lamps be disposed as:

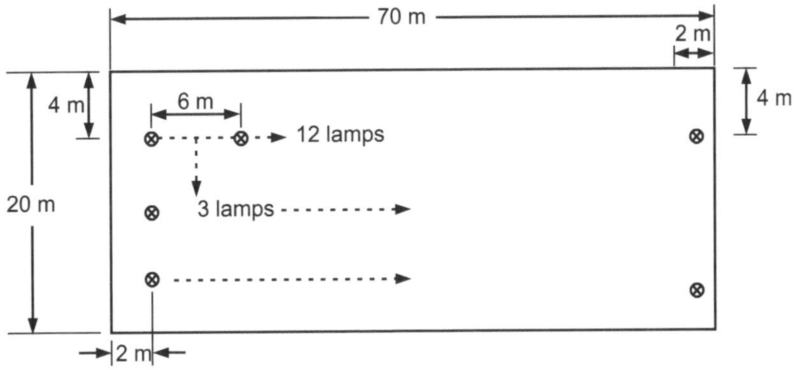

Fig. 1.57

Example 1.19: *A floor lighting is to be provided on the front of a building of 40 m × 25 m for brightness of 18 lumens/square meter. The coefficient of reflection of building surface is 0.21. The lamps of 500 W having lumen output 8400 each are used. If beam factor is 0.65, waste light factor is 1.1, maintenance factor 0.85 then calculate number of lamps for flood lighting.*

Solution: Brightness of incident illumination $= \dfrac{\text{Brightness (in lumens)}}{\text{Reflection factor}} = \dfrac{18}{0.21} = 85.71$ lux

If there are 'N' lamps then total lumen O.P. = 8400 × N

Taking into account the different factors,

$$\therefore \quad 8400\text{ N} \times \text{Beam factor} = \frac{\text{Lux area} \times \text{Waste light factor}}{\text{Maintenance factor}}$$

$$8400\text{ N} \times 0.65 = \frac{85.71 \times (40 \times 25) \times 1.1}{0.85}$$

$$= 22.2 \simeq \mathbf{22\ lamps}$$

Example 1.20: *A hall of 80 m × 40 m with a ceiling height of 5.2 m is to be provided with a general illumination of 150 lumens/sq.m. Assuming coefficient of utilization of 0.6 and depreciation factor of 1.35, determine the number of fluorescent tubes required, the distance between them, mounting height, total wattage. The luminous efficiency of fluorescent tube is 35 lumens per watt for 40 watt tube light.*

Solution: Area = 80 × 40 = 320 m^2

Using standard formula,

$$\text{Total lumens} = \frac{\text{Area} \times \text{Illumination} \times \text{Depreciation factor}}{\text{Utilization factor}}$$

$$= \frac{320 \times 150 \times 1.35}{0.6}$$

$$= 1{,}08{,}000 \text{ lumens}$$

1 tube of 35 lumens/watt.

$$\therefore \quad \text{Total wattage} = \frac{1{,}08{,}000}{35} = 3{,}085 \text{ W}$$

$$\therefore \quad \text{Total tubes} = \frac{3085}{40} = 77 \text{ Numbers}$$

Assume, Space to height ratio = 1

So, Spacing = 5 m, hence tubes = 64

Tubes 16 @ length and $\dfrac{64}{16} = 4$ tubes @ width

Disposition is shown below.

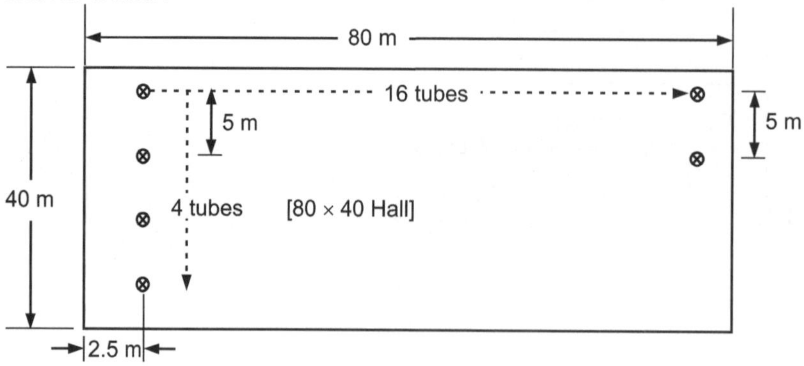

Fig. 1.58

Spacing @ walls = 2.5 m

and tube spacings 5 m space to height = 1

Example 1.21: *An indoor badminton court is accommodated in a hall of 20 m length, 10 m width, and 15 m height. The walls and ceiling of hall are painted black and do not reflect any light. Design a scheme for providing an average illumination of 80 lux at ground surface, using 200 W tungsten filament lamps with suitable fitting. Give reason for your choice. Take coefficient of utilization = 0.5, efficiency of lamps = 15 lumens/watt.*

Solution:

$$\text{Area, A} = 20 \times 10 = 200 \text{ m}^2$$

$$\text{Total lumens required} = \frac{\text{Area} \times \text{Illumination} \times \text{Depreciation}}{\text{Utilization factor}}$$

$$= \frac{200 \times 80 \times 1}{0.5} = 32{,}000 \text{ lumens}$$

$$\text{Total wattage required with 200 watt lamps} = \frac{32000}{15} = 2133.33 \text{ watts}$$

$$\therefore \quad \text{Number of 200 watt lamps} = \frac{2133.33}{200} = 10.6 \text{ numbers} \simeq 10 \text{ lamps}$$

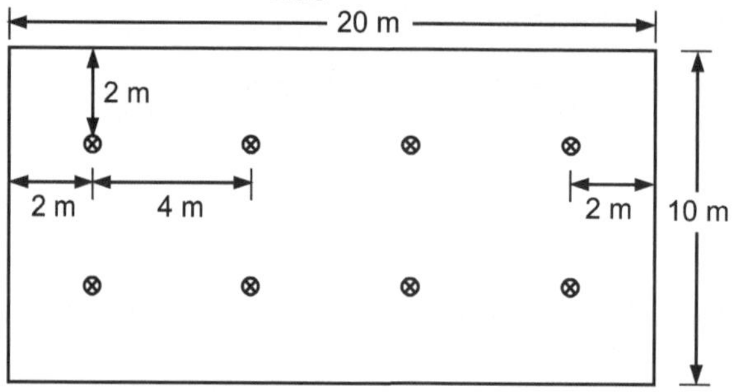

Fig. 1.59: Disposition of 10 lamps

Distance between two lamps = 4 m

Space left @ the walls = 2 m, all sides.

The black painting does not reflect light and glare on the eyes of players is avoided.

The lighting bulbs be suspended to have a space to height ratio ≃ 1.

Example 1.22: *A small assembly shop 16 m long, 10 m wide and 3 m up to trusses is to be illuminated to a level of 200 lux. The utilization and maintenance factors are 0.74 and 0.8 respectively. Calculate number of lamps required to illuminate whole area if the lumen output of the lamp selected is 3000 lumens.*

Solution:

$$\text{Area} = 16 \times 10 = 160 \text{ m}^2$$

$$\text{Gross lumens required} = \frac{\text{Area} \times \text{Illumination}}{\text{Utilization factor} \times \text{Maintenance factor}}$$

$$= \frac{160 \times 200}{0.74 \times 0.8} = 54054 \text{ lumens}$$

$$\text{Number of lamps} = \frac{\text{Total lumens}}{\text{Lumen o.p. of each lamp}} = \frac{54054}{3000} = \textbf{18 lamps}$$

Example 1.23: *A hall of 40 ft by 50 ft is to be illuminated to 45 lumen per sq. ft. on working plane. If utilization factor is 0.6 and depreciation factor is 0.8 and source gives an output of 10 lumens per watt, determine the number of lamps.*

Solution: Given data: E = 45 lumen/sq. ft,

Area of working plane = 45 ft × 50 ft = 2000 sq. ft

Assumed: U.F. = 0.6 and D.F. = 0.8

Wattage of lamps assumed = 100 watt/200/500 watt

Efficiency = 10 lumens/watt assumed

Waste light factor = 1

(i) Total lumens utilized = E × A or 45 × 200 = 90000 lumens

(ii) Total lumens given out by the lamp

$$= \frac{\text{Total lumens utilized}}{\text{U.F.} \times \text{D.F.}} = \frac{90000}{0.6 \times 0.8} = 187500 \text{ lumens}$$

(iii) $$\text{Total wattage} = \frac{\text{Total lumens given out by the lamps}}{\text{Luminous efficiency}}$$

$$= \frac{187500}{10} = 18750 \text{ watts}$$

The wattage of lamps is assumed as 100 watt.

(iv) $$\text{Number of lamps} = \frac{\text{Total wattage}}{\text{Wattage of each lamp}} = \frac{18750}{100} = 187.5 \cong 188 \text{ Nos.}$$

∴ Number of lamps = 188 Nos.

Example 1.24: *A drawing hall 30 m by 15 m with a ceiling height of 5 m is to be provided with a general illumination of 120 lumens per m^2; taking a coefficient of utilization of 0.5 and depreciation factor of 1.4, determine the number of fluorescent tubes required, their spacing, mounting height and total wattage. Take luminous efficiency of fluorescent tube as 40 lumens per watt for 80 watt tubes.*

Solution: Given data: E = 120 lumens, C = 0.5 and D.F. = 1.4

Area of working plane = 30 m × 15 m = 450 m^2

Wattage of each lamp = 200 watt

Efficiency of lamp = 40 lumens/80 watt tube

(i) Total lumens required on working plane

$$= \frac{\text{AIWD}}{\text{C}} = \frac{450 \times 120 \times 1 \times 1.4}{0.5} = 151200 \text{ lumens}$$

(ii) Total number of fluorescent tubes

$$= \frac{\text{Total lumens given out by the lamps}}{\text{Wattage of each lamp} \times \text{Luminous efficiency}}$$

$$= \frac{151200}{80 \times 40} = 47.25 \cong 48 \text{ Number of lamps}$$

The number of lamps can be increased or decreased (46 lamps or 50 lamps) better illumination design.

(iii) Total wattage = Total number of lamps × Wattage of lamp

= 48 × 80 = 3840 watts

(iv) Space and mounting height:

∴ Space = 3.75 m and mounting height 5 m

OR ∴ Space = 3.75 m and mounting height 5 m

$$\frac{3.75}{5} = 0.75$$

Working plane is assumed as a ground surface.

Arrangement of Fluorescent Tubes:

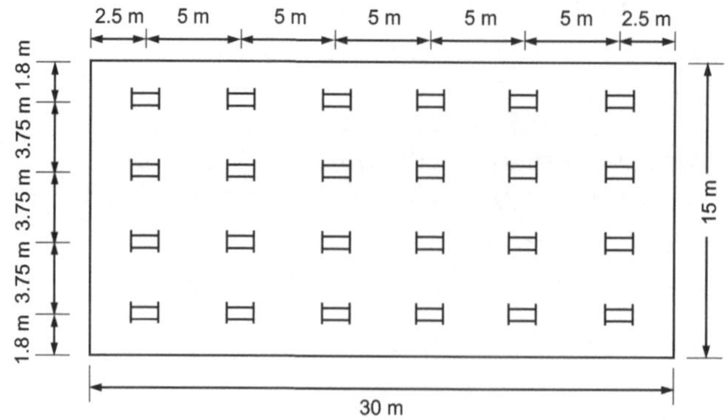

Fig. 1.60

Example 1.25: *An illumination on the working plane of 75 Lux is required in a room 72 m × 15 m in size. The lamps are required to be hunged 4 m above the work bench. Assume a suitable space-height ratio, a utilization factor of 0.5, a lamp efficiency of 14 lumens per watt and a candle power depreciation of 20%. Estimate the number, rating and disposition of lamps.*

Solution: Given Data: E = 75 lux, Area of working plane = 72 m × 15 m = 1080 m^2,
U.F. = 0.5 and D.F. = 0.8 or 1.2, Assume wattage of each lamp = 200 watt,
Efficiency of lamp = 14 lumens/watt, Height = H = 4 m.
Determine: (1) Number of lamps if luminous efficiency of 14 lumens/watt

If student get D.F. = 0.8:

$$\text{Gross lumens} = \frac{A \times E}{U.F. \times D.F.} = \frac{1080 \times 75}{0.5 \times 0.8} = \frac{81000}{0.5 \times 0.8} = 202500$$

$$\text{Number of lamps required} = \frac{\text{Gross lumens}}{\text{Wattage of each lamp} \times \text{Efficiency of each lamp}}$$

$$= \frac{202500}{200 \times 14} = 72.32 \approx 72 \text{ lamps.}$$

Their disposition:

- Assume space to height ratio = 3: 4 = 0.75
- Number of lamps lengthwise (No. of rows) = $\frac{\text{Length (L)}}{\text{Space (S)}} = \frac{75}{3} = 25$
- Number of lamps widthwise (No. of columns) = $\frac{\text{Width (W)}}{\text{Space (S)}} = \frac{15}{5} = 3$

5m	3m	3m	3m	3m	3m	3m	3m	3m	3m	3m	3m	3m	3m	3m	3m	3m	3m	3m	3m	3m	3m	3m	3m	3m
5m																								
5m																								

Fig. 1.61: Disposition of lamps

(at each cross or junction of 3 lines there is one lamp except 1st columns)

Total number of lamps as per illumination design = 24 × 3 = 72 Numbers.

If student write this way D.F. = 1.2:

$$\text{Gross lumens (lumens received)} = \frac{A \times E \times D.F.}{U.F.}$$

$$\text{Gross lumens} = \frac{1080 \times 75 \times 1.2}{0.5} = \frac{97200}{0.5} = 194400$$

Their disposition S.H.R.: 1 Space: 4 m.

- Assume space to height ratio = 1.
- No. of lamps lengthwise (No. of rows) = $\frac{\text{Length (L)}}{\text{Space (S)}} = \frac{72}{4} = 18$
- No. of lamps widthwise (No. of columns) = $\frac{\text{Width (W)}}{\text{Space (S)}} = \frac{15}{4} = 3.75 \approx 4$

Total number of lamps as per illumination design = 18 × 4 = 72 Nos.

Disposition:

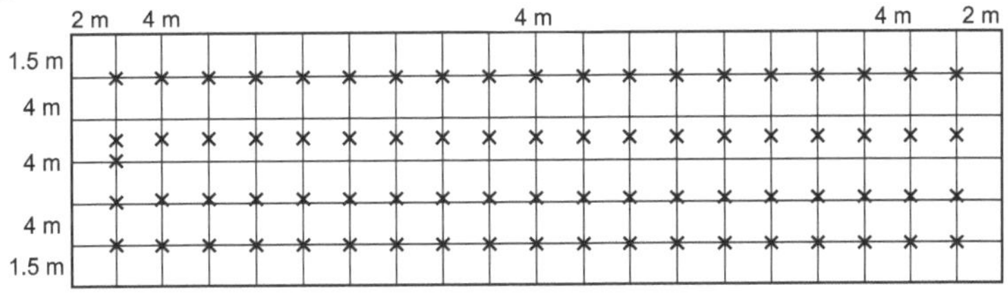

Fig. 1.62

$$\text{Total wattage} = \frac{\text{Total lumens received}}{\text{Luminous efficiency}} = \frac{194400}{14} = 13885.71 \text{ watts}$$

$$\text{Wattage of each lamp} = \frac{\text{Total wattage}}{\text{No. of lamps}} = \frac{13885.71}{72} = 192.85 \approx 200 \text{ watts}$$

Example 1.26: *A minimum illumination of 80 lumens/m² is required in the factory shade of 50 m × 12 m. Calculate the number, location and wattage of the units used. Assume that depreciation factor 0.8, coefficient of utilization is 0.4 and efficiency of lamp units is 14 lumens/watt.*

Solution: Given Data: E = 80 lumens/sq. m.,

Area of working plane = 50 m × 12 m = 600 sq. m., U.F. = 0.4 and D.F. = 0.8,

Wattage of lamps assumed = 100 watt/200/500 watt

Efficiency = 14 lumens/watt, Assumed: Waste light factor = 1.

(i) Total lumens utilized = E × A or 80 × 600 = 48000 lumens

(ii) $$\text{Total lumens given out by the lamp} = \frac{\text{Total lumens utilised}}{\text{U.F.} \times \text{D.F.}} = \frac{48000}{0.6 \times 0.8}$$

$$= 150000 \text{ lumens}$$

(iii) $$\text{Total wattage} = \frac{\text{Total lumens given out by the lamps}}{\text{Luminous efficiency}}$$

$$= \frac{150000}{14} = 10714.285 \text{ watts}$$

The wattage of lamps is assumed - 100 watt

(iv) $$\text{Number of lamps} = \frac{\text{Total wattage}}{\text{Wattage of each lamp}}$$

$$= \frac{10714.285}{100} = 107.142 \cong 107 \text{ Numbers}$$

∴ Number of lamps = 107 Numbers

Location of lamps:

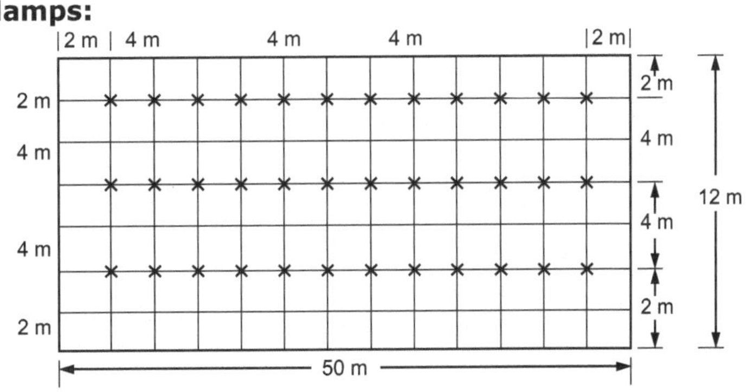

Fig. 1.63

Example 1.27: *A room of size 20 m × 5 m is illuminated by 20 number of 200 watt lamps. The MSCP of each lamp = 250. Assume utilization factor = 0.6 and depreciation factor = 1.2. Find average illumination produced on the floor. State and explain the features of railway platform lighting.*

Solution: (i) Area of room = A = 20 × 5 m = 100 sq. mtr.

(ii) MSCP of each lamp = 250

(iii) Depreciation factor = D.F. = 1.2

(iv) Coefficient of utilization = U.F. = 0.6.

(v) Number of lamps = 20.

(vi) Wattage of each lamp = 200 watts.

Find: Average illumination = E = ?

$$\text{Total lumens given out by all lamps} = (\text{MSCP} \times 4\pi) \times 20 = (250 \times 4\pi) \times 20 = 62831.853 \text{ lumens}$$

$$\text{Total lumens received on the floor} = \text{Total lumens given out by all lamps} \times \frac{\text{U.F.}}{\text{D.F.}}$$

$$\text{Total lumens received on the floor} = 62831.853 \times \frac{0.6}{1.2} = 31415.926 \text{ lumens.}$$

$$\text{Average illumination on the floor} = E_{AV} = \frac{\text{Total lumens received on the floor}}{\text{Area}}$$

$$= \frac{31415.926}{100} = 314.159 \text{ lumens per sq. mtr.}$$

Following are the features for railway platform lighting:

Good platform lighting on all stations is essential for the safety and comfort of passengers and railway staff. The recommended value of illumination is 100 - 150 lux. T-5 fluorescent lamps are used as source of light. The luminaries should be arranged in such a way that the light strikes the platform 'H' straight down and without shadows provided reasonable uniform light across the width of the platform. For non-covered portion of the railway station, street light fitting of T-5 fluorescent lamps with width angle distribution reflectors are suitable. Modern LED street light fitting of 36/40 W are also used.

Example 1.28: *A hall 30 × 12 m is to be illuminated with 50 m candles, DF = 1.3 and UF = 0.5. Calculate space to height ratio and work out the number of lamps from the following table and select suitable wattage lamp for uniform light disposition.*

Watts	*100*	*200*	*300*	*500*	*1000*
Lumens	*1615*	*3650*	*4700*	*9950*	*21500*

Solution: (i) Area of room = A = 30 × 12 m = 360 (ii) MSCP of each lamp = 250

(iii) Depreciation factor = D.F. = 1.3 (iv) Coefficient of utilization = U.F. = 0.5.

(v) Illumination required E = 50 m candle.

Find: Average illumination = E = ?

$$\text{Total gross lumens required} = \frac{A \times E \times \text{D.F.}}{\text{U.F.}} = \frac{360 \times 50 \times 1.3}{0.5} = 46800 \text{ lumens}$$

(1) If 100 watt lamps are used:

$$\text{Number of lamps required} = \frac{\text{Total gross lumens}}{\text{Lumens of 100 watt}} = \frac{46800}{1615} = 29 \text{ Numbers}$$

(2) If 200 watt lamps are used:

$$\text{Number of lamps required} = \frac{\text{Total gross lumens}}{\text{Lumens of 200 watt}} = \frac{46800}{3650} = 13 \text{ Numbers}$$

(3) If 300 watt lamps are used:

$$\text{Number of lamps required} = \frac{\text{Total gross lumens}}{\text{Lumens of 300 watt}} = \frac{46800}{4700} = 10 \text{ Numbers}$$

(4) If 500 watt lamps are used:

$$\text{Number of lamps required} = \frac{\text{Total gross lumens}}{\text{Lumens of 500 watt}} = \frac{46800}{9950} = 5 \text{ Nos.}$$

(5) If 1000 watt lamps are used:

$$\text{Number of lamps required} = \frac{\text{Total gross lumens}}{\text{Lumens of 1000 watt}} = \frac{46800}{21500} = 2 \text{ Numbers.}$$

Space-Height Ratio: Let the mounting height to be 5 m.

Most suitable type of lamps will be 300 watt lamps requiring 10 lamps in two rows, each row having 5 lamps giving spacing of 6 m in length as well as in width and space height ratio of $\frac{6}{5} = 1.2$.

Uniform light disposition:

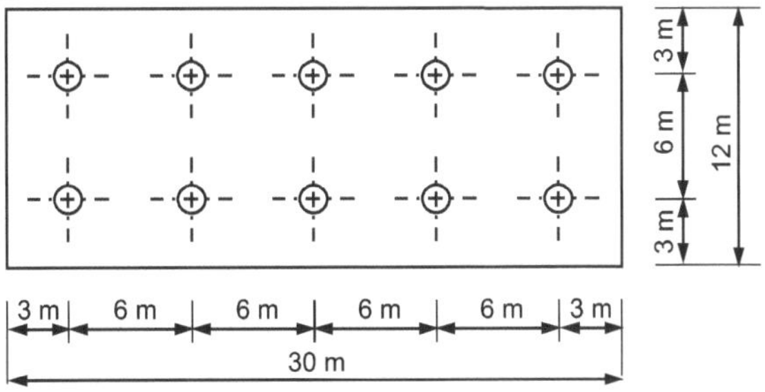

Fig. 1.64

Example 1.29: *A room of 30 m × 10 m is illuminated by 20 numbers of 200 W lamps. The MSCP of each lamp is 250. If utilization factor is 0.4 and depreciation factor 1.2 then find average illumination produced on the surface.*

Solution: (i) Area of room = A = 30 m × 10 m = 300 sq. mtr.

(ii) MSCP of each lamp = 250

(iii) Depreciation factor = D.F. = 1.2

(iv) Coefficient of utilization = U.F. = 0.4

(v) Number of lamps = 20

(vi) Wattage of each lamp = 200 watts.

Find: Average illumination = E = ?

$$\text{Total lumens given out by all lamps} = (\text{MSCP} \times 4\pi) \times 20 = (250 \times 4\pi) \times 20$$
$$= 62831.853 \text{ Lumens}$$

$$\text{Total lumens received on the floor} = \text{Total lumens given out by all lamps} \times \frac{\text{U.F.}}{\text{D.F.}}$$

$$\text{Total lumens received on the floor} = 62831.853 \times \frac{0.4}{1.2} = 20943.951 \text{ Lumens}$$

$$\text{Average illumination on the floor} = E_{AV} = \frac{\text{Total lumens received on the floor}}{\text{Area}}$$

$$= \frac{20943.951}{300} = 69.8131 \text{ Lumens per sq. mtr.}$$

OR

$$E_{AV} = 69.8131 \text{ Lux.}$$

Example 1.30: *A minimum illumination of 80 lux is required in the room of 50 m × 12 m. Calculate the number, location and wattage of the lamps to be used. Assume that depreciation factor is 1.2, utilization factor is 0.4 and efficiency of the lamp is 14 lumens/watt.*

Solution: Given Data: E = 80 lumens/sq. m.,

Area of working plane = 50 m × 12 m = 600 sq. m.

U.F. = 0.4 and D.F. = 1.2

Wattage of lamps assumed = 100 watt/200/500 watt

Efficiency = 14 lumens/watt, Assumed: Waste light factor = 1.

(i) Total lumens utilized = E × A or 80 × 600 = 48000 lumens

(ii) $$\text{Total lumens given out by the lamp} = \frac{\text{Total lumens utilised}}{\text{U.F.}} \times \text{D.F.}$$

$$= \frac{48000 \times 1.2}{0.4} = 144000 \text{ lumens}$$

(iii) Total wattage $= \dfrac{\text{Total lumens given out by the lamps}}{\text{Luminous efficiency}}$

$$= \frac{144000}{14} = 10285.714 \text{ watts}$$

The wattage of lamps is assumed – 100 watt

(iv) Number of lamps $= \dfrac{\text{Total wattage}}{\text{Wattage of each lamp}}$

$$= \frac{10285.714}{100} = 102.85 \cong 103 \text{ Numbers}$$

$\therefore$ Number of lamps = 103 Numbers

Location of lamps:

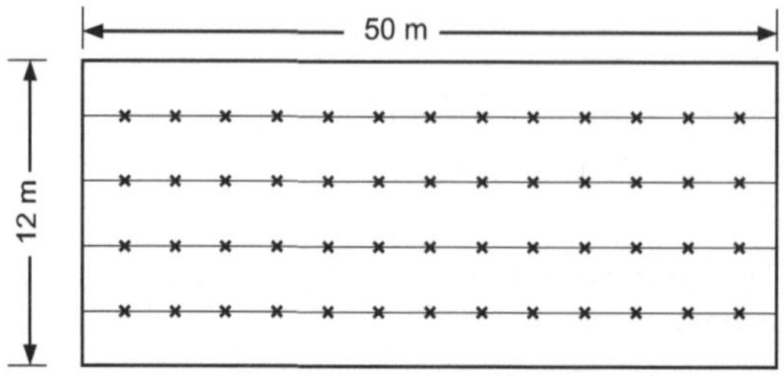

Fig. 1.65

Example 1.31: *A hall 30 metres by 15 metres with a ceiling height of 5 metres is to be provided a general illumination of 120 lumens/m^2 taking a coefficient of utilization of 0.5 and depreciation factor of 1.4. Determine the number of fluorescent tubes required, their spacing mounting height and total wattage. Take luminous efficiency of fluorescent tube as 40 lumens per watt and for 80 watt tubes.*

Solution: Given data: E = 120 lumens, Area of working plane = 30 m × 15 m = 450 m^2, C = 0.5 and D.F. = 1.4, Wattage of each lamp = 200 watt, Efficiency of lamp = 40 lumens/80 watt tube.

(i) Total lumens required on working plane

$$= \frac{AEWD}{C}$$

$$= \frac{450 \times 120 \times 1 \times 1.4}{0.5} = 151200 \text{ Lumens}$$

(ii) Total number of fluorescent tubes

$$= \frac{\text{Total lumens given out by the lamps}}{\text{Wattage of each lamp} \times \text{Luminous efficiency}}$$

$$= \frac{151200}{80 \times 40} = 47.25 \cong 48 \text{ Number of lamps.}$$

As the number of lamps can be increased or decreased (46 lamps or 50 lamps), better will be the illumination design.

(iii) Total wattage = Total number of lamps × Wattage of lamp

$$= 48 \times 80 = 3840 \text{ watts}$$

(iv) Space and mounting height:

$\therefore$ Space = 3.75 metre and mounting height = 5 metre

$$\frac{3.75}{5} = 0.75$$

Working plane is assumed as a ground surface.

Arrangements of Fluorescent Tubes:

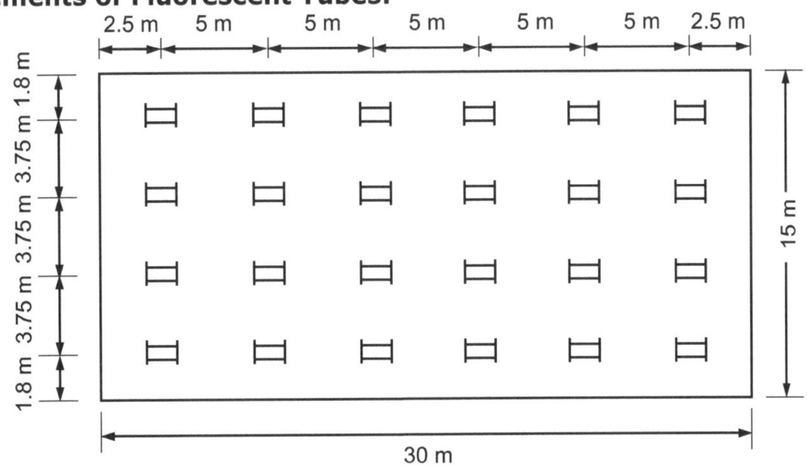

Fig. 1.66

Example 1.32: *The front of building 50 m × 16 m is illuminated by 16 Nos. of 1000 watt lamps arranged so that uniform illumination on the surface is obtained. Assume:*

(i) Luminous efficiency = 17.4 lumens/watt. (ii) Utilization factor = 0.4.

(iii) Depreciation factor = 1.3. (iv) Waste light factor = 1.2.

Determine the illumination on the surface.

Solution: (i) Area of room = A = 50 m × 16 m = 800 m^2

(ii) Wattage = 1000 watt

(iii) Depreciation factor = D.F. = 1.3

(iv) Coefficient of utilization = U.F. = 0.4

(v) Waste light factor = 1.2

(vi) Number of lamps = 16 numbers

Find: Average illumination = E = ?

Average illumination on the floor = E_{AV}

$$= \frac{\text{Number of lamps} \times \text{U.F.} \times \text{Number of wattage} \times \text{Lamp efficiency}}{\text{Area} \times \text{W.L.F.} \times \text{D.L.}}$$

$$= \frac{16 \times 0.4 \times 1000 \times 17.4}{800 \times 1.2 \times 1.3}$$

E_{AV} = **89.23 Lux**

1.7 FACTORY LIGHTING (INDUSTRIAL LIGHTING)

- Knowing the basic terms and their definitions and requirements of illumination, let us apply this knowledge to discuss factory lighting, flood lighting and other places to be illuminated.
- In factory, there are so many provisions for employees and workers such as transport facility, subsidized meals, canteen service, play grounds, common halls for functions, medical facilities, for their motivation towards the work which result in increased production. Providing a good scheme of illumination at the working places in the factories also help to get increased production as it provides healthy surroundings, glareless and fatigueless situations and employees feel friendly. It also eliminates the possibility of accidents due to shadowless illumination. Good illumination also saves time for the same work and increase profitability and employer encourages the workers by giving increased salaries. So a very good lighting scheme in the factory has its own importance.

1.7.1 General Requirements

1. State the general requirements for factory lighting.

- A factory lighting installation, in common with other indoor equipment should provide an adequate illumination on the working place and give a good distribution of light, employ simple and easily cleaned fittings and avoid glare. It is essential not only to avoid glare from the lamp itself but also reflected glare from any polished surface, which may be within the line of vision.

1. **General Lighting:** The usual scheme in factories and workshops is to hung a number of lamps at a sufficient height so that uniform distribution of light over the working plane is obtained. In large machine shops the height is governed by the necessity of keeping the lamps above the travelling crane. In such cases it is often desirable to supplement the main lighting by side lighting in order to give additional illumination on a vertical plane. Since, light coloured walls and ceiling add to the effectiveness of an installation, therefore, it is necessary to paint with white colour.
2. **Local Lighting:** On some points large value of illumination is required. In such case local lighting can be provided by means of adjustable fittings attached to the machine or bench in question or mounted on portable floor standards. Such lamps should be mounted in deep reflectors to avoid glare on the eyes. Portable hand lamps attached to wall plugs by means of trailing leads are used for maintenance work and emergency lighting. Low voltage lamps upto 50 volts are recommended for use us portable hand lamps because such lamps have thicker filament and are, therefore, more robust than those for normal voltage and danger of shock is also avoided. The supply for such lamps can be obtained from a special low voltage distribution system running throughout the factory or by means of a small transformer for each individual lamp provided in the factory.

   Local lighting should never be employed alone, good general lighting is essential so that dark places between the local lighting units are avoided. Dark places between local lighting units cause fatigue to the eyes on account of its continually having to adjust itself to new conditions.
3. **Emergency Lighting:** Some lights, such as for (i) internal pilot lighting required for safe and speedy evacuation of personnel after main lighting circuit is off, (ii) external pilot lighting, provided with careful shades leading to shelters required for evacuation of personnel, (iii) for control posts, first aid centres etc., (iv) dials and gauges in important plants required to be watched regularly are required during an air-raid when all the factory lights are off as a matter of air-raid precaution. The circuit supplying the above emergency lights should be independently controlled. It is very desirable to provide auxiliary lighting from the source other than the main electric supply preferably from batteries or from small petrol driven generator set. If, however, emergency lighting circuits are operated from main electric supply, these should be completely separated from main lighting circuit.

1.7.2 Industrial Light Fittings

- Reflectors are provided for industrial purposes. Their design must be simple so that cleaning is easy. The requirements of most of the installations can be met by one of the following types of fittings.

  Standard Reflectors: These reflectors are made to accommodate lamps of ratings from 40 to 1,500 watts and designed so that they give adequate and uniform illumination when they are mounted at a spacing equal to about 1.5 times their mounting height above the working plane.

  Diffusing Fittings: When more diffused light is required than that given by the standard reflector a diffusing glass screen may be fixed across a standard type of reflector. Such fittings are used where highly polished articles are dealt with.

  Concentrating Reflectors: A reflector with a concentrated beam is employed in large machine shops and foundaries, where the fittings are to be mounted on a considerable height above the working plane. In such places an ordinary reflector would have too wide angle of divergence and would waste a great deal of light on the walls.

  Enclosed Diffusing Fittings: An opal globe completely enclosing the lamp giving a very even and well diffused light is used when light coloured walls and ceiling are there.

  Angle Reflectors: Angle reflectors are used to provide illumination in a vertical plane where concentrating type reflectors are used. These can be mounted on suitable stanchions or the walls.

Maintenance: In order to maintain the fittings in a condition of reasonable efficiency, it is necessary to clean the light fittings periodically. The frequency of cleaning depends on the conditions in the particular factory under consideration and varies from once or twice a week for very dirty surroundings to every four or six weeks under the best conditions.

Types of Lamps: The discharge lamps have been used in where colour rendering is not important. The fluorescent lamps are widely employed on account of its natural day light colour, its even illumination and absence of glare and in some cases, the fact that it gives rise to considerably less heat than filament lamps of the same light output.

1.7.3 Lightning Scheme for Factory/Characteristics of Factory Lighting

1. State any four characteristics of factory lighting.
2. State specific requirements for (i) Factory lighting, (ii) Sports lighting.
3. State general requirements of factory lighting.

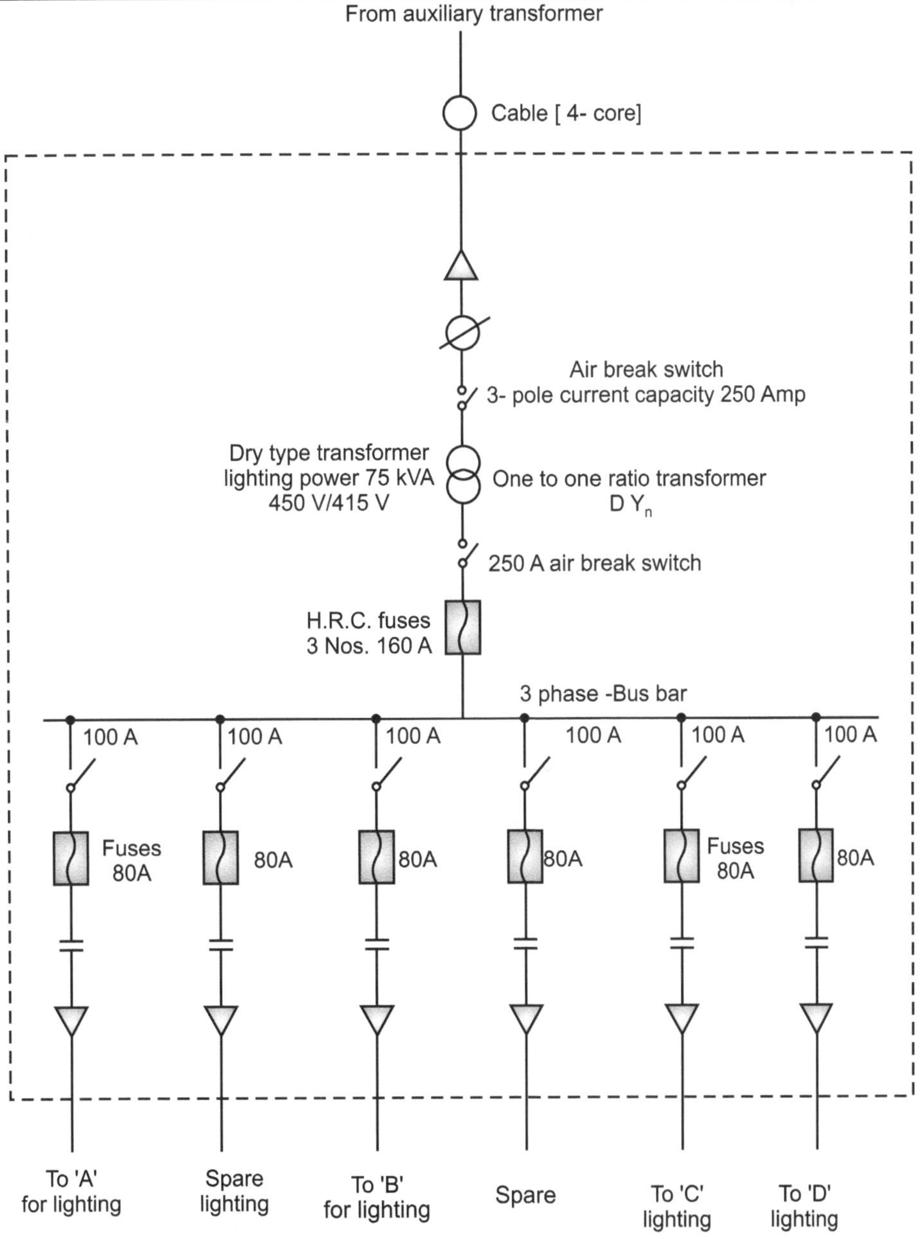

Fig. 1.67: Single line diagram for lighting distribution scheme

- A manufacturing unit has different areas in a big shed where different types of machines are installed for manufacturing products.
- Power wiring system is provided to each and every machine through a common/individual control panel.
- A separate transformer unit is supplying the power to the machines.
- There is a separate provision of illuminating these areas and supply distribution system is altogether separate from the supply system to the various machines.
- The illumination system shall be such that there is a uniform illumination/ adequate illumination and no glares are expected.
- For distribution of electricity the total area is divided into sub-areas like area A, B, C, D etc.
- These areas are illuminated by separate sub-circuits obtained from a common 3-phase bus bar.
- Some spare line provision is also made for future development.
- Switching mechanisms are separate for individual areas.
- Generally a high capacity auxiliary transformer is the starting point of distribution which supply power through cable.
- One more transformer dry type (lightning power transformer) receives electric connection from the auxiliary transformer.
- Connections go to 3-phase bus-bars.
- From bus bar individual areas A, B, C, D etc. tap the electric connections with individual fuse units, contactors.
- All these units are shown in one line diagram.

1.8 LUMEN OR LIGHT LUX METHOD OF LIGHTING CALCULATIONS

Please Refer Numericals in Section 1.6.

1.9 FLOOD LIGHTING

Meaning of Flood Lighting:

- Copious artificial light projected in many directions eliminating all the shadows on the surface illuminated is called as flood light.

1.9.1 Purposes/Applications

- It is used to illuminate various surfaces of various categories, eliminating all the shadows on the surface illuminated. Categories lilke (i) Aesthetic, (ii) Industrial, (iii) Commercial and (iv) Advertising.

(a) Aesthetic flood lighting:

- For enhancing beauty of building at night hours such as:
  - (i) Public places
  - (ii) Ancient buildings
  - (iii) Monuments.
  - (iv) Religious places, buildings, churches, temples
  - (v) Festival occasions
  - (vi) Gardens
  - (vii) Forts, etc.

(b) Industrial and commercial flood lighting:

- (i) Illuminating railway yards
- (ii) Sports stadium
- (iii) Big car park
- (iv) Construction sites
- (v) Queries
- (vi) Industrial campus
- (vii) Commercial important buildings

(c) Advertising:

- (i) To illuminate advertising boards
- (ii) To illuminate show cases
- (iii) Attention seeking item displays
- (iv) Commercial hoarding
- (v) Publicity posters
- (vi) Exhibition

1.9.2 Types of Luminaries Required for Flood Lighting

- 500 to 100 watt lamps with silver or steel reflectors, used in projectors with ventilation, provision, the front of the projector is usually of clear glass, often bowed outwards to protect it from heat of the lamp.
- Use of diffusing glass is made when a diffused beam is required. The projects are so located that they are not visible by persons passing thereby. In some cases, the projectors are housed with ornamental standards.

1.9.2.1 Flood Light Projector

- A particular type of reflector and its housing used for concentrating the light into a "narrow-beam" is known as "flood light projector".

1.9.2.2 Reflecting Surfaces

- Material shape of the reflecting surface of projector is most important.
- Reflecting surface is made up of either of the following:
  (i) Silver glass
  (ii) Chromium plate
  (iii) Stainless steel
- For permanent installation use of cast-metal cases is made to achieve robustness and protection against whether, rain, sunlight conditions.
- For temporary installation, use of sheet metal casing is made.

1.9.3 Classification of Projectors (Tree Pattern)

- The classification is based on beam-spread in degrees and distance.

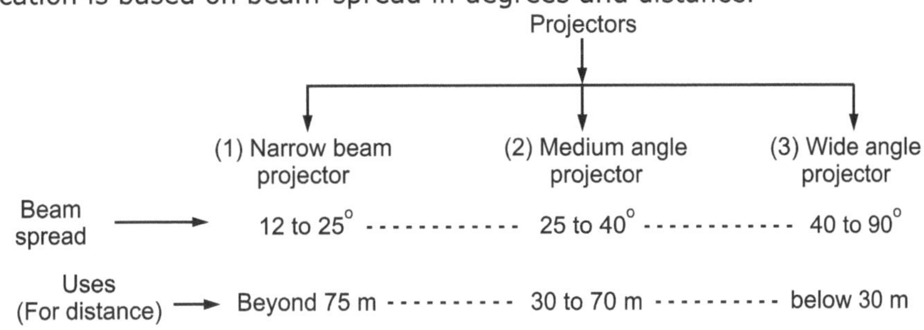

Fig. 1.68

- It is the general practice to use the 3rd type i.e. wide-angle projector with high wattage lamps if preferred over a low wattage lamp as it is more efficient.

1.9.3.1 Lamps Used for Different Projectors

1.	For narrow beam projectors	Lamps having bundled filaments also known specially as projector lamps.
2.	For medium angle projectors	Standard gas filled tungsten filament lamps 250 watt or 500 watt rating.
3.	For wide angle projectors	Standard gas filled tungsten filament lamps but 500 or 1000 watt rating.

1.9.3.2 Location / Mounting of Projectors for Flood Lighting

- Since the projector is costly than its bulb, the number of projectors should be restricted to minimum and as such, use of wide angle projector is encouraged over narrow beam projector. But these projectors suffer from shadows and glare.
- This disadvantage can be reduced if the angle of incidence of beam is less than 45°. If the angle is greater than 45°, then asymmetric projectors are used.

Locations of projectors:

- One of the most important factor which affects the choice of projector is the location of projector lamps. Following are four possible locations of projectors which can be encountered in practical cases.
  - (i) Mounting the projector on the top of an adjacent building beam striking the surface normally [Fig. 1.69 (a)].
  - (ii) Same as (i) but beam striking the surface of an angle other than 90° [Fig. 1.69 (b)].
  - (iii) Mounting the projector in the fore court of the building [Fig. 1.69 (c)].
  - (iv) Mounting the projector at the roof of the building [Fig. 1.69 (d)].
- In first three methods, symmetrical projectors are used whereas in the last case we have to select asymmetrical projector.
- In Fig. 1.69 (a) the beam strikes the surface normally so we get uniform illumination. We have to pay the rent for locating the projectors and access has to be made to them for switching and maintenance purposes.
- The overhead projectors in Fig. 1.69 (b) are desirable for reducing glare and getting soft shadows. These projectors are suitable for yard lighting, quarrying, excavation or construction work.

Projector

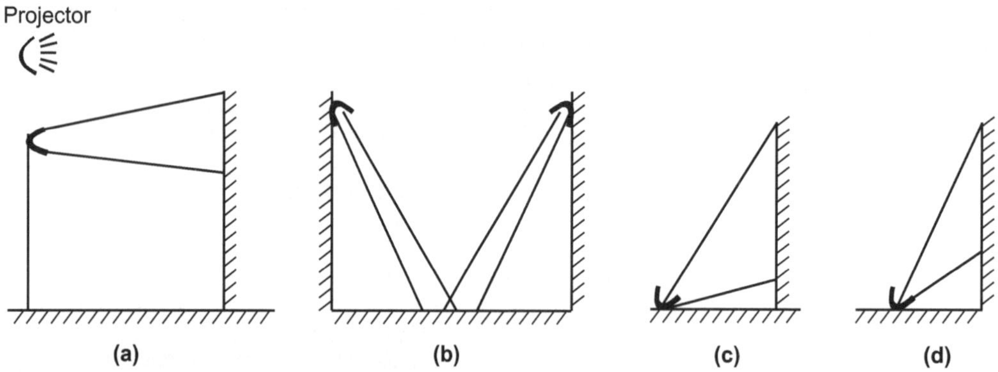

Fig. 1.69

- Location of projector in Fig. 1.69 (c) is possible where there is enough front court yard space. Projectors may be mounted on the pole or kept on the ground. Where there is no enough front court yard projectors are mounted on the brackets fixed to the wall itself and we have to use asymmetric projectors.
- Second important factor affecting the design is the illumination level of the surface. The illumination level of the surface is cancelled with the brightness of the surface. Illumination level when multiplied with the reflection factor of the surface gives us brightness of the surface. The illumination level depends upon the cleanliness of the surface and brightness of surroundings.
- Following are some sketches showing flood lighting, area and luminaires.

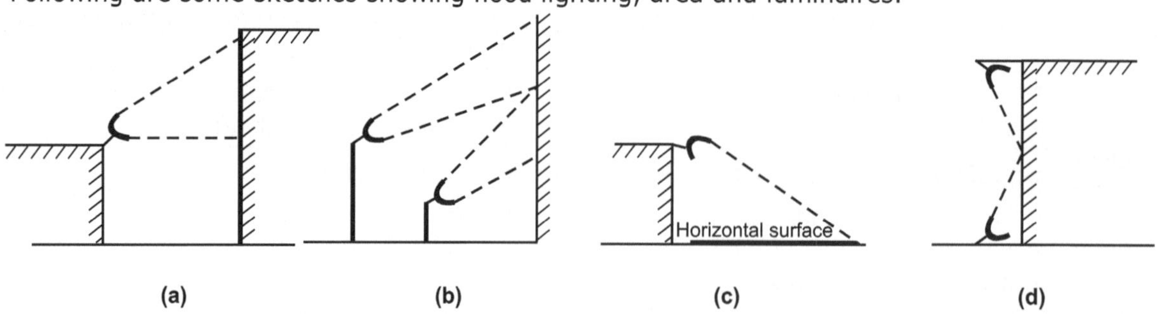

contd. ...

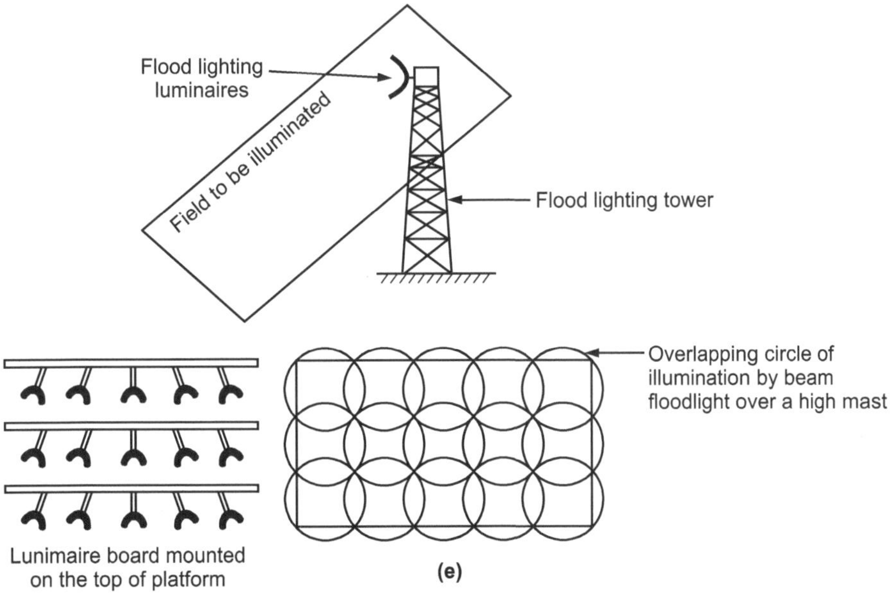

Fig. 1.70: Floodlighting schemes

Example 1.33: *A flood lighting is to be provided on the front of a building of 40 m × 25 m for brightness of 18 lumens/square meter. The coefficient of reflection of building surface is 0.21. The lamps of 500 W having lumen output 8400 each are used. If beam factor is 0.65, waste light factor is 1.1, maintenance factor is 0.85 then calculate number of lamps for flood lighting.*

Solution: To get brightness of incident illumination

$$= \frac{\text{Brightness (Lumens)}}{\text{Reflection factor}} = \frac{18}{0.21} = 85.71 \text{ Lux}$$

Let, Lamps = N

Total lumens output of N lamps = 8400 N

Considering the factors,

$$8400 \text{ N} \times \text{Beam factor} = \frac{\text{Lux} \times \text{Area} \times \text{Waste light factor}}{\text{Maintenance factor}}$$

$$8400 \text{ N} \times 0.65 = \frac{85.71 \times (40 \times 25) \times 1.1}{0.85}$$

∴ Number of lamps, N = 20.2 ≈ **20 Lamps**

1.9.3.3 Some Fitting, Holders, Pictures of Flood Lighting Luminaries

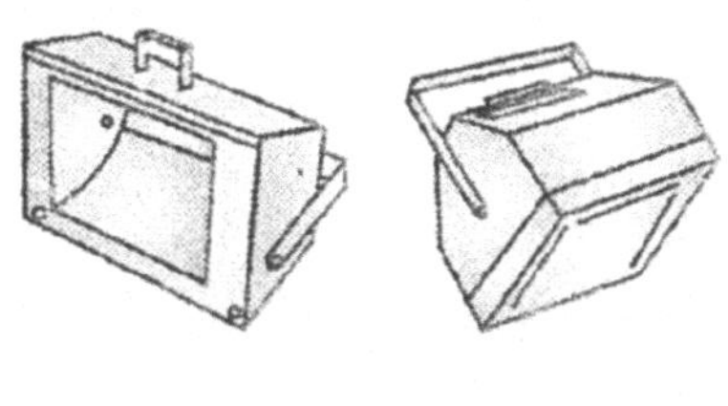

(a) Stage lighting

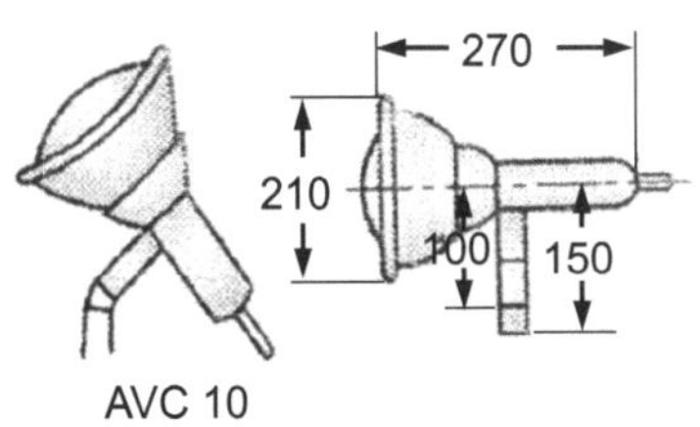

(b) Display lighting

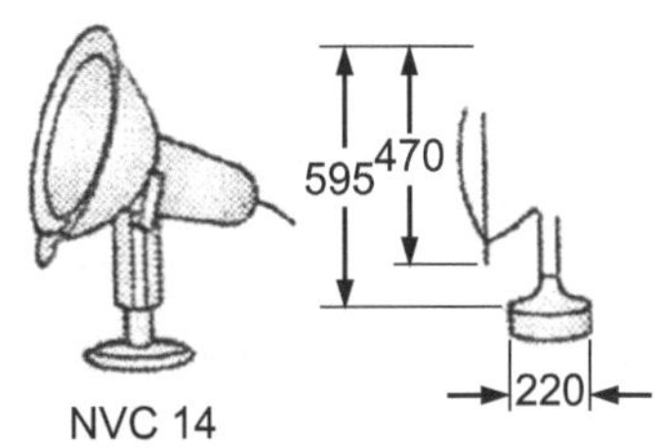

(c) Small flood lighting

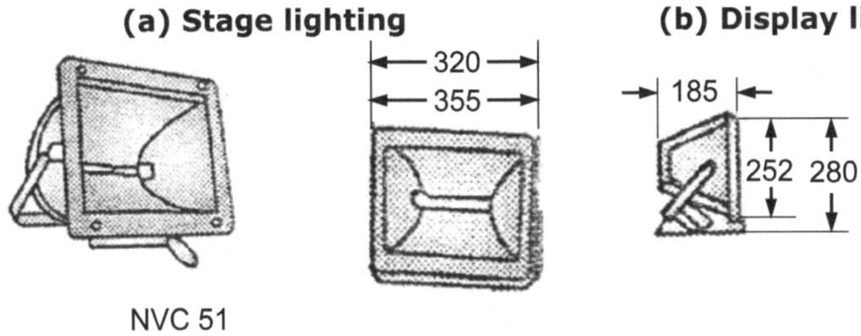

(d) Halogen flood lighting

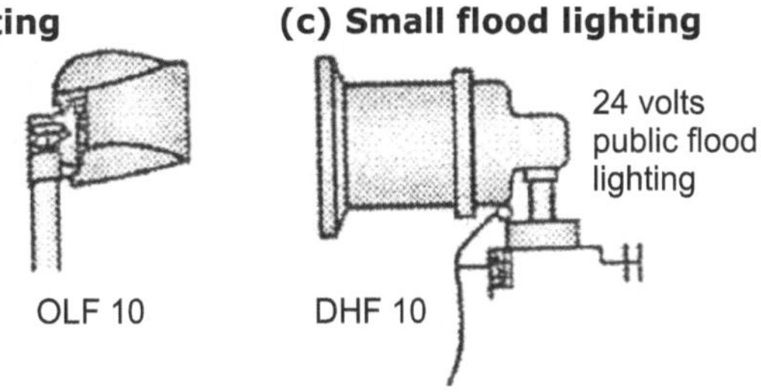

(e) Public flood lighting

contd. ...

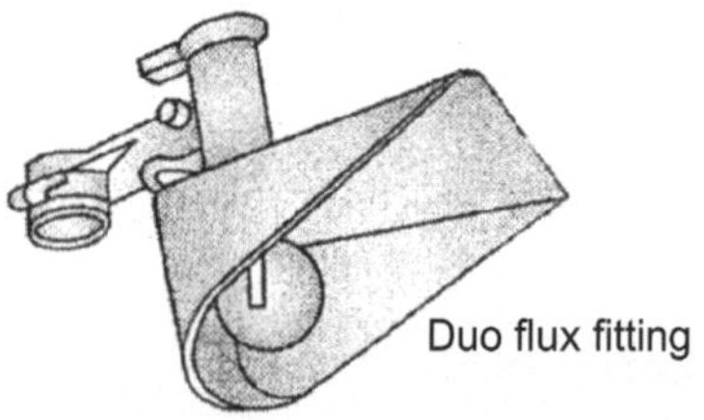

(f) Duo-flux yard flood lighting fitting

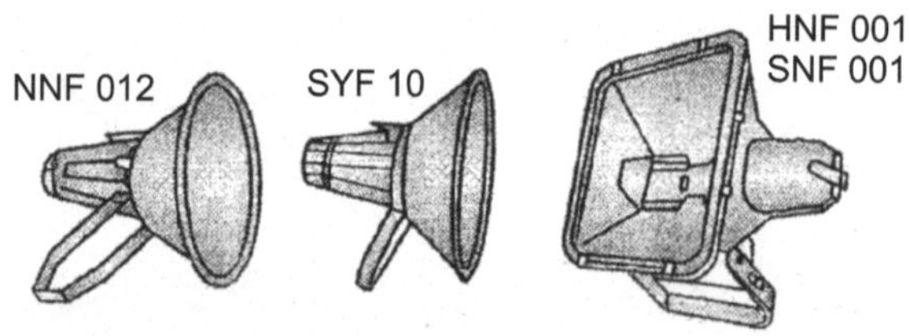

(g) Heavy flood lighting luminaires suitable for HPMV/HPSV lamps of 250 W and above

Fig. 1.71: Various types of flood lighting and display lighting luminaires

1.9.4 Luminaires

- **Basic function:** To provide electrical connections to the light source.
- **Other important functions:** To redirect light in the preferred directions with the minimum of loss, to reduce glare from the light source. Otherwise most of the light is wasted in undesired directions and may cause glare.
- Luminaires also provide acceptable appearance.
- Luminaires must therefore be durable, electrically, mechanically and thermally safe housing for the lighting source.
- It should provide sufficient insulation.

1.9.4.1 Illumination Level for Flood Lighting

Areas	Illumination in Lux
Landscapes	50 - 150
Construction work	40 - 60
Excavation work	10 - 20
Parking	5 - 10
General campus lighting for watching	2 - 5
Quarries	20 - 25
Decorative and specular effects	100, 200
Sign boards advertising	100 - 300
Water tanks	80 - 120
Railway yards	1 - 10
Dredging	40 - 60
Docks and loading platforms	20 - 30
Petrol pumps	20 - 150
Bus depot	20 - 50
Airport Apron	100 - 200
Ship yards	40 - 50
Stadium	100 - 300
Badminton court	200 - 300
Tennis court	150 - 300
Football and hockey	100 - 200
Indoor games	200 - 300

1.9.5 Calculations on Flood Lighting Scheme

Table 1.9: Calculations on flood lighting scheme

Building Surfaces	Reflection factor	Lux recommended for brightness of locality		
		Low	Medium	High
White terra cotta, Cream terra cotta and light marble	60 - 80	50	100	150
Light gray lime stone, Bed ford lime stone, Buff lime stone and Smooth buff face brick	40 – 60	70	120	200
Briar hill sand stone, Smooth gray brick, Medium gray lime stone and Common tan brick	20 – 40	100	150	250
Dark field gray brick, Common red brick and Brown stone	10 – 20	120	180	300

1.9.5.1 Following Steps will Help in Calculations

- **First Step: Illumination Level Required:** The illumination level (lumens/m^2) required depends upon the type of building, the purpose of the flood lighting, the amount of conflicting light in the vicinity, etc.
- **Second Step: Type of Projector:** Two considerations enter into the choice of a projector, viz. beam size and light output. The former determines the area covered by the beam and the later the illumination provided. Beam angle of the projector is decided keeping in view the distance of projector from the surface.
- **Third Step: Number of Projectors:** For any desired intensity over a definite surface, the number of projectors required is obtained from the following relation:

$$N = \frac{\begin{matrix}\text{Area of surface} \\ \text{to be illuminated} \\ \text{in square meters}\end{matrix} \times \begin{matrix}\text{Illumination level} \\ \text{required in} \\ \text{lumens/m}^2\end{matrix} \times \begin{matrix}\text{Depreciation} \\ \text{factor}\end{matrix} \times \begin{matrix}\text{Waste light} \\ \text{factor}\end{matrix}}{\text{Utilization factor} \times \text{Wattage of lamp} \times \text{Luminous efficiency of lamp}}$$

where, N = Number of projectors

A = Area of surface to be illuminated in square meters

E = Illumination level required in lumens/m^2.

- The terms used in flood lighting calculations are described below.
- These terms are same as discussed earlier. If in the example, the values of factors are not mentioned then one has to assume the value from the range given.

1.9.5.2 Factors Considered in Illumination Calculation

1. **Waste Light Factor:** Whenever a surface is illuminated by a number of light sources, there is always a certain amount of waste light on account of overlapping and falling of light beyond the edges of the area to be illuminated. This effect is taken into account by multiplying the theoretical value of lumens required by 1.2 for rectangular areas and 1.5 for irregular areas and objects such as statues and monuments etc.

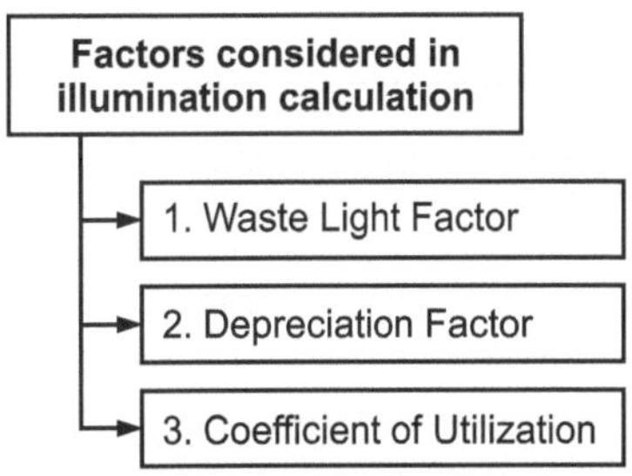

Fig. 1.72: Factors considered in Illumination Calculation

2. **Depreciation Factor:** Dirt and dust depositing on the reflector surface and on the source of light reduce the effectiveness of the projector and 50 to 100% more light must be provided so that the illumination shall be adequate at the end of the interval between the cleaning periods. Depreciation factor is defined *as the ratio of illumination under ideal conditions to the illumination under normal conditions.*
3. **Coefficient of Utilization:** It is also known as beam factor and is defined *as the ratio of beam lumens to the lamp lumens.* Its value lies between 0.3 and 0.5. This factor is due to the fact that all the light emitted by the projector is not along the direction of the beam but some of it is absorbed by the reflector and by the front glass. When the above losses are considerable the value of utilization is low. The coefficient of utilization factor goes up with deep reflectors having a good reflecting surface such as silvered glass.

1.9.5.3 Solved Examples

Example 1.34: *A flood lighting system is to be provided for a building having the surface area on the front side of 800 square metres. The illumination is by the light of total 16 lamps rated for 1000 W each. The arrangement is for uniform spreading of illumination. The luminous efficiency of lamps is 8 lumens/watt. Assume D.F. = 1.3, waste light factor = 1.2 and utilization factor of 0.4. Work out the illumination in lux. The surface measures 50 m × 16 m.*

Solution:

$$\text{Lumens by lamps} = \text{Wattage} \times \text{Number of lamps} \times \text{Efficiency}$$

$$= 1000 \times 16 \times 18 = 2,88,000$$

Lumens actually reaching on the surface < Lumens produced by lamps

$$\text{Lumens on surface} = \frac{\text{Lumens by all lamps} \times \text{Utilization factor}}{\text{Depreciation factor} \times \text{Waste light factor}}$$

$$= \frac{288000 \times 0.4}{1.3 \times 1.2} = 73846 \text{ lumens}$$

$$\text{Illumination } E_m, \text{ Lux} = \frac{73846}{\text{Area}} = \frac{73846}{800} = \mathbf{92.3\ lumens/m^3\ or\ lux}$$

Example 1.35: *A building frontage 50 m × 15 m is to be illuminated by flood lighting projectors situated 25 metres away. If the illumination is 100 lux, coefficient of utilization is 0.5, depreciation factor is 1.5, waste light factor is 1.2, estimate the number and size of projectors. Sketch the projectors recommended indicating the usual adjustments provided.*

Solution: Area to be illuminated,

$$A = 50 \times 15 = 750 \text{ m}^2$$

Total lumens required on the surface = Area to be illuminated × Illumination level required

$$= 750 \times 100 = 75,000 \text{ lumens}$$

As lamp rating and efficiency is not given, let us assume

Lamp rating of 1000 watts with luminous efficiency of 17 lumens/watt.

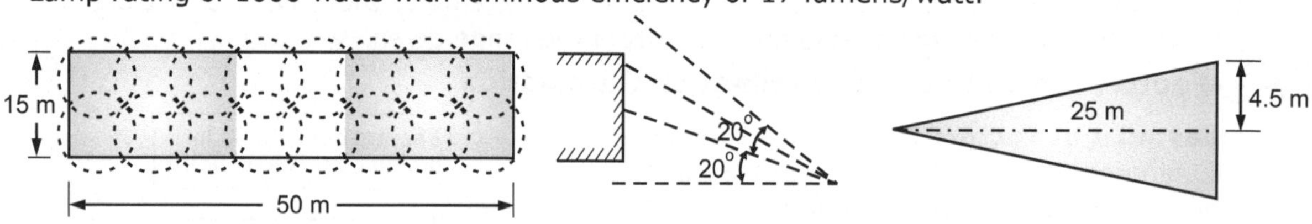

(a) (b)

Fig. 1.73: Location of projectors and illumination

Total lumens required to be given out by lamps,

$$\phi = \frac{\text{Total lumens required on the surface} \times \text{Depreciation factor} \times \text{Waste light factor}}{\text{Utilization factor}}$$

$$= \frac{75000 \times 1.5 \times 1.2}{0.5} = 2,70,000$$

$$\text{Number of projectors required} = \frac{\text{Total lumens required by lamps}}{\text{Wattage of lamp} \times \text{Luminous efficiency of lamp}}$$

$$= \frac{\mathbf{270000}}{\mathbf{1000 \times 17}} = \mathbf{16}$$

Hence, 16 projectors of 1,000 watts each with beam angle of 20° will be required.

Example 1.36: *The front of a commercial building 80 × 10 m in size is to be flood lighted by providing projector lamps placed 8 metres from the frontage. Determine number and size of projector required for flood lighting. Assume the factors: U.F. = 0.4, W.L.F. = 1.2, D.F. = 1.3.*

Select proper lamps from the given table:

Watts	300	500	1000
Lumens	5,000	9,000	18,000

Solution:

Area to be illuminated = 80 × 10 = 800 sq. m.

Intensity of illumination = 50 m candle

Total number of lumens required = 800 × 50 = 40,000

Waste light factor = 1.2

Depreciation factor = 1.3

Utilization factor = 0.4

Total number of lumens to be given out by the projectors,

$$\frac{40000 \times 1.2 \times 1.3}{0.4} = 1.56 \times 10^5$$

Using 1,000 watt lamps,

$$\text{Number of projectors} = \frac{1.56 \times 10^5}{18000} = 8.66 \text{ say } 9$$

But for symmetry use 10 with 1,000 watt projectors fitted with reflectors of 25°.

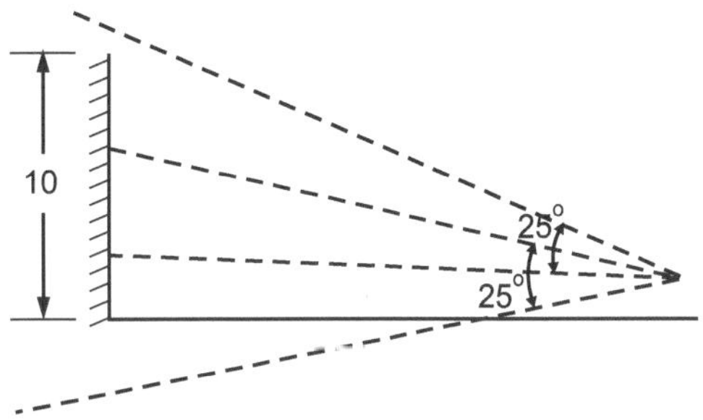

Fig. 1.74: Projector for flood lighting

Example 1.37: *The front of a building 50 m × 16 m is illuminated by 16 number 1000 watt lamps arranged so that uniform illumination on the surface is obtained. Assume:*

1. *Luminous efficiency = 17.4 lumens/watt*
2. *Utilization factor = 0.4.*
3. *Depreciation factor = 1.3.*
4. *Waste light factor = 1.2.*

Determine the illumination on the surface.

Solution: Area 'A' to be flood lighted = 50 × 16 = 800 m^2

Total lumens given out by all the lamps:

$$\phi = \text{Number of lamps} \times \text{Wattage} \times \text{Luminous efficiency}$$
$$= 16 \times 1000 \times 17.4$$
$$= 276400 \text{ lumens}$$

Due to different factors, the total lumens received on the surface will be less than the above calculated value of ϕ.

$$\therefore \quad \text{Lumens actually received} = \frac{\phi \times U{\cdot}F}{D.F. \times \text{Waste light factor}}$$

$$= \frac{278400 \times 0.4}{1.3 \times 1.2} = 71384.61 \text{ lumens}$$

Now area A = 800 m^2

Hence illumination in lumens/m^2 on the surface

$$E = \frac{\phi \text{ received}}{A} = \frac{71384.61}{800} = \mathbf{89.23\ lumens/m^2}$$

Example 1.38: *It is required to flood light the front of building of 30 m × 20 m to give brightness of 20 lumens/m². The co-efficient of reflection of the building surface is 0.2. The lamps of 500 watts having lumen output of 8450 each are used.*

Beam factor = 0.6, Waste light factor = 1.2, Maintenance factor = 0.8.

Calculate number of lamps of flood lighting.

Solution: 20 lumens/sq.m in order to get brightness of incident illumination will be

$$= \frac{\text{Brightness (Lumens)}}{\text{Reflection factor}} = \frac{20}{0.2} = 100 \text{ Lux}$$

$$\text{Total lumen output of N lamps} = 8450 \times N$$

Considering all factors,

$$\therefore \quad 8450 \times N \times \text{Beam factor} = \frac{\text{Lux} \times \text{Area} \times \text{Waste light factor}}{\text{Maintenance factor}}$$

$$\therefore \quad 8450 \times N \times 0.6 = \frac{100 \times (30 \times 20) \times 1.2}{0.8}$$

$$\therefore \quad \text{Number of lamps N} = 17.75 \approx \mathbf{18\ lamps}$$

Practice Questions

1. Give the illumination level required in Lumens/m² (Lux) for the following categories of works:
   (a) Very fine work (b) Fine work
   (c) Medium fine work (d) Plain work
   (e) Very plain work
2. Define the following terms related to illumination:
   (a) Light (b) Luminous flux (ϕ) or (F)
   (c) Luminous intensity (I) (d) Lumen
3. Define luminous flux in lumens and efficiency in lumens per watt of the following:
   (a) Tungsten lamp of input power 40 W; 60 W; 100 W.
   (b) Fluorescent lamp of input power 20 W, 40 W, 100 W.
4. Explain drawing a sketch: (i) Plane angle, (ii) Solid angle.
5. Define following and give their units:
   (a) Illumination/Illuminance (b) Luminous efficiency (Lamp efficiency)
6. Define the following factors:
   (a) Utilization factor (b) Maintenance factor
   (c) Depreciation factor
7. Define the following factors:
   (a) Waste light factor (b) Absorption factor
   (c) Beam factor
   Also give their range values.
8. State and explain inverse square law of illumination.
9. State and explain Lambert's cosine law of illumination.
10. Give features of good illumination scheme.
11. Give any ten advantages of good illumination scheme.
12. State the advantages of electrically produced light.
13. What are ranges of wavelength of (i) visible light, (ii) ultraviolet light, (iii) infra red light?
14. Define radiant efficiency.
15. List different types of lamps.
16. Draw constructional view of incandescent lamp and explain its working. Give its advantages.

17. Explain the working of carbon arc lamp and state its uses.
18. Draw a sketch and explain working of fluorescent lamp. What is the function of choke and starter in it?
19. Compare fluorescent tube with tungsten lamp.
20. What faults can develop in fluorescent tube? What are their reasons and what actions can be taken for repairs?
21. Explain with construction, the working of HPMV lamp.
22. Write a note on mercury iodide lamp.
23. Compare mercury iodide lamp with HPMV lamp.
24. State the uses of HPMV lamp and mercury iodide lamp.
25. Explain the working of sodium vapour lamp.
26. Write a note on Neon lamp.
27. Explain with diagram neon sign tube. Show its circuit arrangement.
28. Explain the working of halogen lamp. State its advantages and applications.
29. Explain the working of CFL.
30. Write a note on metal halide lamp and LED.
31. Give the range of illumination efficiency in lumens/watt of the following lamps:
    - (i) Incandescent lamp
    - (ii) Fluorescent lamp
    - (iii) CFL
    - (iv) HPMV lamp
    - (v) Sodium vapour lamp
    - (vi) Halogen lamp
32. Compare the following lamps on the points of (a) Life, (b) Colour of light:
    - (i) Incandescent lamp
    - (ii) Fluorescent lamp
    - (iii) HPMV lamp
    - (iv) Sodium vapour lamp
33. State the effect of shadows and glare produced by lamps.
34. Explain how uniformity in illumination is obtained.
35. State the importance of mounting heights and spacing between the lamps for illumination control.
36. What different types of dimmers are used for light control ?
37. Explain resistance type dimmer with neat diagram. State its advantages.
38. How salt water dimmer control the light ? Explain its working in brief.
39. Explain autotransformer as a dimmer.
40. Can two winding transformer be used as a dimmer ? Explain in brief.
41. State the types of electronic dimmer.
42. Draw the circuit of thyristor type dimmer and explain its working.
43. Explain TRIAC operated dimmer for light control.
44. Draw the circuit for controlling one lamp by single switch.
45. Draw the circuit for controlling single lamp from two places. Explain the sequence of operation.
46. Draw the circuit to control a single lamp from three places. Which switches are used for the circuit?
47. Can a single lamp be controlled from four places? If yes, then draw the circuit arrangement and explain the control. Which type of switches are needed for this ?
48. Which factors are considered while designing the lighting scheme for internal application ?

49. State the lighting design objectives of internal applications.
50. State and explain with relevant figures
    (i) Choice of general lighting. (ii) Localised lighting. (iii) Local lighting.
51. Write a note on choice of lamps for interior lighting.
52. Write a note on control of lighting.
53. Draw a process (flow) chart expressing lighting objectives.
54. Write a detailed note on home lighting.
55. Explain lighting systems in home lighting, such as:
    (i) General lighting. (ii) Decorative lighting. (iii) Utility lighting.
    (iv) Mood lighting. (v) Spot and focus lighting. (vi) Security and safety living.
56. Write a note on: Lighting sources, Accessories, Fittings, Luminaire for home lighting.
57. Give a chart of illumination level required for different parts of home lighting.
58. What are methods of lighting calculations for house illumination scheme? Explain any one method.
59. What are the different areas in a hotel ?
60. Explain lighting scheme in the following areas of hotel:
    (i) Entrance hall. (ii) Extension to the entrance hall.
61. Explain lighting scheme in the following areas of hotel:
    (i) Reception suits (ii) Dining rooms
62. Explain lighting scheme in the following areas of hotel:
    (i) Corridors, stairs, lift. (ii) Bed rooms.
63. Explain lighting scheme in the following areas of hotel:
    (i) Restaurants. (ii) Pubs and clubs.
64. State the functional requirements of "offices".
65. State the illumination required in the following sections of the office work:
    (i) Conference rooms (ii) General office (iii) Entrance hall
    (iv) Drawing office (v) Corridors/Lift cars (vi) Stairs.
66. State the general rules for interior lighting of office.
67. State the requirements and which luminaires are used and minimum illumination required etc. for a general office.
68. State the meaning of flood lighting and also state the purpose and applications.
69. What Luminaries are required for flood lighting.
70. Show the location and mountings of projectors for flood lighting.
71. Give the illumination levels for:
    (i) Land scapes (ii) Indoor games (iii) Ship yards
    (iv) Parking (v) Bus Depot (vi) Stadium
72. Explain the following factors:
    (i) Waste light factor (ii) Depreciation factor (iii) Utilization factor

■■■

Chapter **2...**

Electric Heating and Welding

Syllabus

2.1 Electric Heating

2.1.1 Advantages of Electric Heating.

2.1.2 Modes of Transfer of Heat:

- Conduction, Convection and Radiation.

2.1.3 Classification of Electric Heating Methods:

2.1.4 Resistance Heating: (Construction and Operation)

- Direct Resistance Heating: Salt Bath Furnace.
- Indirect Resistance Heating: Resistance Ovens, Requirements of Heating Element Material, Causes of Failure of Heating Elements, Methods of Temperature Control.
- Applications of Resistance Heating.

2.1.5 Arc Heating: (Construction and Operation)

- Direct Arc Furnace.
- Indirect Arc Furnace.
- Applications of Arc Heating.

2.1.6 Induction Heating: (Construction and Operation)

- Core Type Induction Furnaces: Ajax Wyatt Furnace.
- Coreless Induction Furnace.
- Applications of Induction Heating (Simple Numericals on Melting Furnaces).

2.1.7 Dielectric Heating:

- Principle of Dielectric Heating.
- Advantages of Dielectric Heating.
- Limitations of Dielectric Heating.
- Applications of Dielectric Heating (Simple Numericals on Dielectric Heating).

2.1.8 Eddy Current Heating

- Principle, Advantages and Applications

2.2 Electric Welding

2.2.1 Methods of Electric Welding: Electric Arc Welding, Resistance Welding.

2.2.2 Resistance Welding:

- Principle of Resistance Welding.
- Advantages of Resistance Welding.
- Types of Resistance Welding - (Only List)

2.2.3 Spot Welding Machine.

2.2.4 Electric Arc Welding:

- Formation and Characteristics of Electric Arc.
- Effect of Arc Length.
- Arc Blow.

2.2.5 Polarity in DC Welding

2.2.6 Electrodes for Metal Arc Welding

2.2.7 V-I Characteristics of Arc Welding

2.2.8 Arc Welding Machines: [Only List]

- DC Welding Machines – MG Set, AC Rectified Welding Unit.
- AC Welding Machines – Welding Transformer.

Objectives

- Explain with sketches and broad specifications working principle of the specified electrical heating system.
- Describe with sketches construction of the specified electric furnace.
- Recommend the relevant heating system for the given application with justification.
- Design the heating element of the given type of furnace for the specified data.
- Select the relevant welding system for a specified application with justification.
- Describe the procedure to maintain the given type of heating and welding system.

2.1 ELECTRIC HEATING

2.1.1 Concept of Electric Heating

- Fire (heat) and electricity are the basis of present day civilization. Man used fire to cook food, to warm himself during cold weather, to make tools and weapons. Later in modern times he used fire to extract metals from earth's crust, to raise steam from water, to move turbines of generator to produce electricity.
- With great discovery of electricity, it was found that many things are common with electricity and heat. The various effects of electrical current are utilized for heating purpose. When current is passed through a conductor, heat is produced (Resistance heating). If a magnetic material is subjected to alternating flux, heat is produced in the material (induction heating). Similarly, if a insulating or dielectric material is subjected to alternating electric field, the dielectric loss is produced causing heating (dielectric heating).
- Thus, effective use of effects of electricity – I^2R effect, induction effect, dielectric loss etc. can be utilized effectively for heating of different category of materials.

2.2 CLASSIFICATION OF ELECTRIC HEATING METHODS

1. Give classification of electric heating.

- Fig. 2.1 shows tree pattern for classification of electric heating.

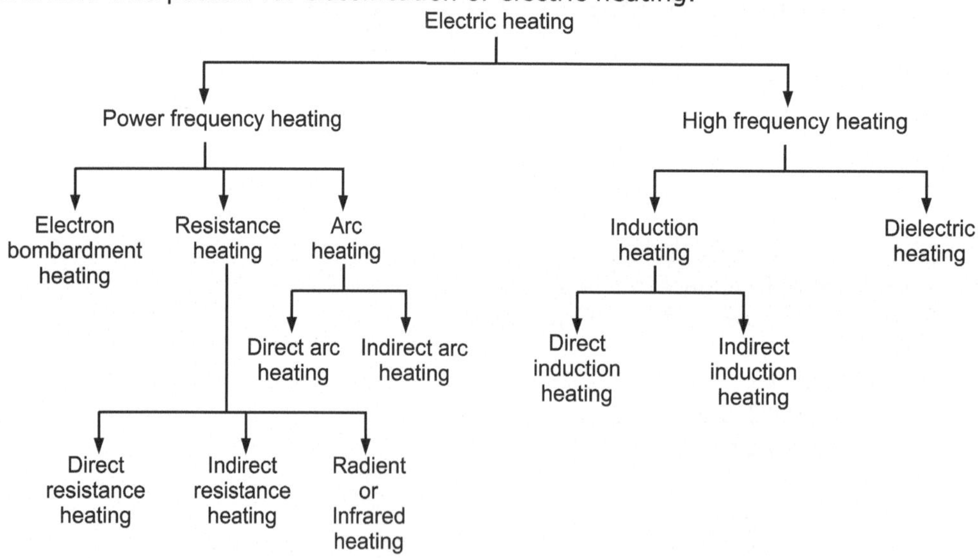

Fig. 2.1: Classification of Electric Heating

- Electric heating can be broadly classified under two categories:
  (i) Power frequency heating in which the furnace operates with 50 Hz A.C. supply.
  (ii) High frequency heating in which special high frequency generators are essential, e.g. dielectric heating, induction heating.

Electron bombardment heating: Bombardment of electron causes heating.

Resistance heating:

(i) Direct resistance heating: The electric current is made to pass through the substance to be heated or charged itself. The electric current while passing through charge produces I^2R loss which appears in the form of heat, thus charge is heated up.

Few examples of this heating method are: resistance welding, electrode boiler for heating water, salt bath furnace.

(ii) Indirect resistance heating: Current is passed through heating element and I^2R loss is produced which appears in the form of heat. This heat is passed on to the substance or charge to be heated by radiation and convection.

e.g. room heater, hair drier, soldering iron, immersion water heater, hot plate, frying pan, electric oven etc.

Arc heating: The heat energy of arc produced by electric means is utilized for heating the substance or charge. The arc temperature is about 3000 to 3500°C.

(a) Direct arc heating: An arc is made to struck between electrode and charge itself. The heat energy of arc is absorbed by the charge and thus heating is done.

(b) Indirect arc heating: An arc is made to struck between two electrodes. The heat of arc is then passed on to the charge through radiation.

(iii) Radiant or Infrared heating: Heat energy from electric lamp is focussed on the charge to be heated. The heat energy is transferred through electromagnetic radiations. This is used for drying the paint on objects.

1. Differentiate direct and indirect arc furnace.

Sr. No.	Item	Direct Resistance Heating	Indirect Resistance Heating
1.	Current	Directly passed through substance to be heated.	Current is passed through heating element.
2.	Mode	Direct conduction.	By the mode of radiation and convection.
3	Examples	Boiler, salt bath furnace, resistance welding.	Room heater, hair dryer, soldering iron, hot plate, electric oven.

High Frequency Heating:

- **Induction heating:** In this method, currents are induced in the charge by electromagnetic induction action and circulation of these currents in the charge causes its heating. This is used in metallurgical industries for melting the metals.
  - **(i) Direct induction heating:** Currents are induced in the charge by electromagnetic induction action, in case of steel and other metals, these currents are sufficient enough to melt the metals. The equipment used for melting is known as induction furnace and process used for general heat treatment of metals is known as eddy current heating.
  - **(ii) Indirect induction heating:** Eddy currents are induced in the heating element by electromagnetic induction, the heat so produced is transferred to the charge by radiation and convection, certain heat treatment methods for metals make use of this method.
- **Dielectric Heating:** Dielectric loss is produced in the charge itself when it is subjected to alternating electric field. This dielectric loss appears in the form of heat, thus charge is heated up.

2.3 ADVANTAGES AND DISADVANTAGES OF ELECTRIC HEATING

1. State advantages of electric heating.

(i) Clean operation: There is no coal ash or smoke etc. hence the operation is clean, cost of cleaning is very less.

(ii) No pollution: Absence of flue gases, smoke, dust hence it does not produce any contamination or pollution in atmosphere.

(iii) Ease of control: The temperature can be controlled easily over a wide range just by switching OFF or ON the number of heating elements manually or automatically. So possibility of overheating/under heating is avoided. Temperature can be maintained constant.

(iv) Low ambient temperature: The temperature around an electric furnace is much lower as compared to other heat treatment methods which may be troublesome for the workers/operators.

(v) Localised application: A workpiece can be heated to the required depth for heat treatment, this is not possible with non-electrical heating methods.

(vi) Uniform heating: Uniform heating is possible with induction heating.

(vii) Heating of bad conductor of heat and electricity: Wood, plastic and bakery items can be heated uniformly and suitably with dielectric heating.

(viii) Highest efficiency of utilization: Due to less losses of heat the efficiency (η) of operation is very high.

(ix) Less attention: Electrical heating equipment requires less attention in comparison to other types of heating.

(x) Cheapness: Electrical furnace does not require big space as other furnaces. Also other accessories such as coal storage, fire wood, chimney are absent. Hence, cost is reduced to great extinct. Running cost is also very less.

(xi) Quick heating: Time required for increasing the temperature is less as compared to other methods of heating.

(xii) More suitable for large scale production.

(xiii) Mobility of job: The job to be heated can be passed through heating furnace by conveyor belts.

2.4 ELECTRODES USED FOR ARC FURNACE

1. Carbon
2. Graphite
3. Soderberg are the various materials used for electrodes in arc furnace.

Properties of electrode materials:

(i) Good electric conductivity

(ii) Insolubility

(iii) Infusibility

(iv) Chemical inertness

(v) Good mechanical strength

(vi) Better resistance to thermal shock.

- **Comparison of graphite electrode over carbon electrode:**

Sr. No.	Point of Comparison	Carbon Electrode	Graphite Electrode
1.	Cost	Cheaper.	Cost is more double of carbon.
2.	Nature, obtained from	These are amorphous.	Obtained by heating carbon electrode at high temperature.
3.	Specific resistance	More.	Lesser (0.25 of carbon).
4.	Current carrying capacity	Less (as specific resistance is more).	More (as specific resistance is lesser).
5.	Use	For small capacity furnace.	Used for large capacity furnace.

2.5 MODES OF TRANSFER OF HEAT

- Heat is always transferred from hot body to cold body. e.g. from the heating element to the body to be heated. This transfer of heat is due to conduction, convection, radiation or by a combination of these methods.

1. **Conduction:** In this mode of heat transfer, heat is transferred through a substance from one part to another i.e. one molecule of substance gets heated and transfers the heat to adjacent molecule and so on.
   The rate at which heat is conducted through a substance depends on the temperature gradient. The heat flow takes place till there is difference in temperature and continuity.
   This conduction mode of heat transfer is important in case of heating insulating materials, refractory etc.
   The conductivity is expressed in MJ per hour per square meter per meter or watts per square centimeter per centimeter.
2. **Convection:** In this mode, heat transfer takes place by actual motion of the molecules of the substance and it mostly takes place in liquids.
   e.g. immersion water heater, when the heater is immersed in water, a continuous flow of water is initiated in the vicinity of water heater (rods). This flow is in upward direction. Because of this heated water goes in upward direction and during its motion, the heat is also passed on to the entire mass of the water. Thus, entire mass of water is heated up.
   For vertical heating surface in air, the heat dissipated due to convection is given by:
   $$H_c = 3.875\,(T_1 - T_2)^{1.25} \text{ watts/m}^2$$
   where, 3.875 = Constant (value depends on heating surface arrangement i.e. shape and size).
   T_1 = Absolute temperature (in ^{o}C) of heating surface
   T_2 = Absolute temperature (in ^{o}C) of air

- **Radiation:** In this mode, heat transfer takes place by means of heat waves. These waves do not heat the medium in between two bodies and heat the body which intercepts these waves.
- The rate of heat dissipation by Stefan's law is given by,

$$H = 5.72 \text{ K.e.} \left[\left(\frac{T_1}{100}\right)^4 - \left(\frac{T_2}{100}\right)^4\right] \;\ldots \text{ watts/m}^2$$

where, K = Constant (radiant efficiency)
e = Emissivity
T_1 = Temperature of source in Kelvin
T_2 = Temperature of substance to be heated in Kelvin

The heat transfer rate by these conventional methods is as follows:

(i) Conduction – 20 watts/cm^2
– 1000 watts/cm^2 (for intense flame)
(ii) Convection – 0.5 watts/cm^2
(iii) Radiation – 8 watts/cm^2

2.6 RESISTANCE HEATING

- The resistance heating process is based on heating effect of electric current i.e. I^2R effect. This method is widely used in industrial applications such as heat treatment of metals, stoving of enameled wave, drying and baking of potteries.

2.6.1 Properties of Material Used for Heating Element

1. What are the desired properties of heating materials?
2. What are the requirements of heating materials? Give minimum four names of heating materials.

- The material used for heating elements should have following properties:
  (i) Low temperature coefficient of resistance: The variation in resistance with temperature should be as small as possible, so that resistance should not change much more when temperature is increased.

(ii) **High melting point:** Melting point should be sufficiently higher than operating temperature.

(iii) **High specific resistance:** The resistivity of material used should be high due to which length will be smaller.

(iv) Its oxidising temperature should be higher.

(v) **Ductile:** To have convenient shapes and sizes the material should be ductile.

(vi) **Mechanical strength:** It should withstand vibrations so it should be mechanically strong.

(vii) **It should be anti-corrosive:** It should not be affected due to chemical fumes.

(viii) **Cost:** Cost should be minimum.

- Table 2.1 shows properties of some commercial heating elements.

Table 2.1: Properties of some Commercial Heating Elements

Type of alloy	Ni-Chromium	Nickel-Cr-Fe	Ni-Cu	Fe-Cr-Al
Composition	80% Ni, 20% Cr.	60, 16, 24	45, 55	70, 25, 5
Commercial name	Nichrome		Eureka or Constantan	Kanthal
Maximum working temperature	1150°C	950°C	400°C	1200°C
Specific resistance at 20°C	109 $\mu\Omega/cm^3$	110 $\mu\Omega/cm^3$	49 $\mu\Omega/cm^3$	140 $\mu\Omega/cm^3$
Specific gravity	8.36	8.28	8.88	7.2

2.6.2 Causes of Failure of Heating Elements

1. Describe any four causes of failure of heating element.
2. State the causes of failure of heating element.
3. What are the causes of failure of heating element?

- Following are the causes:
  1. Formation of hot spots
  2. Oxidation
  3. Corrosion
  4. Mechanical failure.

1. **Formation of hot spots:** The filament may break where it shines brightest during its operation that means the temperature at that particular spot is higher compared to rest of the filament. This is called formation of hot spots. It may be created due to following causes:

(i) **Unequal spacing:** If spacing between the heating element is non-uniform then the temperature will be maximum where spacing is minimum. And thus hot spot may be formed.

(ii) **Supporting structure:** If supporting structure is a bad conductor of heat then it will not transfer any heat, hence temperature of the heating element will be higher near the supports resulting in formation of hot spots.

(iii) **Excessive oxidation:** The portion where the filament is oxidized offers comparatively higher resistance for the same current so heat produced in that portion will be more resulting in formation of hot spots.

2. **Oxidation:** The outer surface of the heating element which is open to atmosphere gets oxidized due to high temperatures. During switching operation (i.e. ON or OFF) the oxide layer gets flicked off and due to this the inner surface is now open to atmosphere. This procedure continues due to which that particular portion of heating element gets thinner and thinner i.e. its cross-sectional area reduces. Thus, hot spots are formed. Further, the heating element may break.

3. **Corrosion:** Chemical fumes produced during industrial operations corrode the surface of the heating element where the actual contact of fumes with the heating element occurs. Due to this failure of the heating element occurs.
4. **Mechanical Failure:** During alloying, a portion of the heating element may have a higher content of higher resistivity material so this portion will produce more amount of heat for the same current. Thus, the heating element may be damaged.

2.6.3 Design of Resistance Heating Element

- When the heating elements reach steady-state temperature, all electrical input is given out as heat. According to Stefan's law the heat radiated to the charge is given by:

$$H = 5.72\ K.e. \left\{ \left[\frac{T_1}{100}\right]^4 - \left[\frac{T_2}{100}\right]^4 \right\} \text{ watt/m}^2.$$

where, T_1 = Temperature of source in ^oC absolute
T_2 = Temperature of object absorbing the heat in ^oC absolute
K = Constant called radiating η
(1 for single element and 0.5 to 0.8 for several elements)
e = Emissivity
(1 for black body and 0.9 for resistance heating element)

$$\boxed{^oC \text{ abs} = {^oC} + 273}$$

Let V = Applied voltage to heating element in volt
R = Resistance of heating element in Ω

$$R = \rho \frac{l}{a}$$

If circular conductor is used, $a = \frac{\pi d^2}{4}$

(d = diameter of circular conductor, l = length of conductor, ρ = resistivity of conductor material)

$$\therefore \quad R = \rho \frac{l}{\frac{\pi d^2}{4}}$$

$$R = \frac{4\rho l}{\pi d^2}$$

Now, $$P = \frac{V^2}{R} = \frac{V^2}{\frac{4\rho l}{\pi \rho l}}$$

$$P = \frac{\pi d^2 V^2}{4\rho l}$$

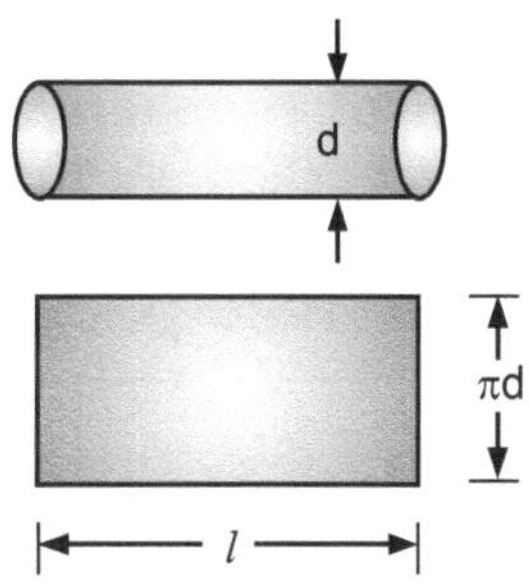

Fig. 2.2: Circular Heating Elements

$$\therefore \quad \boxed{\frac{l^2}{d^2} = \frac{\pi V^2}{4\rho P}} \quad \dots (2.1)$$

Surface area = S = $\pi\ dl$ and Heat dissipated / Unit area = H

$\therefore$ Total heat dissipated = $\pi\ dl \cdot H$

At steady-state temperature,

Power input = Total heat dissipated

$$\boxed{P = \pi\ dl \times H}$$

$$\therefore \quad \frac{\pi d^2 V^2}{4\rho l} = \pi\ dl \times H$$

$$\frac{d^2}{l} = \frac{\pi dl \times H \times 4\rho}{\pi V^2}$$

$$\boxed{\frac{d}{l^2} = \frac{4\rho H}{V^2}} \quad \dots (2.2)$$

Solving equations (2.1) and (2.2), length and diameter of wire are found.

If flat rectangular cross-section conductor of width 'w' and thickness 't' is used. So area 'a' = w.t.

$$P = \frac{V^2}{R} = \frac{V^2}{\rho \frac{l}{a}} = \frac{V^2}{\rho \frac{l}{w \cdot t}}$$

$$P = \frac{V^2 w \cdot t}{\rho l}$$

$$\therefore \quad \boxed{\frac{l}{wt} = \frac{V^2}{\rho P}} \quad \dots (2.3)$$

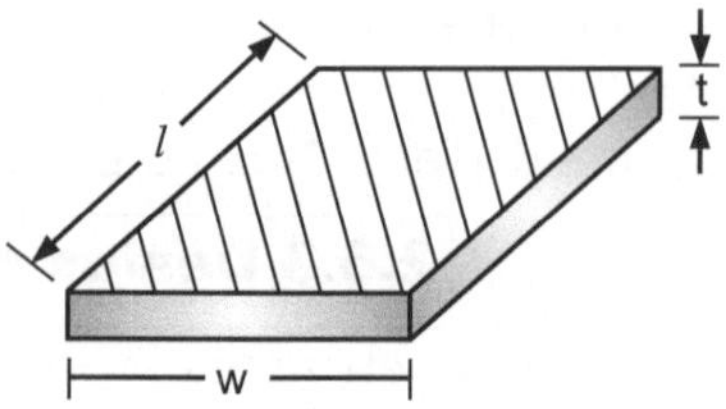

Fig. 2.3: Rectangular heating element

At steady-state temperature,

Power input = Heat dissipated

$P = 2wlH$ (neglecting side area $2tl$ because thickness is very very small)

or $$P = \frac{V^2 w \cdot t}{\rho l} = 2wlH$$

$$\boxed{\frac{t}{l^2} = \frac{2H\rho}{V^2}} \quad \dots (2.4)$$

Solving equations (2.3) and (2.4), l, t, w are found out.

2.7 METHODS OF TEMPERATURE CONTROL OF RESISTANCE FURNACES

- In electric furnace the constant voltage energises the heating element. Now, even if the voltage is constant the temperature will increase with the passage of time and it may cross the desired level of temperature. So to avoid this controlling mechanism is required.
- In large furnaces the control may be electronic control or electromagnetic control involving servomechanism.
- Following methods are used for temperature control:

  (i) Variation of Number of Heating Elements: In this type, instead of cutting off the supply of heating element the number of heating elements are changed as per the requirements.

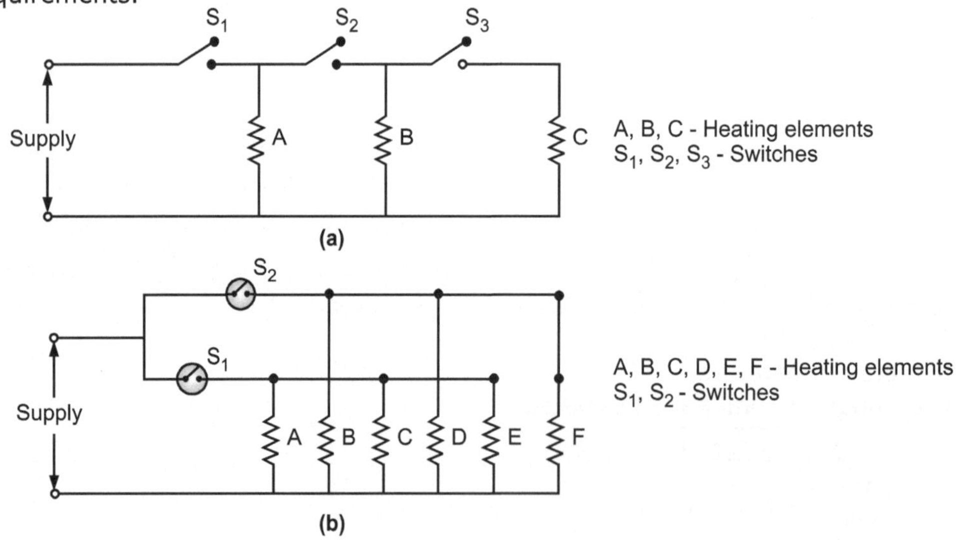

Fig. 2.4: Temperature Control by Variation of Number of Heating Elements

In case of a 3ϕ circuit one has to take care that equal number of heating elements should be switched 'ON' or 'OFF' (as per requirement) to maintain the balanced condition.

The uniform spacing of heating elements also plays an important part as it provides informality of operating temperature.

(ii) Change of Applied Voltage: This is achieved by following methods.

(a) By providing Tappings on Transformer: The power dissipation voltage is less if the magnitude of applied voltage is less and hence naturally temperature produced is also less. In case of furnace transformer tappings are provided on the H.V. side. It is easy to provide the tappings on the H.V. side because the magnitude of current is lesser.

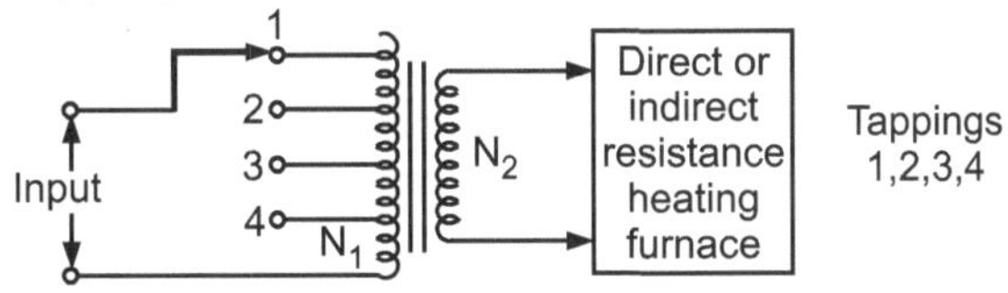

Fig. 2.5: Temperature Control by Change of Applied Voltage

(b) Bucking-Boosting the Secondary Voltage: In this case, the secondary is having two sections, one having small number of turns and the other having more number of turns. Now, if the connections are made as shown in Fig. 2.6 (a); the direction of current in the secondary sections is same, hence there is addition of emfs. So the voltage fed to furnace is more. This action is boosting action.

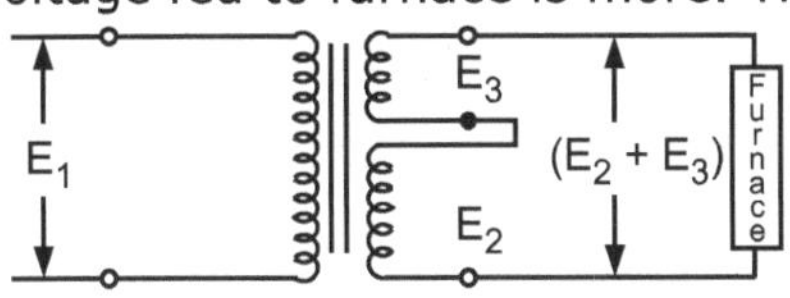

(a) Boosting connection

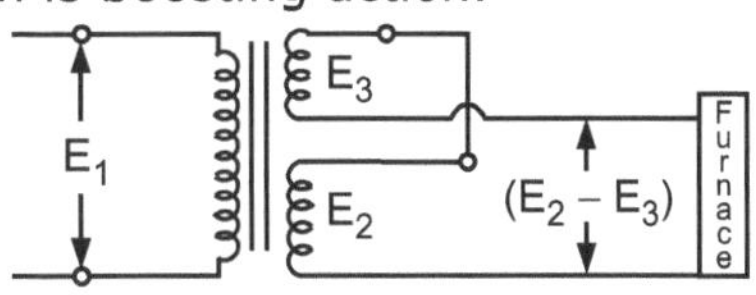

(b) Bucking connection

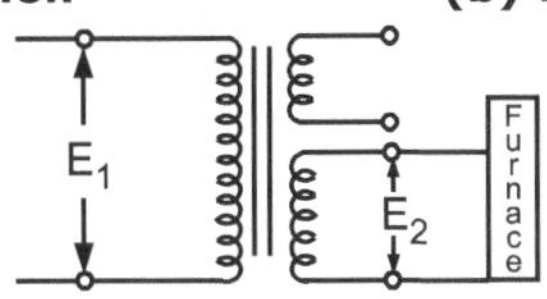

(c) Without boosting/bucking

Fig. 2.6: Temperature Control by Bucking-boosting

Now when the connections are made as shown in Fig. 2.6 (b) the induced emfs in both the windings come in opposition. This is Bucking effect i.e. the voltage of secondary is reduced and hence the voltage fed to the furnace also decreases.

The third possibility is to provide the voltage of extra secondary to the furnace. Thus, the three voltages available are:

(i) $E_2 + E_3$ – boosting

(ii) $E_2 - E_3$ – bucking

(iii) E_2 only – no bucking or boosting

(c) Auto Transformer Control: This is used for small resistance furnaces. The voltage supplied to the resistance furnace is controlled by an auto transformer. Thus, temperature control is obtained. The circuit used is as shown in Fig. 2.7.

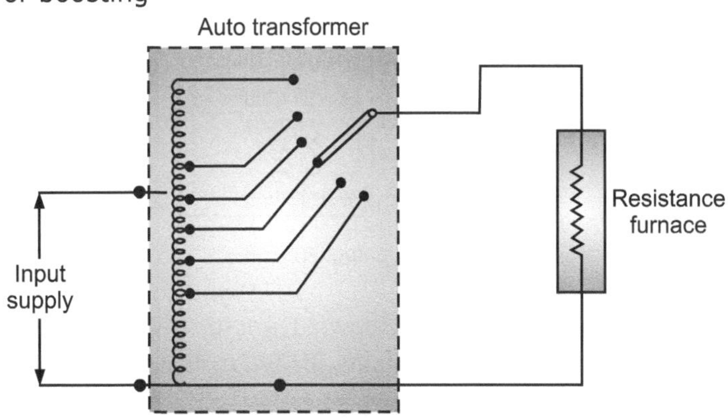

Fig. 2.7: Temperature Control by Autotransformer

Applications of Resistance Heating:

(a) For metal processing – annealing, general heat treating, heating for case hardening, melting – lead, tin, zinc etc., enameling.

(b) For ceramic – treating china earthenware.

(c) For chemical purpose – cracking of gases.

(d) For miscellaneous uses e.g. cremation furnaces.

2.8 DIRECT RESISTANCE HEATING

1. Explain the operation of direct resistance furnace with a neat diagram.
2. Draw the sketch of salt bath furnace. Name the various parts and explain its operation.

- As said earlier, in case of direct heating the current is passed through charge which produces I^2R loss in the charge and thus charge is heated up.
- Simple arrangement of direct resistance heating is shown in Fig. 2.8.

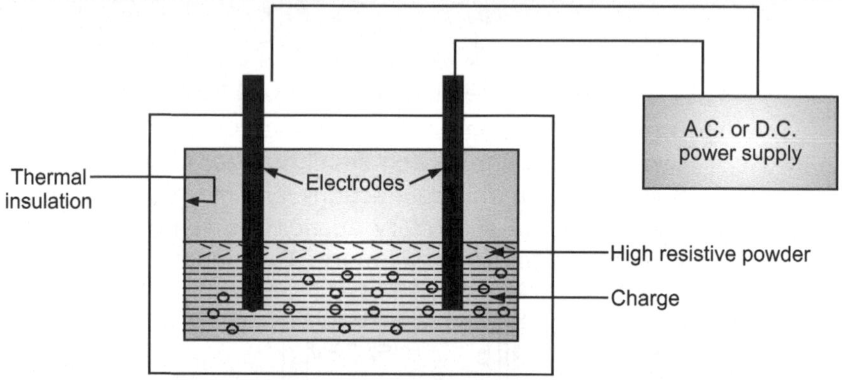

Fig. 2.8: Direct Resistance Heating

- The charge may be in solid or liquid form. Two electrodes are immersed in the charge. If the charge has a very low resistance the two electrodes may be cause of short circuit, so a high resistance powder is generally sprinkled over the charge. The current from power supply is passed to the charge through the electrodes thus the charge is heated up. This method is used in salt bath furnace and electrode boiler for heating water.

Salt Bath Furnace

- Salt can conduct electrical current. Electrodes are immersed in salt (sodium chloride). The fusing point of salt is 1000°C and it can be heated to 1500°C or more. There is circulation of salt after it is heated up. This can provide uniform heating for the charge. Fig. 2.9 shows electrode type salt bath furnace.

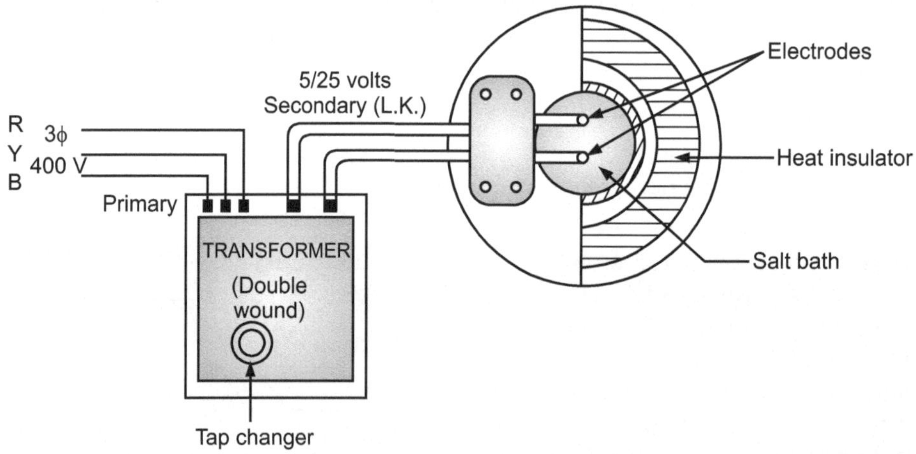

Fig. 2.9: Salt Bath Furnace

- **Construction:** A step-down transformer provided with tappings 400/5 to 20 volt is used for this furnace, it has a 3ϕ primary and 1ϕ secondary. The electrodes are connected to the secondary as shown in Fig. 2.9. These electrodes are put in the bath and current is circulated through the electrodes. The salt in solid state is put in the bath. In the solid state, it is insulator so initially some molten salt is poured to start conduction of current. If molten salt is not available then a separate arrangement is used to heat the salt in solid state between the electrodes. Now the circuit is charged with a lower voltage. Electrodynamic force is produced due to which stirring action occurs resulting in circulation of molten salt.

- The *UP and DOWN movement* of the electrodes for effective control is provided by a *separate motor*. Thermocouple parameters are used to measure the temperature inside the bath. The temperature of the bath is between 500 to 1400°C.

Use: For heat treatment such as hardening, annealing, tempering etc.

Precautions:

(i) The poisonous gases which are produced during stirring action of molten salt must be thrown out. This is done with the help of exhaust fans.

(ii) The bath material should not be too conductive in its solid state because of this short circuit may take place. So to prevent this it is desirable to sprinkle some non-conducting powder between the two electrode tips.

2.9 INDIRECT RESISTANCE HEATING (RESISTANCE OVEN)

- In indirect heating the current is passed through some heating elements and this heat is passed to the charge through radiation, convection.
- Bath type of resistance furnace (resistance oven) is an example of indirect resistance heating.

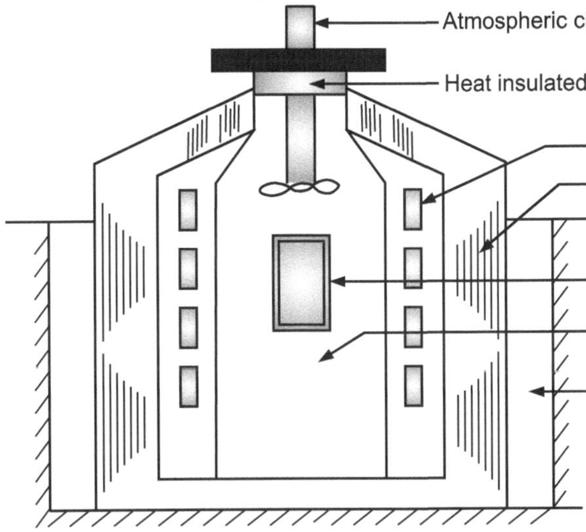

Fig. 2.10: Bath Type Resistance Furnace

- **Construction:** A typical bath type resistance furnace is shown in Fig. 2.10. The inside chamber in which the body to be heated is kept which is surrounded with heating element made up of nichrome. The chamber and the door are made air tight with the help of rubber gasket. The chamber again is surrounded by thermal insulation so that, the heat is retained inside and it will not pass out. An atmospheric circulating fan unit is provided at the top so that heat is transferred to the body by convection.

  For automatic temperature control, a suitable thermocouple is used. The low temperature furnace (about 300°C) is generally called as oven. But these high range temperatures between 300°C to 1400°C are called as furnaces.

- **Applications:**

  (i) Small ovens for drying the coatings like varnish.

  (ii) High temperature furnaces for annealing, hardening, melting of non-ferrous metals etc.

SOLVED EXAMPLES

Example 2.1: *A resistance oven employing nichrome wire is to be heated from 220 V, 1ϕ supply and is rated at 16 kW. If temperature of element is to be limited to 1170°C and average temperature of charge is 500°C, find diameter and length of wire.*

$$\text{Radiating efficiency } K = 0.57,$$

$$\text{Emissivity } e = 0.9,$$

$$\rho \text{ of nichrome} = 109 \times 10^{-6}\ \Omega\,\text{cm}$$

Solution: Given: $K = 0.57$, $e = 0.9$, $\rho = 109 \times 10^{-6}\ \Omega\ cm = 1.09 \times 10^{-6}\ \Omega m$

$T_1 = 1170°C = 1170 + 273 = 1443$ abs.

$T_2 = 500°C = 500 + 273 + 773$ abs.

From Stefan's formula, we have,

$$H = 5.72\ K.e \left\{\left(\frac{T_1}{100}\right)^4 - \left(\frac{T_2}{100}\right)^4\right\}$$

$$= 5.72 \times 0.57 \times 0.9 \left\{\left(\frac{1443}{100}\right)^4 - \left(\frac{773}{100}\right)^4\right\}$$

$$\boxed{H = 11.6749 \times 10^4\ W/m^2}$$

We have,

$$\frac{l}{d^2} = \frac{\pi V^2}{4\rho P} \quad \text{...(1)}$$

$$\frac{l}{d^2} = \frac{\pi \times (220)^2}{4 \times 1.09 \times 10^{-6} \times 16 \times 1000}$$

$$\frac{l}{d^2} = 2179660 \quad \text{...(2)}$$

$$\text{Heat dissipated} = \text{Electrical power input}$$

$$\therefore \quad \pi\, dl \times H = P$$

$$\therefore \quad \pi \times d \times l \times 11.6749 \times 10^4 = 16000$$

$$dl = 0.043623$$

$$d^2l^2 = 1.90297 \times 10^{-3}$$

$$d^2 = \frac{1.90297 \times 10^{-3}}{l^2}$$

Putting value of d^2 in (1),

$$\frac{l}{\frac{1.90297 \times 10^{-3}}{l^2}} = 2179660$$

$$l^3 = 2179660 \times 1.90297 \times 10^{-3}$$

$$\boxed{l = 16.06\ m}$$

Putting this value in (1), we will find d.

$$\frac{16.06}{d^2} = 2179660$$

$$\boxed{\mathbf{d = 2.714\ mm}}$$

Example 2.2: *A 20 kW, 1ϕ, 220 V resistance oven employes a circular nichrome wire for its heating element. If wire temperature not to exceed 1170°C and temperature of charge is to be 500°C, calculate diameter and length of wire. Take K = 0.57, e = 0.95 and ρ = 1.09 $\mu\Omega$m. What would be the temperature of the element when charge is cold at 20°C?*

Solution: $T_1 = 1170°C = 1170 + 273 = 1443$ Abs., $T_2 = 500°C = 500 + 273 = 773$ Abs., $K = 0.57$, $e = 0.95$, $\rho = 1.09 \times 10^{-6}\ \Omega m$

From Stefan's formula, we have,

$$H = 5.72\ K.e. \left\{\left(\frac{T_1}{100}\right)^4 - \left(\frac{T_2}{100}\right)^4\right\}$$

$$= 5.72 \times 0.57 \times 0.95 \left\{\left(\frac{1443}{100}\right)^4 - \left(\frac{773}{100}\right)^4\right\} \text{ watt/m}^2$$

$$H = 12 \times 10^4\ W/m^2$$

Now,

$$\frac{l}{d^2} = \frac{\pi V^2}{4\rho P} = \frac{\pi \times (220)^2}{4 \times 1.09 \times 10^{-6} \times 20 \times 10^3}$$

$$\frac{l}{d^2} = 1740000 \quad \text{... (1)}$$

$$\text{Heat dissipated} = \text{Electrical power input}$$

$$\pi d l \times H = P$$

$$\pi \times d \times l \times 12 \times 10^4 = 20 \times 10^3$$

$$ld = 0.053$$

$$l^2 d^2 = (0.053)^2$$

$$\therefore \quad d^2 = \frac{(0.053)^2}{l^2} \quad \ldots (2)$$

Putting value of d^2 in (1), we get,

$$l = 16.69 \text{ m} \simeq 17 \text{ m and } d = 3.09 \text{ mm}$$

Now the charge is remaining cold at 20°C.

$$\therefore \quad T_2 = 20°C$$

$$\therefore \quad 12 \times 10^4 = 5.72 \times 0.57 \times 0.95 \left[\left(\frac{T_1}{100}\right)^4 - \left(\frac{273 + 20}{100}\right)^4\right]$$

$$\therefore \quad T_1 = 1403 \text{ K}$$

$$\therefore \quad T_1 = 1403 - 273$$

$$\boxed{T_1 = 1130°C}$$

Example 2.3: *A 30 kW, 3ϕ, 400 volt resistance oven is nichrome strip 0.254 mm thick for the three λ-connected heating elements. If wire temperature is to be 1100°C and that of charge to be 700°C, estimate suitable width for strip. Assume emissivity = 0.9 and radiating efficiency to be 0.5. Take specific resistance = 101.6 × 10^{-8} Ωm*

Solution:

$$\text{Power / Phase} = \frac{30}{3} = 10 \text{ kW} = 10{,}000 \text{ watt}$$

$$\text{Voltage/Phase} = \frac{400}{\sqrt{3}} = 231 \text{ volt}$$

$$P = \frac{V^2}{R}$$

$$\therefore \quad R = \frac{V^2}{P} = \frac{(231)^2}{10{,}000}$$

$$\therefore \quad R = 5.34\ \Omega$$

Let l = length of strip

w = width of strip

Heat dissipated from Stefan's formula:

$$H = 5.72 \text{ K.e.} \left\{\left(\frac{T_1}{100}\right)^4 - \left(\frac{T_2}{100}\right)^4\right\}$$

$$= 5.72 \times 0.5 \times 0.9 \left\{\left(\frac{1100 + 273}{100}\right)^4 - \left(\frac{700 + 273}{100}\right)^4\right\}$$

$$H = 68{,}600 \text{ watt/m}^2$$

$$\text{Total surface area available} = 2W \times l$$

$$\text{Total heat dissipated} = \text{Power input}$$

$$\therefore \quad 68{,}600 \times [(W.l.) \times 2] = 10{,}000$$

$$W.l \times 2 = 0.1457 \text{ m}^2$$

$$\therefore \quad W.l = 0.07288 \text{ m}^2 \quad \ldots (1)$$

$$R = \rho \frac{l}{a} = \rho \frac{l}{Wt}$$

$$\therefore \quad 5.34 = \frac{101.6 \times 10^{-8} \times l}{W \times 0.254 \times 10^{-3}}$$

$$\frac{l}{W} = 1333.33 \quad \ldots (2)$$

$$\therefore \quad l = W \times 1333.33$$

Putting it in equation (1),

$$W \times (W \times 1333.33) = 0.07288$$

and $l = 9.85$ m

$$\therefore \quad \boxed{W = 7.38 \text{ mm}}$$

Example 2.4: *A 40 kW, 3-phase, 400 V resistance oven uses nickel-chromium strip of 0.3 mm thickness. The heating elements are star connected. If the wire temperature is to be 1127°C and that of charge is to be 727°C, estimate suitable width and length of the wire required.*

Given: radiation efficiency = 0.6, specific resistance of Ni-Cr = 1.03×10^{-6} Ωm, emissivity = 0.9.

Solution:

$$\text{Power/phase} = \frac{40}{3} = 13.333 \times 10^3 \text{ watt}$$

$$\text{Voltage/phase} = \frac{400}{\sqrt{3}} = 230.94 \text{ V}$$

Let, l = Length of strip

w = width of strip

Heat dissipated from Stefan's formula,

$$H = 5.72 \text{ Ke}\left[\left(\frac{T_1}{100}\right)^4 - \left(\frac{T_2}{100}\right)^4\right]$$

$$= 5.72 \times 0.5 \times 0.9 \left[\left(\frac{1127 + 273}{100}\right)^4 - \left(\frac{727 + 273}{100}\right)^4\right]$$

$$= 87771.34 \text{ W/m}^2$$

Thickness t = 0.3 mm = 0.3×10^{-3} m (given)

Now,

$$\frac{l}{wt} = \frac{V^2}{P\rho}$$

$$\frac{l}{w \times (0.3 \times 10^{-3})} = \frac{(230.94)^2}{(13.333 \times 10^3) \times (1.03 \times 10^{-6})}$$

$$\therefore \quad \frac{l}{w} = \frac{1.5999 \times 10^{-3}}{0.013733}$$

$$\therefore \quad \frac{l}{w} = 1165.044 \qquad \text{... (1)}$$

$$\therefore \quad \frac{t}{l^2} = \frac{2\rho H}{V^2} = \frac{2 \times (1.03 \times 10^{-6}) \times 87771.34}{(230.94)^2}$$

$$\therefore \quad \frac{t}{l^2} = \frac{0.1808}{53333.28}$$

$$\therefore \quad \frac{t}{l^2} = 3.3901 \times 10^{-6}$$

$$\therefore \quad \frac{t}{3.3901 \times 10^{-6}} = l^2$$

$$\therefore \quad \frac{0.3 \times 10^{-3}}{3.3901 \times 10^{-6}} = l^2$$

$$\therefore \quad l^2 = 88.491$$

$$\therefore \quad \boxed{l = 9.406 \text{ m}}$$

Substituting the value of l in equation (1), we get

$$\frac{9.406}{w} = 1165.044$$

$$\therefore \quad \boxed{\text{Width w} = 0.008074 \text{ m}}$$

2.10 ARC HEATING

1. Explain principle of working of indirect arc furnace.
2. Explain with neat sketch principle of operation of indirect arc furnace.

- The heat energy of electric arc is utilized for heating.

Arc Furnaces:

- For melting of ferrous and non-ferrous materials, a high temperature is required (3000°C to 3500°C) which is achieved by electric arc furnaces.

  Methods of striking the arc: (i) H.T. strike, (ii) L.T. strike.

(i) H.T. Strike:

In this method, a constant air gap is maintained between two electrodes which are connected to secondary of step-up transformer. By gradually increasing the primary voltage with the help of potential divider, arc is made to struck between electrodes. The arc is struck when voltage across the section is sufficient to ionize the air gap between electrodes.

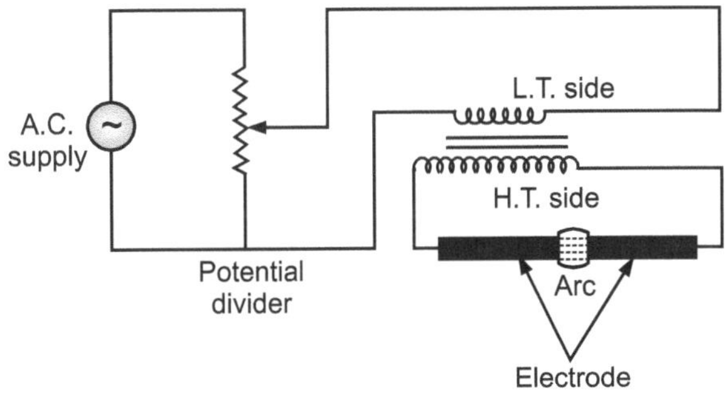

Fig. 2.11: H.T. Strike

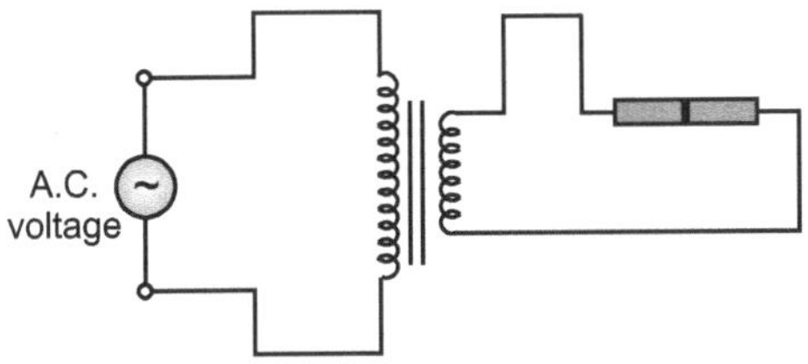

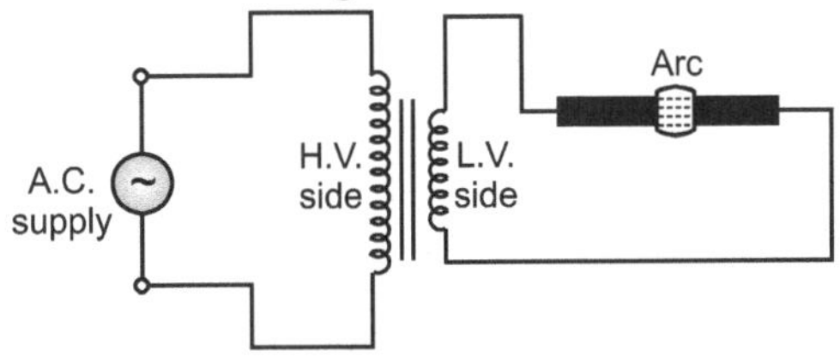

Fig. 2.12: L.T. Strike

In this method, the electrodes are connected across L.V. side of transformer. First they are shorted and then immediately separated resulting in production of arc.

This method is widely used in different arc furnaces.

2.11 DIRECT ARC FURNACES

1. Draw the sketch of direct arc furnace. Explain construction, operation and applications.

- The arc is formed between electrode and charge.
- **Construction:** It consists of arc chamber with refractory lining. Total structure is supported on metal frame work. Carbon or graphite electrodes are used.

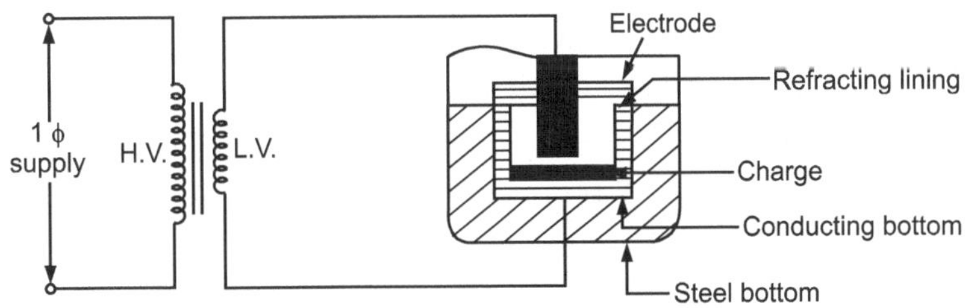

Fig. 2.13: Direct Arc Furnace

- **Operation:** First the electrode is lowered and then separated from the charge. An arc gets strucked and a very high temperature is produced. Arc is maintained due to ionization of air, ionized air acts as conductor. Current passing through the charge causes heating of charge. Temperature of arc is between 3000°C to 3500°C. But losses take place and actual temperature obtained is 1500°C to 2000°C.
- **Applications of direct arc furnace:** The main application is found in production of steel. The steel obtained by this method is more pure than other method (cupola method).

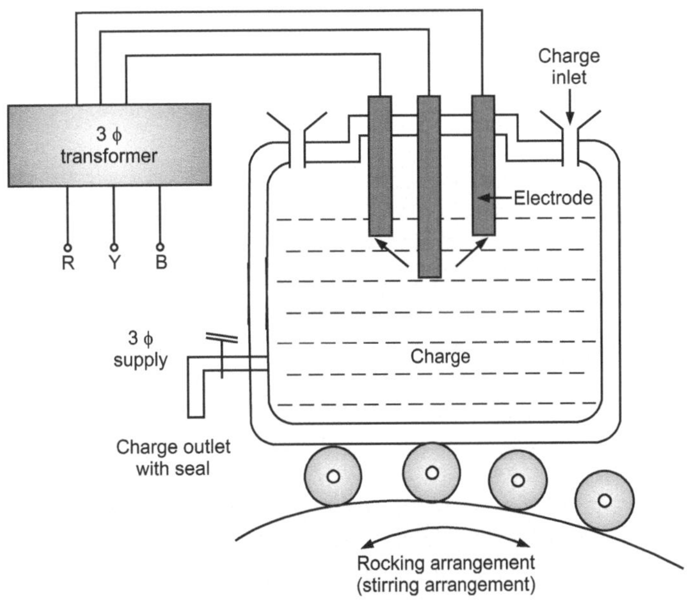

Fig. 2.14: Submerged Furnace

Non-conducting bottom type direct arc furnace/submerged furnace:

- This furnace is used for ferro alloys. The electrodes are submerged in charge flows, current is proportional to resistance offered by the charge. The stirring arrangement is also provided.

2.12 INDIRECT ARC FURNACE (ROCKING FURNACE)

- **Construction:** Graphite electrodes are projected from side walls. These electrodes are at certain height from the charge. Separate stirring arrangement is provided.
- **Operation:** 1ϕ A.C. supply is applied across the electrodes. The arc is struck by separating the electrodes manually or automatically.
- The heat is transferred to the charge top layer and refractory lining through radiation and from top layer of charge to other portion of charge through conduction. To distribute the heat uniformly the furnace is rocked continuously. This type of construction limits the number of electrodes to two and therefore this furnace is available for single phase supply only. Capacity of the furnace is limited to 1.5 tonnes only. It is used for melting *non-ferrous materials.*

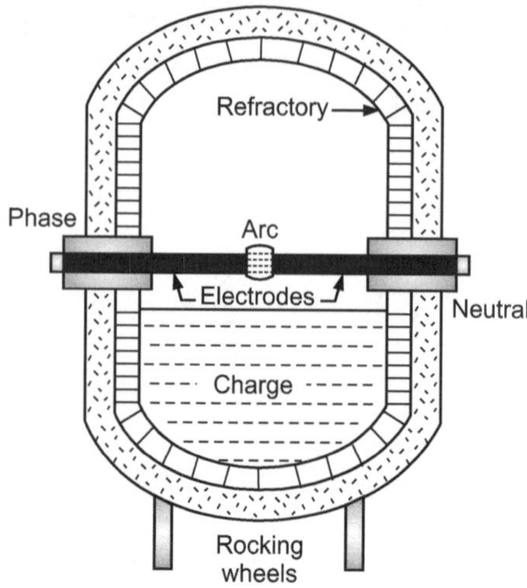

Fig. 2.15: Indirect Arc Furnace

- **Electrodes used for arc furnace:** Three types of electrodes are used: Carbon, Graphite, Soderberg.
- **Properties of carbon and graphite:** Infusibility, good electrical conductivity, insolubility, chemical inertness, mechanical strength, resistant to thermal shock.
- **Comparison between Carbon and Graphite Electrodes:**

Carbon	Graphite
• These are amorphous.	• Obtained by heating carbon electrodes at high temperature.
• For same size, current carrying capacity is less because specific resistance is more.	• For same size, current carrying capacity is more because specific resistance is less $\left(\frac{1}{4}^{th} \text{ as that of carbon}\right)$.
• It is cheap (mostly it is $\frac{1}{2}$ of graphite electrodes).	• It is costly.
• Used for small capacity furnaces.	• Used for large capacity furnaces.

- The basic arrangement for arc furnace consists of a transformer, electrode, heating chamber. The arc is formed either between two electrodes (indirect arc furnace) or between one electrode and charge (direct arc furnace).
- An electrical circuit with constant voltage supply and an arc load is basically unstable. This is because once arc is established, current increases, arc diameter increases, arc resistance decreases, therefore, current again increases. In this way, cause helps the effect which in turn helps the cause. Thus arc is said to have negative volt-ampere characteristics i.e. voltage drop in arc decreases with increase in arc current.

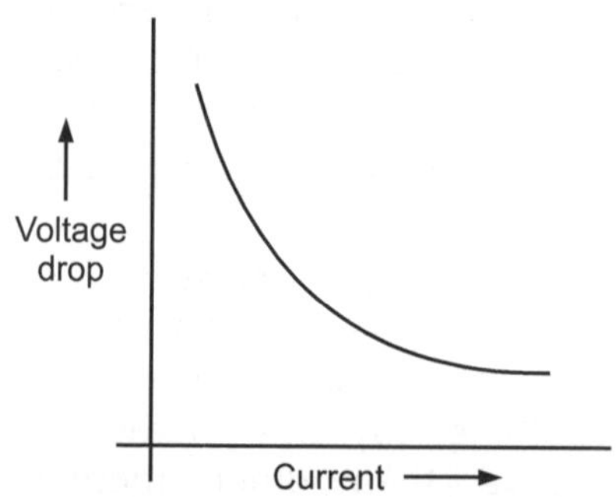

Fig. 2.16: Arc Characteristics

- So some stabilizing device is essential that will help to maintain stable arc. "Reactor" is used for this purpose. The reactor is inserted in series with the electrode. The basic property of reactor (to oppose charge) is utilized for arc stability.

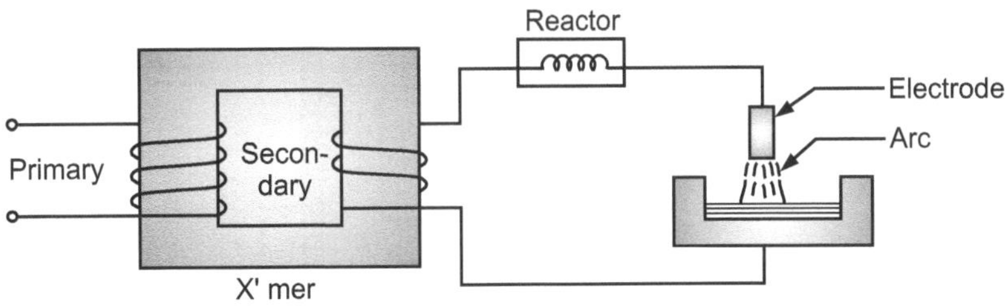

Fig. 2.17: Series Reactor to Stabilize Arc

- The volt-ampere characteristics of reactor is as shown in Fig. 2.18.

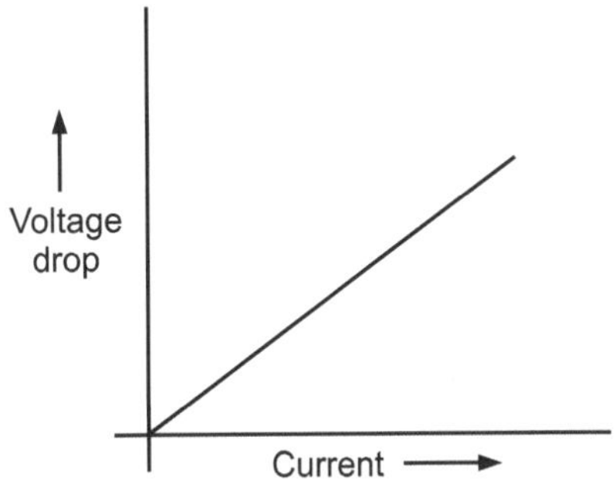

Fig. 2.18: Volt-ampere Characteristics of Reactor

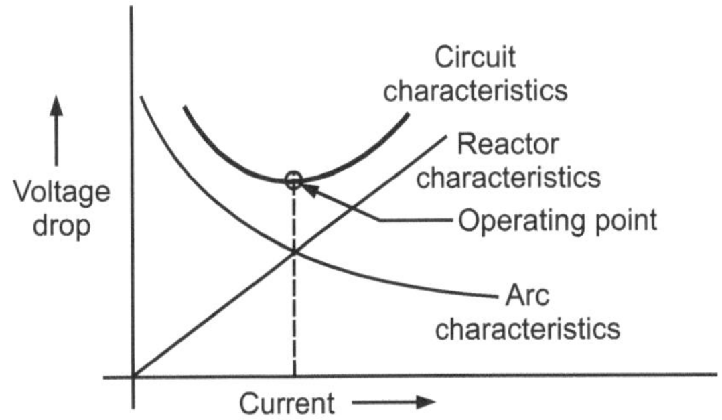

Fig. 2.19: Volt-ampere Characteristics of Arc

- Merging the two characteristics we get the total characteristics for the circuit. (Refer Fig. 2.19)
- From Fig. 2.19 it is seen that after operating point the characteristics is linear.

Comparison between Direct Arc Furnace with Indirect Arc Furnace

Sr. No.	Direct Arc Furnace	Indirect Arc Furnace
1.	Temperature achieved is more. This is due to touching of arc between electrode and charge directly.	Temperature achieved is less. This is due to touching between electrodes and heat transfer by radiation.
2.	Size is much more than 5 to 10 tonnes.	Size is lesser than 0.5 to 3 tonnes.
3.	Power requirement is lesser i.e. 200 kWh/tonne.	Power requirement is more i.e. 400 kWh/tonne.
4.	Uses: For steel production.	Uses: Melting metals.
5.	Power factor: 0.8 lagging.	Power factor: 0.85 lagging.

Power Supply for Arc Furnaces:

- It should have low voltage and high current characteristics because of the following reasons:
  (i) $H \propto I^2$, therefore heavy current is needed to produce large amount of heat.
  (ii) If voltage between electrode and charge is high, the voltage gradient will be high. The nitrogen gas present in furnace atmosphere gets ionized and it is absorbed by the charge. The nitriding effect on charge produces embrittlement. So to avoid it, voltage gradient and hence voltage should be low.
  (iii) High currents and low voltages keep the electrode very near to the charge because arc is small in length. Thus arc is away from the roof and therefore, life of the roof refractory is increased.

Conditions for maximum output from arc furnace:

- Arc furnace consists of a transformer, series reactor, electrodes and connection leads (Refer Fig. 2.20).

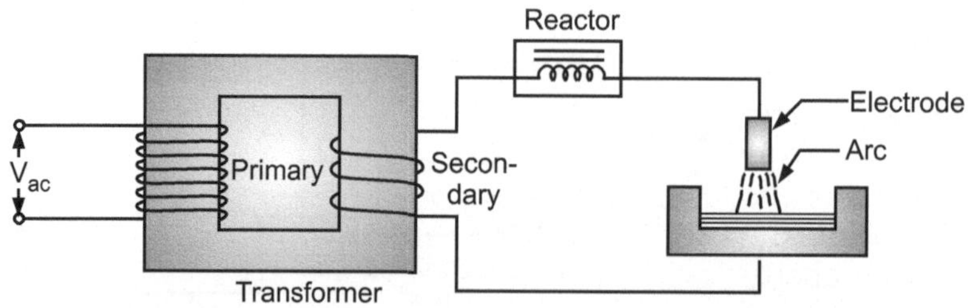

Fig. 2.20: Representation of Arc Furnace Arrangement

Equivalent Circuit Development:

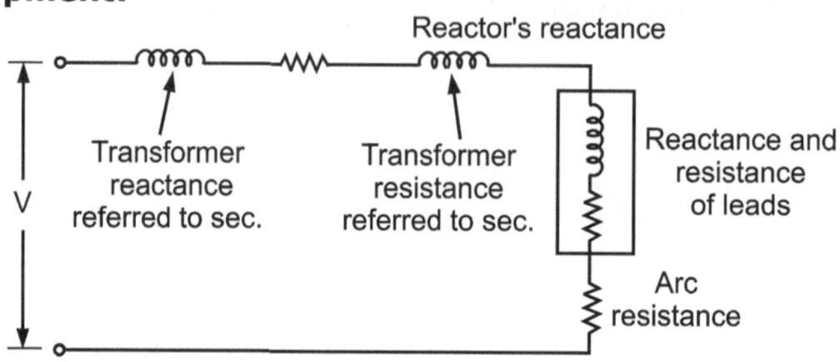

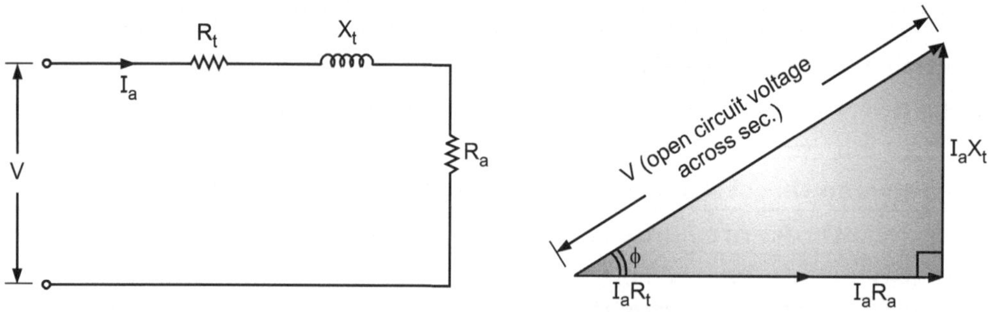

Fig. 2.21: Equivalent Circuit and Phasor Diagram for Arc Furnace

R_t = (Total resistance of transformer referred to secondary side) + (Resistance of leads and electrode)

X_t = (Total reactance of transformer referred to secondary side) + (Resistance of stabilizing reactor and reactance of leads)

R_a = Resistance of arc

The phasor diagram is drawn as per the equivalent circuit.

Arc current,
$$I_a = \frac{V}{\sqrt{(R_a + R_t)^2 + X_t^2}}$$

Power loss in the arc 'P' $= I_a^2 R_a$

$$= \left[\frac{V}{\sqrt{(R_a + R_t)^2 + X_t^2}}\right]^2 \cdot R_a$$

$$P = \frac{V^2}{(R_a + R_t)^2 + X_t^2} R_a$$

$$P = \frac{V^2 R_a}{R_a^2 + 2R_aR_t + R_t^2 + X_t^2}$$

$$P = \frac{V^2}{R_a + 2R_t + \left(\frac{R_t^2 + X_t^2}{R_a}\right)}$$

The power loss 'P' will be maximum when denominator is minimum.

$$\therefore \quad \frac{d(\text{Denominator})}{dR_a} = 0$$

$$\therefore \quad \frac{d\left[R_a + 2R_t + \left(\frac{R_t^2 + X_t^2}{R_a}\right)\right]}{dRa} = 0$$

$$\therefore \quad 1 + 0 - \left(\frac{R_t^2 + X_t^2}{R_a^2}\right) = 0$$

$$1 - \left(\frac{R_t^2 + X_t^2}{R_a^2}\right) = 0$$

$$\therefore \quad \boxed{R_a = \sqrt{R_t^2 + X_t^2}}$$

Thus power in furnace is maximum when arc resistance R_a = impedance of the circuit referred to secondary (excluding arc resistance).

Power Factor at Maximum Power:

$$\cos\phi = \frac{R_a + R_t}{\sqrt{(R_a + R_t)^2 + X_t^2}} = \frac{R_a + R_t}{\sqrt{R_a^2 + R_t^2 + 2R_aR_t + X_t^2}} = \frac{R_a + R_t}{\sqrt{\left(R_t^2 + X_t^2\right) + R_a^2 + 2R_aR_t}}$$

For maximum power $\quad R_a = \sqrt{R_t^2 + X_t^2}$

$$\therefore \quad R_a^2 = R_t^2 + X_t^2$$

Putting this condition in above equation,

$$\cos\phi = \frac{R_a + R_t}{\sqrt{R_a^2 + R_a^2 + 2R_aR_t}} = \frac{R_a + R_t}{\sqrt{2R_a^2 + 2R_aR_t}}$$

$$= \frac{\sqrt{R_a + R_t} \cdot \sqrt{R_a + R_t}}{\sqrt{2R_a\,(R_a + R_t)}} = \frac{\sqrt{R_a + R_t} \cdot \sqrt{R_a + R_t}}{\sqrt{2R_a} \cdot \sqrt{R_a + R_t}}$$

$$= \frac{\sqrt{R_a + R_t}}{\sqrt{2R_a}} = \sqrt{\frac{R_a + R_t}{2R_a}} = \frac{1}{\sqrt{2}}\sqrt{\frac{R_a + R_t}{R_a}} = \frac{1}{\sqrt{2}}\sqrt{1 + \frac{R_t}{R_a}}$$

$$= \frac{1}{\sqrt{2}} \ \ldots \text{ neglecting resistances } \boxed{\cos\theta = 0.707}$$

Thus, maximum useful heat power loss/power consumption occurs when p.f. of secondary is 0.707. (Refer Fig. 2.22).

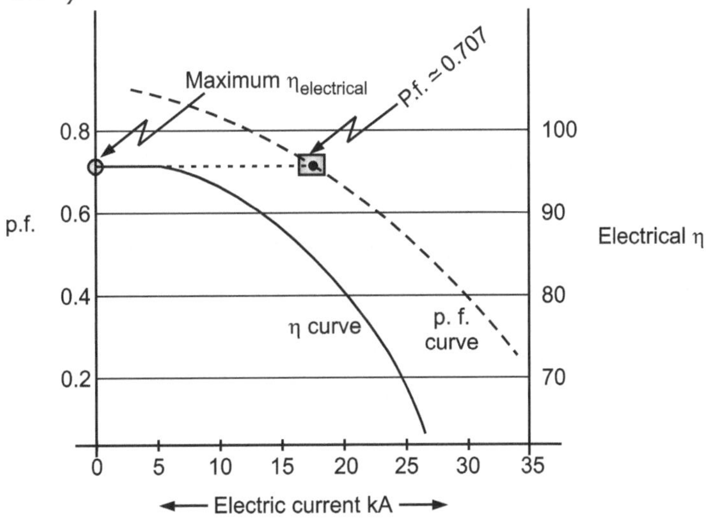

Fig. 2.22: Arc Furnace Performance Curves

$$\text{Electrical } \eta = \frac{\text{Power of arc}}{\text{Power input}} \times 100 = \frac{V_{arc}\, I_{arc}}{V \cdot I} \times 100$$

SOLVED EXAMPLES

Example 2.5: *Calculate kVA and kW drawn from supply, p.f., electrical efficiency for and 3 phase electric arc furnace with the following data:*

Current drawn = 4500 amp.

Arc voltage = 50 volt

Resistance of transformer referred to secondary = 0.002 Ω

Reactance of transformer referred to secondary = 0.004 Ω

Solution:

'V' drop in R_t = 4500×0.002

= 9 volt

V drop in X_t = 4500×0.004

= 18 volt

The V drop in arc = 50 volt, it is resistive drop

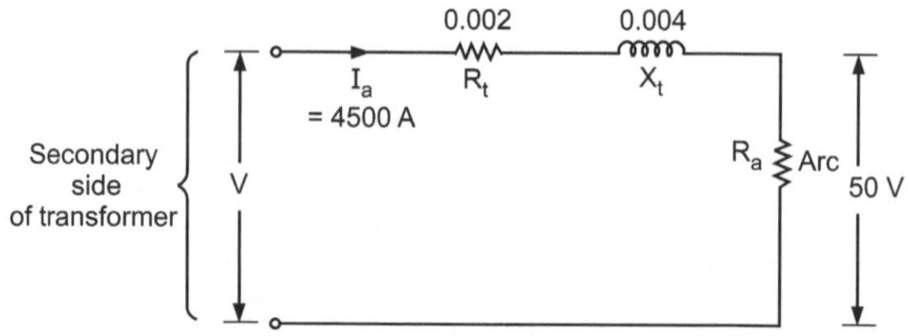

Fig. 2.23

The secondary voltage

$$\bar{V} = \overline{IR_t} \times \overline{IX_t} + \overline{V_{arc}}$$

$$\therefore \quad V = \sqrt{59^2 + 18^2}$$

$$\boxed{\mathbf{V = 62\ volt}}$$

$$\cos\phi = \frac{59}{62} = 0.952$$

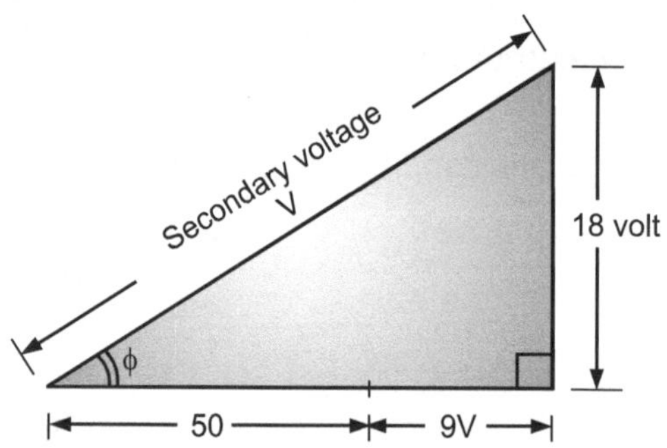

Fig. 2.24

The furnace is 3 phase furnace of voltage 62 volt is phase value.

$$\therefore \quad \text{Line voltage} = \sqrt{3} \times 62 = 107 \text{ volt}$$

$$\text{kVA I/P} = \frac{\sqrt{3}V \cdot I_a}{1000} = \frac{\sqrt{3} \times 107 \times 4500}{1000} = 834$$

$$\text{p.f. of supply} = 0.952$$

$$\therefore \quad \text{kW} = \text{kVA} \cos\phi$$

$$\therefore \quad = 837 \times 0.952 = 794$$

$$\text{kW of arc} = 3 \times \frac{4500 \times 50}{1000} = \boxed{\mathbf{675}}$$

$$\text{Electrical } \eta = \frac{675}{796.5} \times 100 = \boxed{\mathbf{85.01\%}}$$

Example 2.6: *A 3 phase arc furnace has to melt 10 tonnes of steel in 2 hours. Estimate the average input to the furnace if overall efficiency is 50%. If current input is 9000 amp. with the above kW input and the resistance and reactance of furnace leads (including transformer) are 0.003 Ω and 0.005 Ω respectively, estimate the arc voltage and total kVA taken from the supply.*

Given: Specific heat of steel = 444 J/kg/°C

Latent heat of fusion = 37.25 kJ/kg

Melting point of steel = 1370°C

Initial temperature = 20°C

Solution: Sensible heat required = m.s. $(t_2 - t_1) = 10000 \times 444 \times (1370 - 20) = 5.99 \times 10^9$ J

Latent heat required = (Mass) × (Latent heat of fusion)

$= 10000 \times (37.25 \times 1000) = 3.725 \times 10^8$ J

$$\text{Total heat required} = (\text{Sensible heat}) + (\text{Latent heat})$$
$$= 6.3625 \times 10^9 \text{ joule} = 6.3625 \times 10^9 \text{ watt sec.}$$

The overall furnace efficiency is 50%.

$$\therefore \quad \text{Heat to be produced} = \frac{6.3625 \times 10^9}{0.5} = 1.2725 \times 10^{10} \text{ watt. sec.}$$

The furnace works for 2 hrs. i.e. for $(2 \times 60 \times 60)$ sec. so heat to be produced in kW.

$$= \frac{1.2725 \times 10^{10}}{(2 \times 60 \times 60)} \times \frac{1}{1000} = 1767.36 \text{ kW}$$
$$= 1767.36 \times 10^3 \text{ watt (Total power input)}$$

Now consider arc furnace circuit.

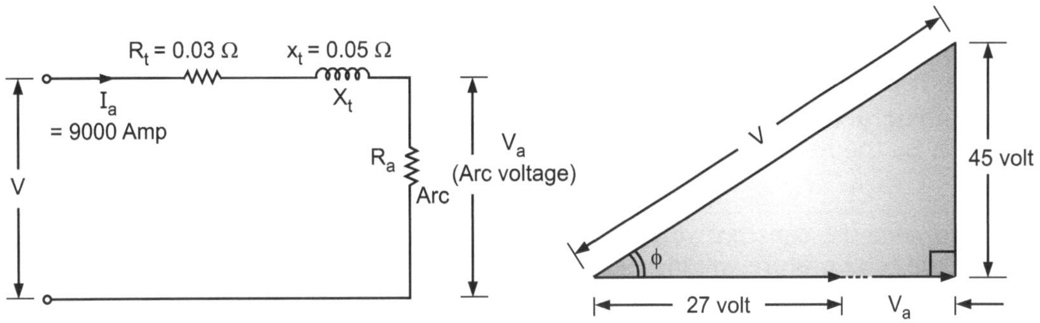

Fig. 2.25

$$\text{'V' drop in } R_t = 9000 \times 0.003 = 27 \text{ volt}$$
$$\text{'V' drop in } X_t = 9000 \times 0.005 = 45 \text{ volt}$$

The open-circuit voltage or secondary voltage is given by

$$V = \sqrt{(27 + V_a)^2 + (45)^2}$$

$$\text{Power factor } \cos\phi = \frac{V_a + 27}{V} = \frac{V_a + 27}{\sqrt{(27 + V_a)^2 + (45)^2}}$$

$$\text{The total power input} = 3\, V_{ph}\, I_{ph} \cos\phi$$

$$1767.36 \times 10^3 = 3 \cdot \sqrt{(27 + V_a)^2 + (45)^2} \times 9000 \times \frac{V_a + 27}{\sqrt{(27 + V_a)^2 + (45)^2}}$$

Solving we get, arc voltage $\boxed{V_a = 38.5 \text{ volt}}$

$$\therefore \quad \text{Secondary voltage 'V'} = \sqrt{(27 + V_a)^2 + (45)^2}$$
$$= \sqrt{(27 + 38.5)^2 + (45)^2} = 79.5 \text{ volt}$$

$$\text{Total kVA drawn from the supply} = \frac{3 \times V_{ph} \times I_{ph}}{1000} = \frac{3 \times 79.5 \times 9000}{1000} = \boxed{2146 \text{ kVA}}$$

Basic Electrical Requirements of Arc Furnace:

1. **Furnace Transformer:** The arc voltage is about 50 to 150 volts. The secondary of transformer is provided for low voltage and a very current carrying capacity. Multiple tappings are also provided on the secondary in order to get variable voltage. The transformer is oil immersed and cooling is done with the help of water.
2. **Electrodes:** Generally, these are madeup of graphite or carbon. The carbon electrodes are cheap and their current density is 8 to 15 A/cm^2 whereas the graphite electrodes have current density of 82-60 A/cm^2. So carbon electrodes are suitable for low capacity arc furnaces whereas the graphite electrodes are preferred in case of higher capacity arc furnaces.
3. **Furnace leads:** Several flexible cables are used for connection from transformer to furnace. The cables are used in parallel because the furnace current is very heavy. These are arranged such that the distance between them is kept minimum in order to reduce the inductance. Their run should be straight and short as far as possible.

4. **Arc stability:** The electrodes may burn gradually after the arc is struck. This causes increase in arc length and the arc may extinguish because the gap is not maintained properly. Moreover, the resistance of the charge which is melting changes because of temperature variations. These two factors change the magnitude of arc current.

   In order to get stability of arc the effect is transmitted with the help of C.T. to an automatic regulator circuit which adjusts the height of electrodes during melting process or a series of reactor is used.

5. **Overcurrent protection:** This is obtained by using C.T., relays and circuit breaker.

Basic mechanical requirements of arc furnace:

1. **Charging:** Proper arrangement for pouring the charge for heating should be provided.
2. **Removal of melt and slag:** When the melt is ready the front door opens and the furnace is tilted by 45°C and thus the melt is taken out. For slag removal, it is tilted 15°C and the slag door opens.
3. **Shell:** Madeup of steel and cylindrical shape. It has two doors one for tapping and slag removal. It is air leak proof.
4. **Lining structure:** The inner surface of the steel shell has to be lined with something heat resistant. So the heat is not allowed to leak.

2.13 TEMPERATURE CONTROL

- In electric furnace the constant voltage energises the heating element. Now, even if the voltage is constant the temperature will increase with the passage of time and it may cross the desired level of temperature. So to avoid this controlling mechanism is required.
- In large furnaces the control may be electronic control or electromagnetic control involving servomechanism.
- Following methods are used for temperature control:

  (i) Interminant Switching (i.e. ON-OFF Control): The temperature rise beyond the predetermined value (beyond a certain value) can be prevented by switching the supply OFF. Here care has to be taken that the switching has to be done before the temperature actually reach it's maximum value because even if the switch is kept open the temperature increases for some time in the furnace. Now afterwards the temperature begins to fall and when it reaches the predetermined value the supply is again switched ON. Due to continuous 'ON-OFF' action of switches there is a lot of arcing at the switching contacts and wear and tear of the contacts also occurs. That is why such a method is suitable for small furnace only.

  (ii) Variation of Number of Heating Elements: In this type, instead of cutting OFF the supply of heating element the number of heating elements are changed as per the requirements.

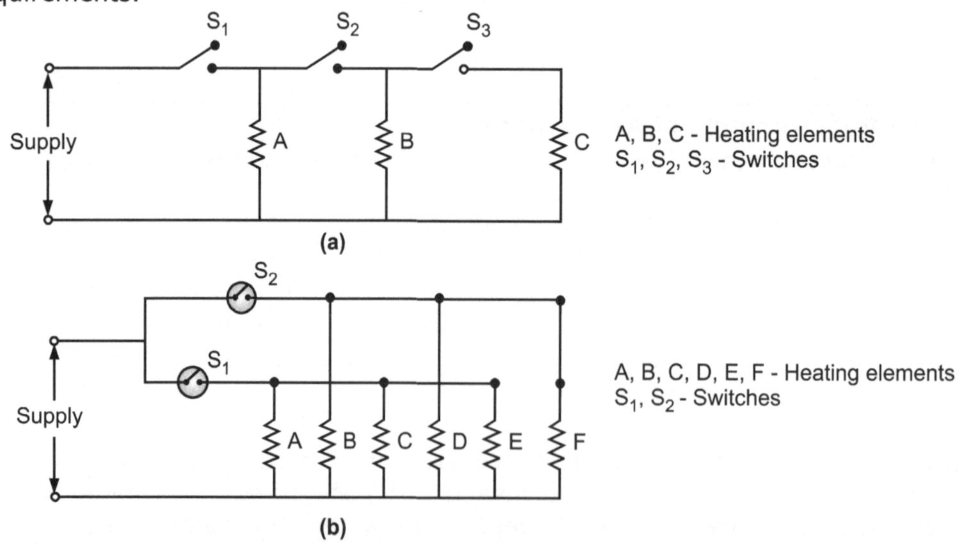

Fig. 2.26: Temperature Control by Variation of Number of Heating Elements

In case of a 3ϕ circuit one has to take care that equal number of heating elements should be switched 'ON' or 'OFF' (as per requirement) to maintain the balanced condition.

The uniform spacing of heating elements also plays an important part as it provides informality of operating temperature.

(iii) Variation in Circuit Configuration:

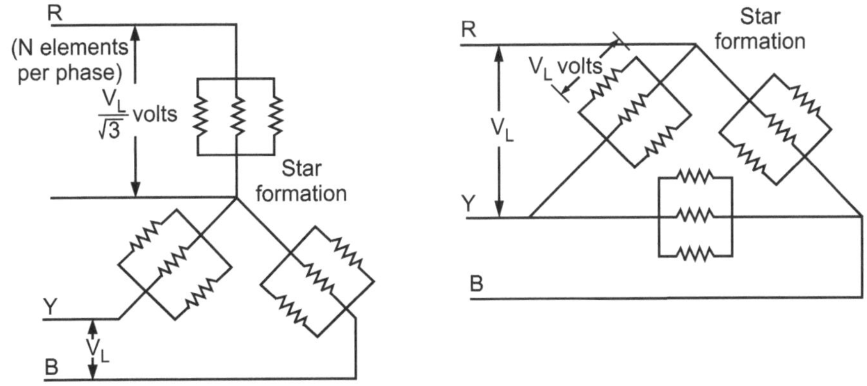

(a) Star delta configuration of heating elements

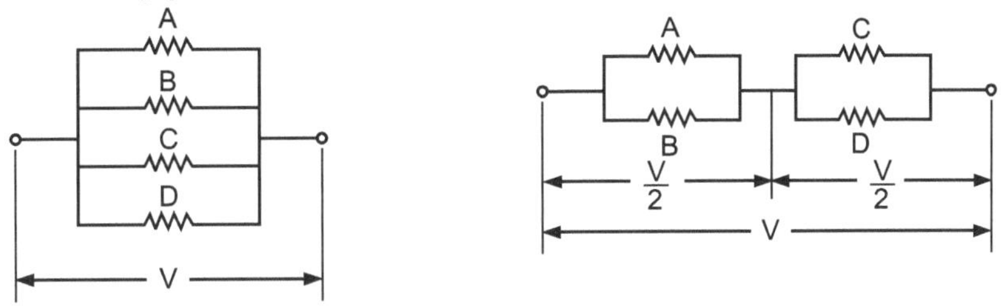

(b) Series parallel configuration of heating elements

Fig. 2.27: Temperature Control by Variation of Circuit Configuration

In case of three phase supply the heating element will give different heats when connected in star or delta formation.

Now consider Fig. 2.27 (a) when 'N' numbers of heating elements are connected in star, the voltage across each element is $\frac{V_L}{\sqrt{3}}$ volts. This power will be $1/3^{rd}$ of the power when the same element is connected in delta formation. (Power in delta = $\frac{1}{3}^{rd}$ power in star).

Thus by using the star-delta formation, amount of heat produced can be varied. Consider Fig. 2.27 (b) we have used series parallel grouping for temperature variation (control) purpose. In the first figure, the voltage received by each element is V and in the second figure voltage received is $\frac{V}{2}$. Power dissipated in the second case will be one fourth of the previous power.

(iv) Change of Applied Voltage: This is achieved by following methods.

(a) By providing Tappings on Transformer: The power dissipation voltage is less if the magnitude of applied voltage is less and hence naturally temperature produced is also less. In case of furnace transformer tappings are provided on the H.V. side. It is easy to provide the tappings on the H.V. side because the magnitude of current is lesser.

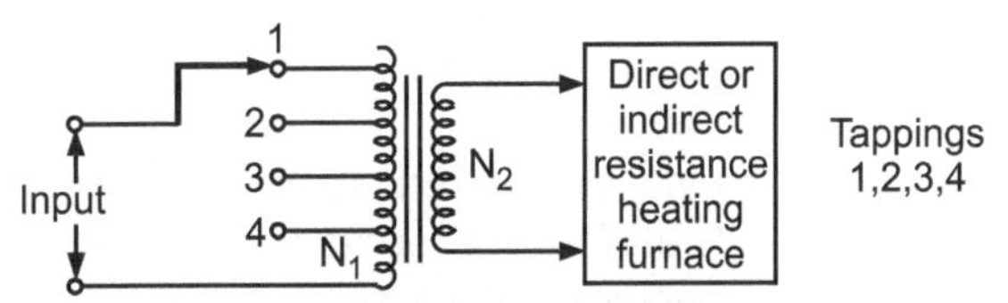

Fig. 2.28: Temperature Control by Change of Applied Voltage

(b) Bucking-Boosting the Secondary Voltage: In this case, the secondary is having two sections, one having small number of turns and the other having more number of turns. Now, if the connections are made as shown in Fig. 2.29 (a); the direction of current in the secondary sections is same, hence there is addition of emfs. So the voltage fed to the furnace is more. This action is boosting action.

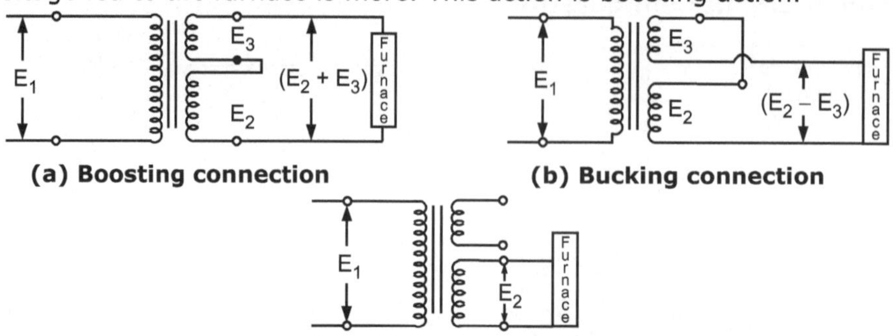

(a) Boosting connection (b) Bucking connection

(c) Without boosting/bucking

Fig. 2.29: Temperature Control by Bucking-boosting

Now when the connections are made as shown in Fig. 2.29 (b) the induced emfs in both the windings come in opposition. This is Bucking effect i.e. the voltage of secondary is reduced and hence the voltage fed to the furnace also decreases.

The third possibility is to provide the voltage of extra secondary to the furnace.

Thus, the three voltages available are:

(i) $E_2 + E_3$ – boosting

(ii) $E_2 - E_3$ – bucking

(iii) E_2 only – no bucking or boosting

(c) Auto Transformer Control: This is used for small resistance furnaces. The voltage supplied to the resistance furnace is controlled by an auto transformer. Thus, temperature control is obtained. The circuit used is as shown in Fig. 2.30.

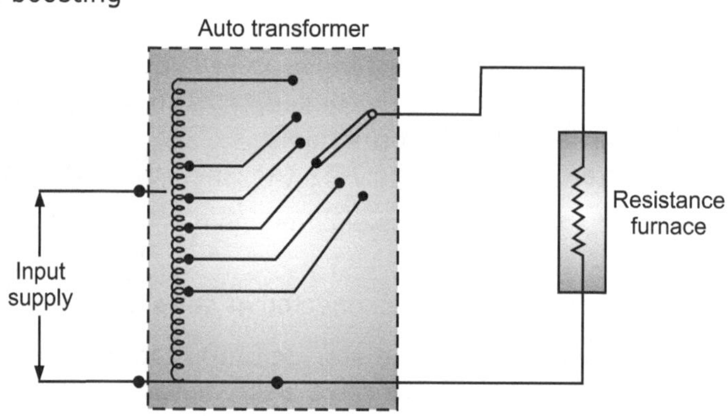

Fig. 2.30: Temperature Control by Autotransformer

(d) Series Reactor Control: The arrangement is shown in Fig. 2.31. When switch is in on position (1), the heating furnace receives rated voltage. When switch is put in on position (2), the reactor gets connected in series with the furnace and it receives less voltage. Inclusion of reactor causes lagging power factor, hence a capacitor is also connected in parallel so that furnace operates with a better power factor.

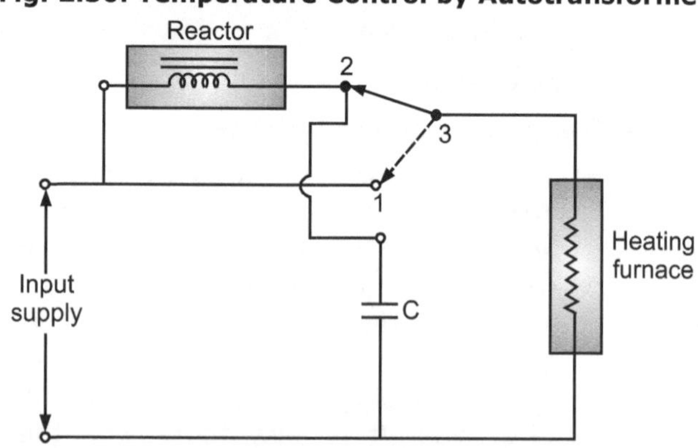

Fig. 2.31: Temperature Control by Series Reactor Method

2.14 LIST OF EQUIPMENTS USED IN ARC FURNACE AND THEIR APPLICATIONS

- **Basic Electrical Requirements of Arc Furnace:**
  1. **Furnace Transformer:** The arc voltage is about 50 to 150 volts. The secondary of transformer is provided for low voltage and a very current carrying capacity. Multiple tappings are also provided on the secondary in order to get variable voltage. The transformer is oil immersed and cooling is done with the help of water.
  2. **Electrodes:** Generally, these are madeup of graphite or carbon. The carbon electrodes are cheap and their current density is 8 to 15 A/cm^2 whereas the graphite electrodes have current density of 82-60 A/cm^2. So carbon electrodes are suitable for low capacity arc furnaces whereas the graphite electrodes are preferred in case of higher capacity arc furnaces.
  3. **Furnace leads:** Several flexible cables are used for connection from transformer to furnace. The cables are used in parallel because the furnace current is very heavy. These are arranged such that the distance between them is kept minimum in order to reduce the inductance. Their run should be straight and short as far as possible.
  4. **Arc stability:** The electrodes may burn gradually after the arc is struck. This causes increase in arc length and the arc may extinguish because the gap is not maintained properly. Moreover, the resistance of the charge which is melting changes because of temperature variations. These two factors change the magnitude of arc current.

     In order to get stability of arc the effect is transmitted with the help of C.T. to an automatic regulator circuit which adjusts the height of electrodes during melting process or a series of reactor is used.
  5. **Overcurrent protection:** This is obtained by using C.T., relays and circuit breaker.
- **Basic mechanical requirements of arc furnace:**
  1. **Charging:** Proper arrangement for pouring the charge for heating should be provided.
  2. **Removal of melt and slag:** When the melt is ready the front door opens and the furnace is tilted by 45°C and thus the melt is taken out. For slag removal, it is tilted 15°C and the slag door opens.
  3. **Shell:** Madeup of steel and cylindrical shape. It has two doors one for tapping and slag removal. It is air leak proof.
  4. **Lining structure:** The inner surface of the steel shell has to be lined with something heat resistant. So the heat is not allowed to leak.

2.15 INDUCTION HEATING

1. State the principle of induction heating.

- Principle of electromagnetic induction is used for heating the charge. The heat energy is transferred to the charge at extremely rapid rate much faster than other conventional methods. This is because heat is developed within the charge and whole resistance of charge is utilized to produce heat. The heat is generated within the charge without any physical contact between the source and the charge. The medium of energy transfer is electromagnetic field.

2.15.1 Core Type Induction Furnace

- **Construction:** It consists of magnetic core having thin laminations of silicon steel. The primary is wound on one limb and a crucible type non-conducting ring is provided on the secondary. The crucible provides housing for the metal to be heated. The crucible is made-up of ceramic material so that it can withstand high temperatures. The core of the transformer is designed such that it does not saturate even if high current is passed.

* **Operation:** First of all the charge in the form of different sizes is poured into the crucible. Due to these different sizes of the material there is a possibility that the secondary circuit may not complete, to avoid it, this more conductive material of powdered form is put on the pieces of material.

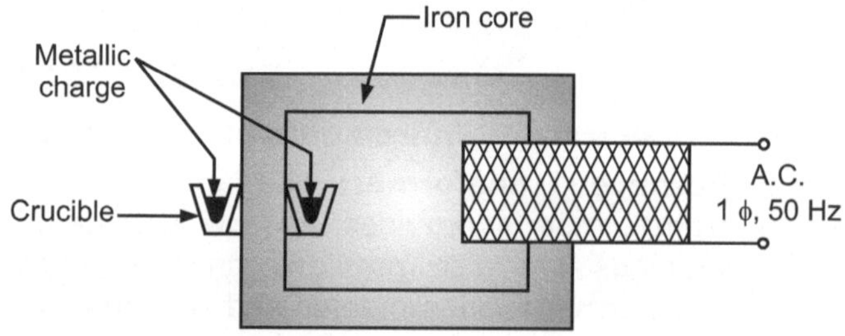

Fig. 2.32: Core Type Induction

* When A.C. supply is given to the primary, alternating flux is produced which links with the secondary. So e.m.f. is induced in the charge itself. Heavy current is circulated through the top powdered layer of conductive material and heat is produced in the top layer of charge which is transferred to the bottom layer of charge.

Disadvantages:

(i) Due to irregular shape of the charge, magnetic coupling between primary and secondary is poor.

(ii) High leakage reactance due to poor magnetic coupling causes poor power factor.

(iii) To avoid low p.f. low frequency supply is necessary. So low frequency generator is required to be installed. The overall cost of the furnace is increased.

(iv) Pinch effect may occur which causes open circuit to secondary side.

(v) Crucible is of odd shape and it is difficult to manufacture.

Pinch-off effect in core type induction furnace:

* The magnetic coupling between primary and secondary is very poor, resulting in high leakage reactance and low p.f. To overcome this difficulty frequency is kept small (as low as 10 Hz).
* The electromagnetic forces produce turbulence of the molten metal. This tubulance effect is dependant on frequency and current density. The turbulence effect is severe at high values of frequency and current density.

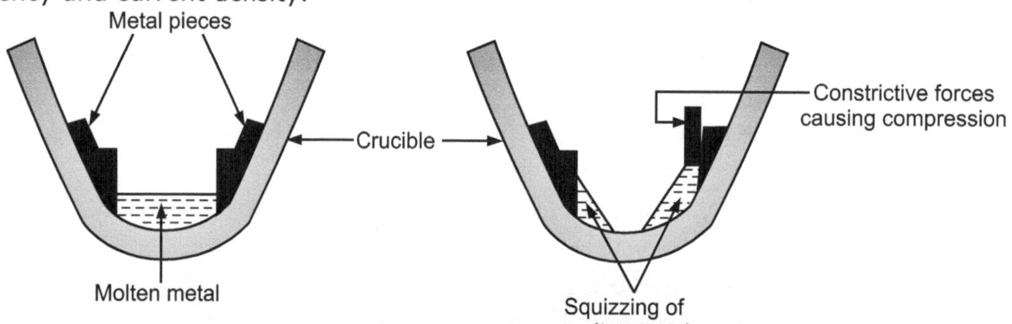

Fig. 2.33: Pinch-off Effect

* Hence, frequency is kept upto 10 Hz and current density is 5 amp/mm^2. The current flowing around the melt interacts with the alternating flux and produces *constrictive* forces on a section of the metal. This squeezes the melt and results in complete interruption of the section circuit. This effect is known as Pinch-off effect.

2.15.2 Vertical Core Type Induction Furnace / Ajax-Wyatt Furnace

1. Draw the sketch of Ajax Wyatt Furnace. Explain its operation.
2. Explain with neat sketch construction of 'Ajax Wyatt' vertical core furnace. List any two applications of it.
3. Describe core type induction furnace with neat sketch and give applications (uses).

- **Construction:** It consists of a transformer with horizontal core and vertical channel for the charge. Secondary is formed by the charge itself. Comparatively better magnetic coupling between primary and secondary is obtained due to 'V' shape. So the odd shaped crucible is avoided. In the upper portion, inlet is provided for the charge. Outlet for pouring the molten metal is also provided. Mechanical geared fitting is provided at the bottom of the furnace. Refractory lining is provided to reduce heat loss.

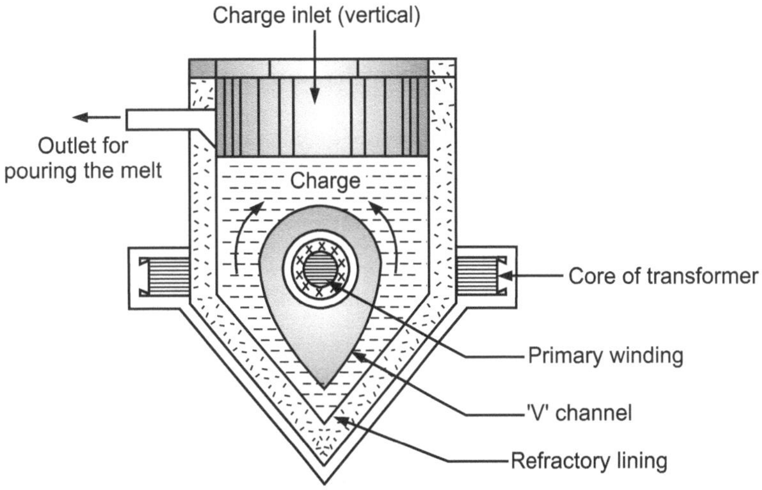

Fig. 2.34: Ajax-Wyatt Induction Furnace

- **Operation:** The 'V' shaped channel is always filled with molten metal. The total melt is never drained out from the furnace. Even if the furnace is not functioning a low voltage is maintained across the primary with the help of auto transformer which circulates currents and thus keeping the charge in the molten stage. Now the additional charge to be melted is poured in the furnace. Then primary is given A.C. supply causing heavy current in the material housed in 'V' channel. Thus heat is produced which is utilized for melting the charge. The circulation of molten metal takes place in 'V' channel due to convection. The high temperature metal goes up and low temperature metal comes down. Thus, the stirring action is automatically obtained. This causes homogeneous heating.
- **Uses:** Upto 2 tonnes commercially used for melting non-ferrous metals like copper, brass etc.

Advantages:

(i) Magnetic coupling between primary and secondary is better.

(ii) Leakage flux is less, hence p.f. is high (about 0.8 to 0.85).

(iii) It is not necessary to operate this furnace on low frequency supply, hence no alternator set is required, so overall cost is less.

(iv) Odd shaped crucible is not required.

(v) Pinch-off effect is avoided.

2.15.3 Coreless Induction Furnace (High Frequency Induction Furnace)

- It works on the principle of transformer. The primary winding is wound on furnace and is connected to A.C. supply. The charge inside the furnace acts as secondary. Induced currents heat up the charge.
- Due to skin effect, primary winding gets heated up. To keep the coil temperature within limits, these coils are made hollow and water is circulated through it. Heat in the charge is produced by eddy currents flowing concentrically. High frequency supply is preferred 500 to 1000 Hz because eddy current $\propto B^2f^2$.
- With this method it is possible to employ vacuum heating required for heating precious materials. p.f. of this furnace is very low. Efficiency is also low.

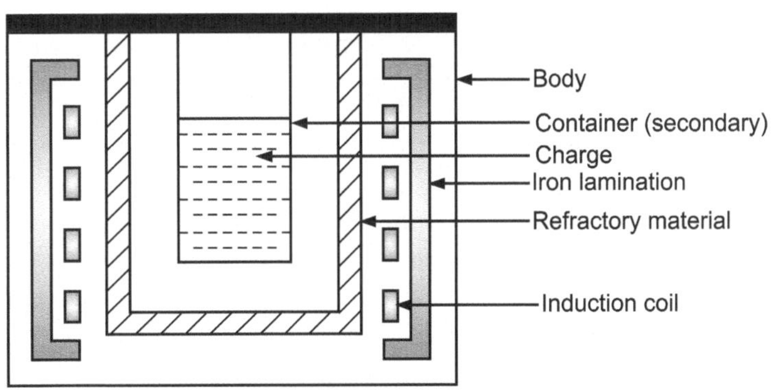

Fig. 2.35: Coreless induction furnace

- Stray magnetic field produced by coil outside the heating coil may link with supported steel structure and causes serious heating to supporting structure. To avoid this external packets of laminated core are provided which provide less reluctance path for magnetic field. This also decreases magnetizing current and increases p.f. Sometimes capacitors are used to improve p.f.

Advantages:

1. Time required for melting is less.
2. Temperature control is easy.
3. Shape of container and crucible is simple and suitable.
4. Automatic stirring action is obtained due to eddy currents.
5. Overall cost is less.

Comparison between High Frequency and Power Frequency Coreless Furnace:

1. High frequency furnace requires a separate high frequency source. So capital cost is more than coreless furnace of power frequency.
2. **Maintenance cost** of power frequency coreless furnace is less compared to high frequency coreless furnace. Because it does not require separate high frequency motor generator set.
3. **Losses:** Losses are more in high frequency coreless furnace due to converter. These losses are not present in power frequency coreless furnace.
4. **Efficiency:** Efficiency of power frequency coreless furnace is greater than high frequency coreless furnace.
5. Power frequency coreless furnace has tremendous tubular and stirring effect which is more than high frequency coreless furnace. Thus a better melt is obtained.
6. Power frequency coreless furnace requires special procedure for starting and is used generally for long production schemes of one alloy and it is not a simple thing to switch from one type of alloy to the other.

   High frequency coreless furnaces on the other hand does not require any special procedure for starting and can be drained after each heat cycle.
7. **Capacity:** Power frequency coreless furnace in size ranging from one tonne upwards. High frequency furnace from few kg to 15 tonne capacity.
8. In power frequency furnace the lining gets damaged due to turbulence of metal washed against it. So life of basic lining is less than high frequency furnace.

2.15.4 Applications of Induction Heating

1. Write applications of induction heating.

- Induction heating is employed for various industrial applications such as soldering, brazing, heat treating including hardening, annealing, melting, sintering powdered metals, drying paints, sterilizing surgical instruments, bonding, clutch facing.

2.15.5 Simple Numericals on Melting Furnaces

- Some heat equations are to be studied before solving the numericals on melting furnaces.

1. **Sensible heat:**

   It is the quantity of heat gained or lost by a mass of a substance when a temperature change takes place without change in its state.

$$\text{Sensible heat} = m.s.\,(t_2 - t_1)$$

where,

m = mass of substance in kg

s = specific heat of substance in J/kg/°C

t_2 = final temperature in °C

t_1 = initial temperature in °C

2. Latent heat:

It is the quantity of heat required to change the state of a given mass of substance without change in temperature.

$$\text{Latent heat} = (\text{Mass}) \times (\text{Specific latent heat})$$

where, m = mass in kg or gram

s = specific latent heat in J/kg or cal/gm

$$\text{Total heat required} = (\text{Sensible heat}) + (\text{Latent heat})$$

$$\eta_{\text{thermal}} = \frac{\text{Total heat required}}{\text{Total heat generated by the furnace}} \times 100 = \text{m.s}$$

Following conversions are important for problem solving:

$$1 \text{ watt sec.} = 1 \text{ joule} = 0.24 \text{ cal} = 1 \text{ Nm}$$

or $$1 \text{ kW sec} = 1 \text{ kJ} = 0.24 \text{ kcal}$$

$$1 \text{ kg} = 2.205 \text{ lb} \rightarrow (\text{pound})$$

$$1 \text{ tonne} = 1000 \text{ kg}$$

It is quiet possible that in the problems, units for specific heat, specific latent heat are in different way. So proper conversion is necessary.

2.16 EDDY CURRENT HEATING

2.16.1 Principle

1. Explain high frequency eddy current heating. Where it is applied?
2. State the principle and applications of eddy current heating.
3. State the principle of eddy current heating.

- Heating effect is produced due to eddy currents induced in the mass to be heated. Alternating current cutting the conducting substance induces eddy currents in it and the loss of energy due to eddy currents is "eddy current loss" which appears to be in the form of heat.
- This eddy current loss $W_e = K \cdot B^2 \times f^2 \times \text{Volume}$. So $W_e \propto B^2$ and also $W_e \propto f^2$.
- So eddy current's principle is used in heating a conducting substance.

2.16.2 Nature of Supply Used

- Very high frequency (about 10,000 Hz to 40,000 Hz) A.C. supply is used with the frequency control and flux density control equipment.

2.16.3 Description of Eddy Current Heating

- Charge or load to be heated is placed inside the coil arrangement (coil surrounds the charge).
- The heat, so produced, by eddy currents in the articles or charge penetrates the charge upto a sufficient depth and outside on the surface of the charge.

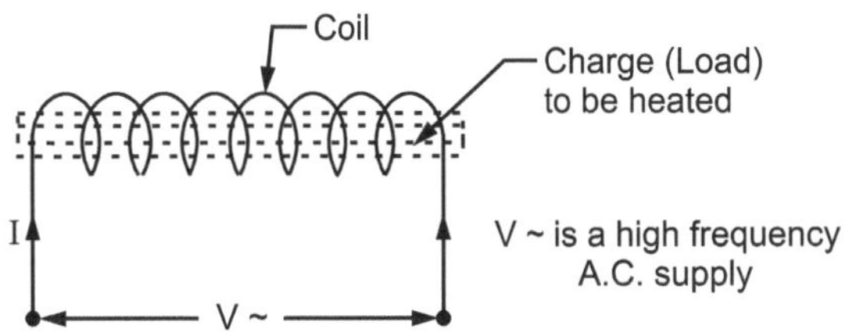

Fig. 2.36

2.16.4 Advantages of Eddy Current Heating

1. State advantages and disadvantages of eddy current heating.

1. Control is smooth and very easy.
2. Heat can penetrate to desired depth.
3. Method is very quick and convenient.

4. Process is clean.
5. Heat loss is minimum as it is produced in the material to be heated.
6. Process can be carried out in ordinary place, no effect of surroundings/ atmosphere etc.
7. Process is very easy to be handled, hence semi-skilled person can carry the work.
8. Temperature can be controlled by simple method.
9. Automatic control can be employed.
10. Coil spacing need not be charged for different objects. Air gap between coil and object does not affect the process.

2.16.5 Disadvantages of Eddy Current Heating

1. The system of heating is comparatively costly than other types of heating.
2. Initial cost of equipment is more.
3. Efficiency is less upto 40 to 50% only.
4. It is difficult to heat to a great depth of the material.
5. Requirement of frequency is very high from 10 kHz to 400 kHz.

2.16.6 Applications of Eddy Current Heating

- Heating effect is produced due to eddy currents induced in the mass to be heated. Alternating current cutting the conducting substance induces eddy currents in it and the loss of energy due to eddy currents is "eddy current loss" which appears to be in the form of heat.
- This eddy current loss $W_e = K \cdot B^2 \times f^2 \times$ Volume. So $W_e \propto B^2$ and also $W_e \propto f^2$.
- So eddy current's principle is used in heating a conducting substance.

2.17 DIELECTRIC HEATING

2.17.1 Principle of Dielectric Heating

- An atom of any material consists of nucleus having net positive charge at the centre and electrons carrying equal negative charge surrounding the nucleus. The configuration of such an atom is concentric. Refer Fig. 2.37 (a).

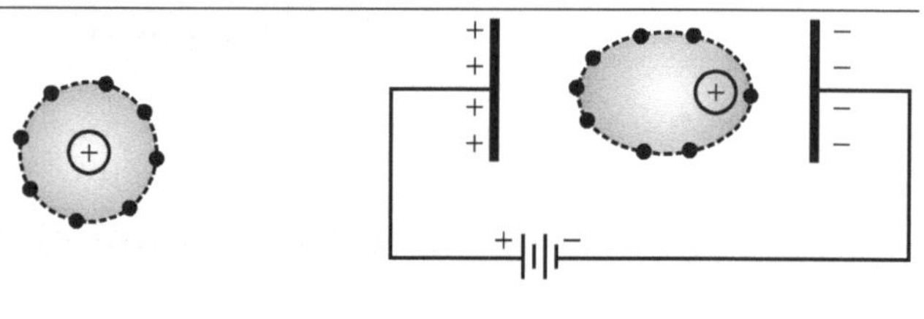

(a) (b)

Fig. 2.37: Dipole Formation

- When such atom is subjected to the influence of electric field, the positive plate repeals the positively charged nucleus and the symmetry is now disturbed (Refer Fig. 2.37 (b)). Such atom in stretched form is known as dipole or atom in this state is said to be polarized.
- Now for a substance, there are number of such atoms which will be polarized. If the electric field strength is increased, the degree of polarization will go on increasing i.e. more and more number of atoms will be polarized. There will be certain value of electric field for which all the electric dipoles will align themselves as shown in Fig. 2.38.

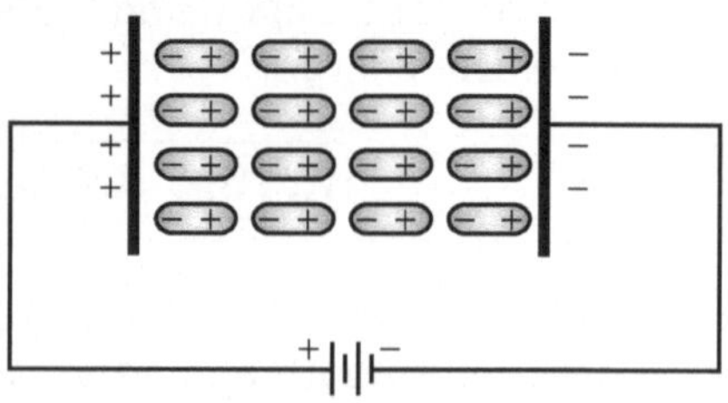

Fig. 2.38: Dipole Alignment

- The dielectric material in this state is said to be saturated.

- If a high frequency A.C. supply is applied across the plates, the dipoles will change their alignment after each half cycle (Fig. 2.39). During this process, some energy will be wasted as interautomatic friction which is called dielectric loss. Thus heat is generated within the substance which is indeed utilized for heating the substance itself.
- The electric field applied should not be too large otherwise, it will dislodge the electrons from the orbit, thereby ionizing the dielectric i.e. breakdown occurs. Thus, the value of electric field applied should be always less than the breakdown value. The dielectric heating is also known as high frequency capacitance heating.

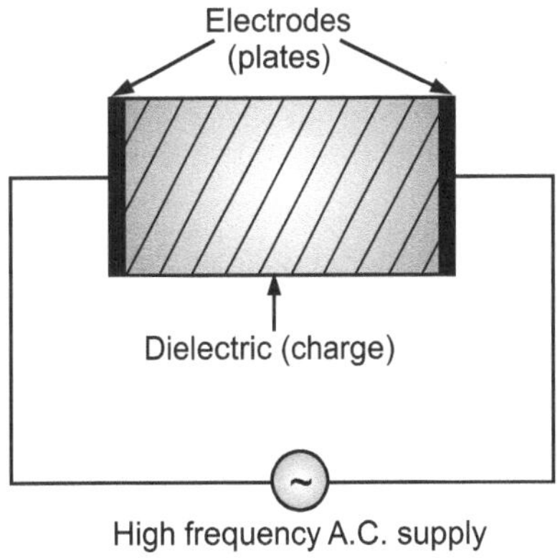

Fig. 2.39: Dielectric Heating Arrangement

- The dielectric loss taking place in insulating material is analogous to hysteresis loss taking place in magnetic materials, hence it is called as dielectric hysteresis.

Choice of frequency

- $\varepsilon_r.\delta$ is called loss factor. For a given material, it does not vary greatly with frequency. To obtain given amount of power loss, it is possible to use high voltage and medium frequency supply and vice versa. However, breakdown strength of specimen puts upper limit on voltage, generally the maximum voltage is limited to 20 kV.
- The frequency source has got some inductance and it forms resonant circuit when connected to the dielectric. For maximum power output from the oscillator, its impedance should match with the load circuit. A special matching circuit is required at higher frequencies. The frequency range used is 10 to 40 mega cycles/sec. are used.
- The maximum frequency that can be used is given by,

$$f = \frac{18.75}{l\sqrt{\varepsilon_r}}$$

where, l = Length of electrode

ε_r = Relative permittivity of the dielectric material

Because of leakage current (I_r), the current (I) flowing through the dielectric does not lead the applied voltage exactly by 90°.

The phase angle 'ϕ' is always less than 90°.

Let d = Distance between the electrodes or plate

A = Area of plate $\simeq$ Area of dielectric medium

ε_r = Relative permittivity of the dielectric medium

ε_o = Absolute permittivity (8.854×10^{-12})

- Power consumed is given by,

$$P = VI\cos\phi$$

But $\phi = (90 - \delta)$

and $\cos\delta = \frac{I_c}{I}$

$$\therefore \quad I = \frac{I_c}{\cos\delta}$$

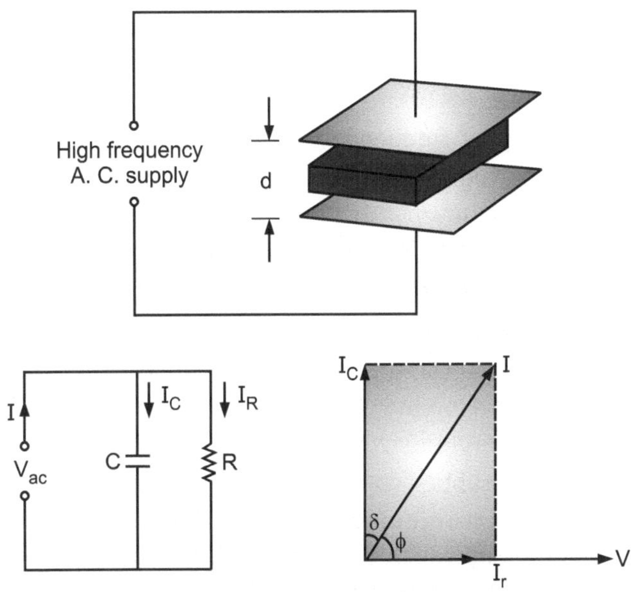

Fig. 2.40: Dielectric Heating Equivalent Circuit

$$\therefore \quad P = V \cdot \frac{I_c}{\cos\delta} \cos(90 - \delta) = V \cdot \frac{I_c}{\cos\delta} \sin\delta$$

$$= V \cdot I_c \tan\delta = V \frac{V}{X_c} \tan\delta = \frac{V^2}{X_c} \tan\delta$$

$$P = V^2\, 2\pi f c \tan\delta$$

'δ' is called dielectric loss angle.

$$\boxed{P = V^2\, 2\pi f\, \frac{A\, \varepsilon_0\, \varepsilon_r}{d} \tan\delta}$$

where, c = Capacitance of dielectric = $\frac{A\, \varepsilon_0\, \varepsilon_r}{d}$

Dissipation factor (tan δ):

- It is defined as the dielectric loss expressed as tangent of the angle between reactance and impedance vectors of a capacitor.

2.17.2 Advantages of Dielectric Heating

1. Uniform heating is achieved because heat is produced throughout the whole mass of the material.
2. Bad conductors of heat can also be heated. This is impossible to achieve specially in bad conductors of heat, by employing any of the conventional source of heat where heat is received by conduction, convection and radiation. This is because heat conductors from surface to interior will cause temperature gradient.
3. Non-conducting materials are heated speedily.
4. The materials heated by this method are combustible which cannot be heated by flame.

2.17.3 Limitations of Dielectric Heating

1. To get desired amount of heat or dielectric loss, high voltage medium frequency or vice versa is to be used. But there is limit on voltage because it should not exceed breakdown voltage of the medium. The voltage applied is upto 200 kV. It is therefore necessary to use high frequency supply. Hence, capital cost is more.
2. The frequency source impedance and local impedance should be matched to get maximum output. A special matching circuit is required at higher frequencies.
3. A definite length of electrode is to be maintained for a particular range of frequency, otherwise there is variation in voltage across the electrodes which results into uneven heating.
4. Higher frequencies may cause interference with radio stations.
5. At higher frequencies, uniform voltage distribution across the electrodes is difficult to obtain.

Applications of Dielectric Heating:

1. For food processing like drying of foods.
2. In wood processing industries for providing joints in the plywood.
3. In tobacco manufacturing industries for baking the tobacco.
4. In cotton industries for drying or heating of cloth for different processes.
5. In terylene cloth industry for producing terylene threads.
6. For rubber vulcanizing.
7. **Electronic sewing machine:** The films of plastic to be joined are rolled in between cold rollers which are connected to high frequency voltage supply.
8. **Diathermy:** The bones and tissues of our body are heated with this method.

2.18 ELECTRIC WELDING

2.18.1 Requirements of Good Welding

(i) A weld should be within the prescribed dimensions and free from deviations.

(ii) It should be free from undercuts slag on the surface.

(iii) It should be free from unfilled creators on the surface.

(iv) Free from visible cracks.

(v) Outer layer shall be free from porosity.

(vi) It should be free from external defects like irregular heights and width.

(vii) It should be free from poor fusion between layers.

2.18.2 Advantages of Electric Welding

1. Give advantages of electric welding.

- In comparison with other types of welding, the electric welding has following advantages:

  (i) The electric welding process and welding work is very clean.

  (ii) Weld joint and work piece has the same material.

  (iii) Time required is less, hence cheaper.

  (iv) Time and current can be very easily varied and controllable.

  (v) The "weld" by this process is very strong and homogeneous.

  (vi) Welding work can be done at the workshop as prefabricated or at site.

2.19 CLASSIFICATION OF ELECTRIC WELDING

1. Write classification of electric welding.

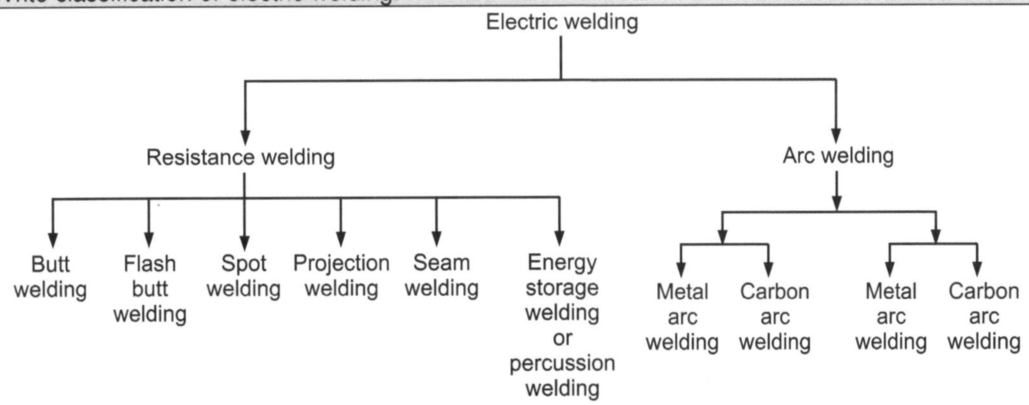

2.20 FACTORS CONSIDERED WHILE SELECTING WELDING SYSTEM FOR PARTICULAR JOB

- Welding means joining of two metals placed together by heating them to melting points.
- In electrical welding, this phenomenon is obtained by the use of electrical energy.
- There are different materials (metal) to be welded and different types of welds are required for the same.
- Hence depending upon the metal, its dimensions, requirements, different systems are selected.

- Table 2.2 shows the selected systems for a particular job.

Table 2.2

Sr. No.	Nature of Job	System
1.	Two thick materials of low conductivity.	Resistance welding system by controlling the contact resistance and suitable electrodes.
2.	Thick materials of high conductivity.	
3.	Two materials of different conductivity.	
4.	Thin metals.	
5.	Localised welding (not air tight or water proof).	Spot welding system.
6.	Punched, form or stamped parts.	Projection welding system.
7.	Leak proof joints of tanks, transformers, refrigerators, gasoline tanks, air-craft tanks.	Seam welding system.
8.	End to end joints for roads, tubes, wires etc.	Butt welding system.
9.	Two metallic parts, the electrode of the same material.	Metal arc welding system.
10.	For aluminium, magnesium and alloys, stainless steel joints.	Helium or Argon arc welding system.
11.	Joining copper metal and its alloys.	Carbon arc-welding system.
12.	Welding work of alloy steel, carbon steel, stainless steel, aluminium.	Atomic hydrogen arc welding system.

2.21 WAYS AND MEANS OF AVOIDING WELDING DEFECTS

2.21.1 Various Defects in Welds

(i) Blow hole
(ii) Slag inclusion
(iii) Undercut
(iv) Incomplete penetration
(v) Cracks
(vi) Distortion
(vii) Overlap
(viii) Bad appearance
(ix) Pit

2.21.2 Reasons for Welding Defects and How to Avoid Them

Sr. No.	Type of Defect	Why? Reasons	What to do to avoid?
1.	Blow hole 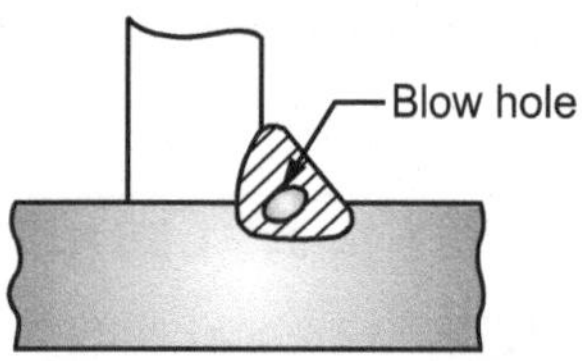 **Fig. 2.41**	(i) Wet, uncleared damaged electrodes. (ii) Too fast cooling rate. (iii) Too thick base metal. (iv) Long/Short arc length. (v) Over travelling speed. (vi) Base metal covered with oil rust etc. (vii) Rusted wire or flux.	(i) Maintain properly. (ii) Proceed proper pre-heating. (iii) Proceed proper pre-heating. (iv) Maintain proper arch length. (v) Reduce travel speed. (vi) Clean properly before welding. (vii) Store wire and flux properly.

contd. ...

2.	Slag inclusion 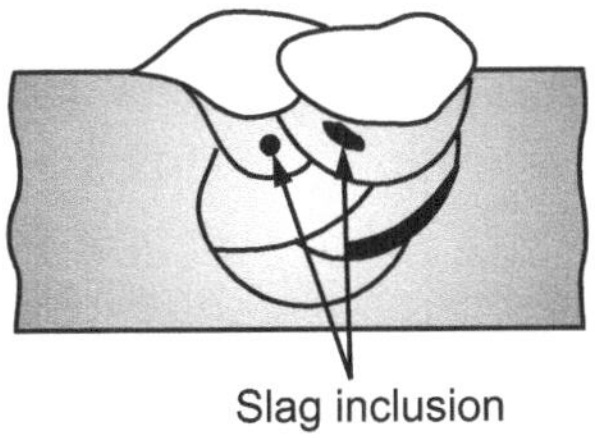 **Fig. 2.42**	(i) Welding current is too low. (ii) Slow travel speed. (iii) Improper groove shape. (iv) Slag removal incomplete from previous bead. (v) Too large weaving width.	(i) Use higher welding current. (ii) Use higher travel speed. (iii) Modify groove angle and root gap. (iv) Remove completely the slag from previous bead. (v) Reduce weaving width.
3.	Undercut 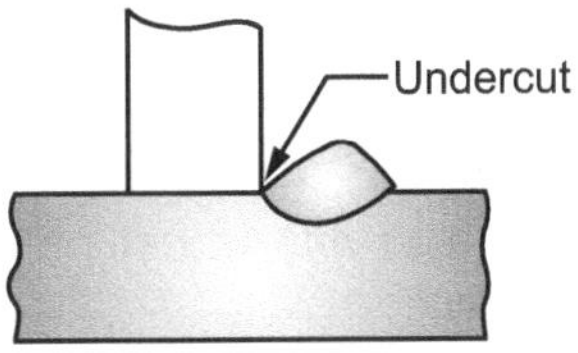 **Fig. 2.43**	(i) Welding current too high. (ii) Unclean base metal surface. (iii) Improper manipulation. (iv) Arc length too excessive. (v) Travel speed too fast. (vi) Base metal is overheated. (vii) Incorrect aiming position for wire when fillet welding.	(i) Reduce welding current. (ii) Clean it properly. (iii) Use proper electrode angle and pause at each side of weld bead when using a weaving technique. (iv) Reduce arc length. (v) Reduce travel speed. (vi) Use small size electrode. (vii) Maintain properly the position of aiming for wire.
4.	Incomplete penetration 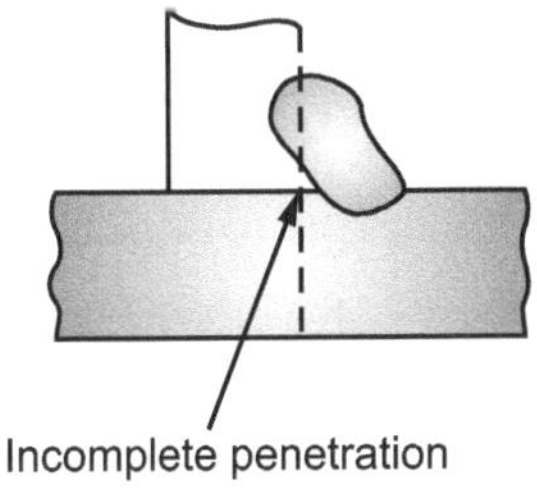 **Fig. 2.44**	(i) Improper electrode. (ii) Travel speed too fast/slow. (iii) Improper grooving and fixturing. (iv) Arc voltage is too high. (v) Improper weaving skill. (vi) Arc length too high. (vii) Improper welding skill.	(i) Select the electrode with better penetration. (ii) Adjust speed properly. (iii) Increase groove angle and root gap. (iv) Reduce arc voltage. (v) More practice. (vi) Reduce arc length. (vii) Practice more.
5.	Cracks 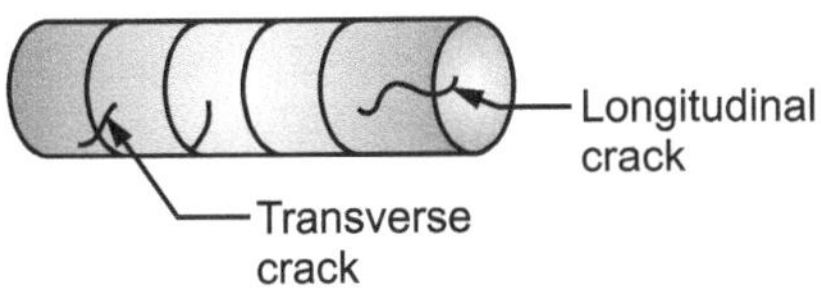 **Fig. 2.45**	(i) Base metal with over carbon and manganese contents. (ii) Electrode not of good quality or may be damp. (iii) Base metal having higher contents of sulphur. (iv) Without pre or post heating. (v) Too large welding current. (vi) Base thickness is much more. (vii) Too strong restraint for groove.	(i) Use low hydrogen type electrode with higher basicity. (ii) Use proper electrode. (iii) Avoid such metal. (iv) Proceed preheating. (v) Reduce welding current. (vi) Preheat base metal. (vii) Modify groove design.

contd. ...

6.	Distortion 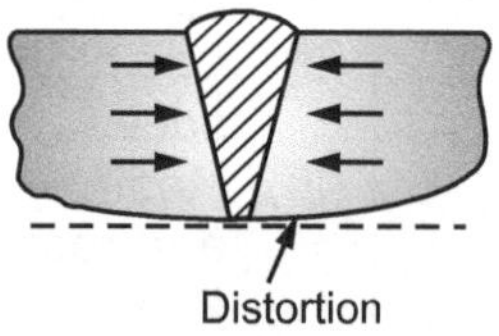 Fig. 2.46	(i) Excessive weld size. (ii) Improper set-up and fixture. (iii) Improper bead sequence. (iv) Overheating base plate because of less thickness. (v) Excessive layers and or faulty joint penetration. (vi) Wornout tip. (vii) Wire feeder works unsmoothly. (viii) Feed roller not fasten properly.	(i) Make welds to specify size. (ii) Tack or clamp parts securely. (iii) Use proper bead sequencing. (iv) Use thicker base plate. (v) Take weld-parts with allowance for distortion. (vi) Replace it. (vii) Grease wire feeder. (viii) Adjust properly.
7.	Overlap	(i) Too low welding current. (ii) Too slow travel speed.	(i) Properly use welding current. (ii) Use proper speed.
8.	Bad appearance	(i) Bad electrode (ii) Improper manipulation (iii) Tip worn out. (iv) Base metal overheated. (v) Lack of fusion in the weld bead.	(i) Choose proper electrode. (ii) Choose proper travel speed and welding sequence. (iii) Replace by new. (iv) Reduce welding current. (v) More practice.
9.	Pit	(i) Unclean electrode and segregation of base metal (ii) Weld metal contains high carbon/manganese (iii) Damp electrode (iv) Base metal cooling rate is too fast.	(i) Use good low-hydrogen type electrode. (ii) Use higher basicity electrode. (iii) Use dried electrode. (iv) Reduce travel speed and proceed pre and post heating.

2.22 RESISTANCE WELDING

2.22.1 Principle of Resistance Welding

1. State the principle of resistance welding.

- A heavy current is made to flow through the joint to be welded. The resistance of the joint generates I^2R loss causing heat which melts the metal joint and causes fusion of the joint.

2.22.2 Types of Resistance Welding

1. State the various types of resistance welding.

1. **Butt Welding:** Rods, pipes and wires are welded by this method. As shown in Fig. 2.47 there is a transformer, across secondary of the transformer two pieces of metal to be welded are held. The two pieces are held in clamps. One clamp is fixed and other is movable.

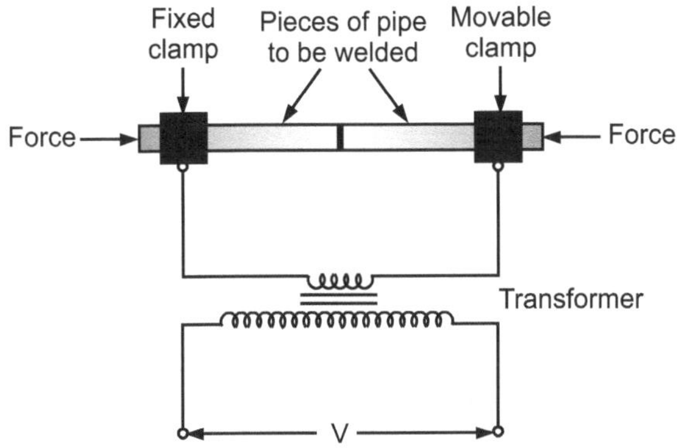

Fig. 2.47: Butt Welding

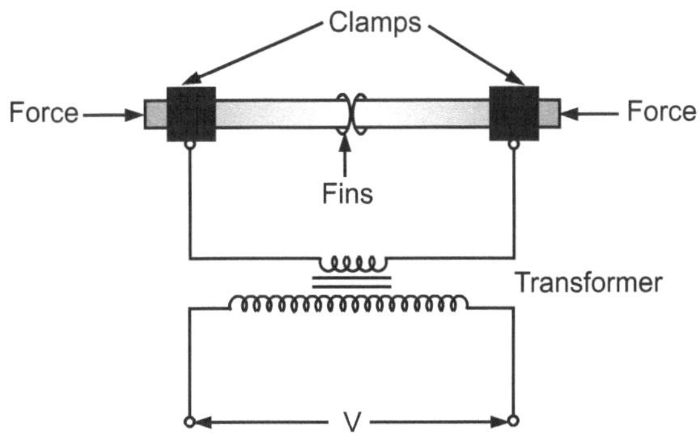

Fig. 2.48: Flash Butt Welding

The pieces of pipe are made to touch each other and force/pressure is applied through springs. When supply is given to transformer, current passes through the pieces of pipes and sufficient heat is developed at the joint causing welding at the joint.

2. **Flash Butt Welding:** This is a modification of butt welding. The ends of the pipe to be welded are held together with a little force/pressure applied to them. Arcing at the joint is allowed to take place. This removes any unevenness at the joint. A weld having a fin around the joint is obtained (Refer Fig. 2.48).

   The fin can be easily removed. Two dissimilar metals can be effectively welded with this method. This welding method is faster than other methods.

3. **Spot welding:** In spot welding, the welding is done at certain points on metallic sheets. The pieces to be welded are held between the two electrodes as shown in Fig. 2.49.

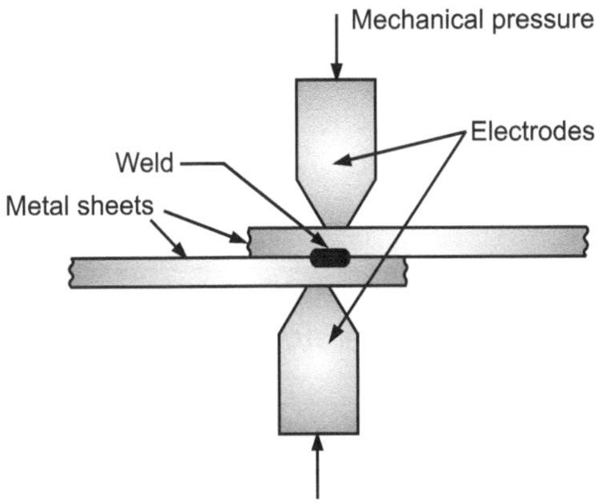

Fig. 2.49: Spot Welding

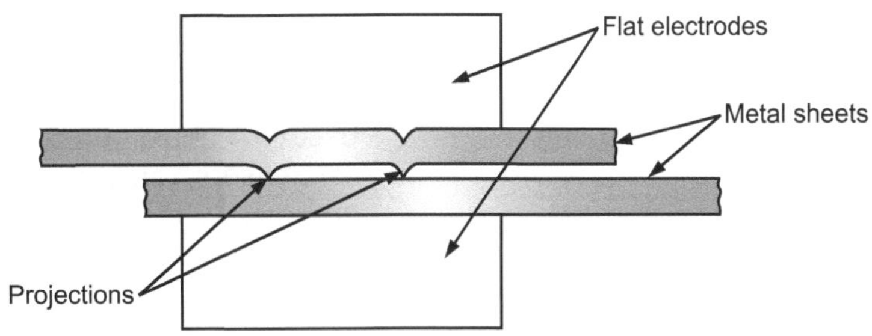

Fig. 2.50: Projection Welding

   When current passes through the electrodes, a spot weld is produced between the sheets.

4. **Projection Welding:** It is a modified form of spot welding. The electrodes used are flat and not pointed as in the case of spot welding. (Refer Fig. 2.50).

   Projections are made on one sheet so that heat is concentrated in that region. When current is passed through the electrodes, the current density and pressure required is less, hence electrode life is more. The welding automatically takes place only at the projections.

1. Describe with sketch seam welding.

5. **Seam Welding:** It is similar to spot welding. In seam welding, a series of spots is produced by roller electrodes, as shown in Fig. 2.51.

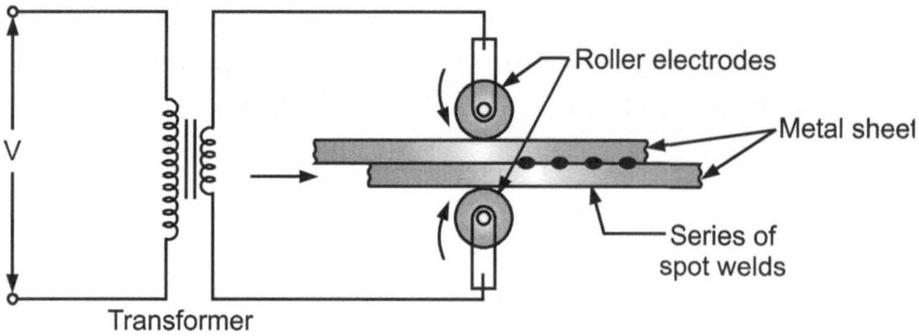

Fig. 2.51: Seam Welding

Two sheets are passed under the two rolling electrodes. Depending upon the number of welding current pulses per second and speed of electrodes, a series of spot welds is obtained.

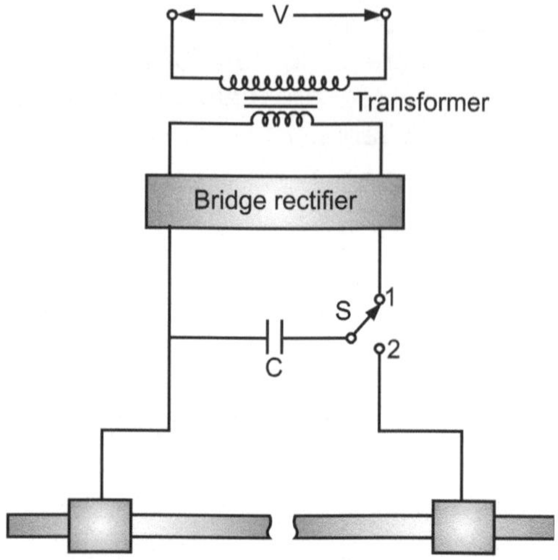

Fig. 2.52: Energy Storage Welding

6. **Energy Storage Welding / Percussion Welding:** Welding is obtained by discharging the stored energy in the capacitor. As shown in Fig. 2.52, it consists of a bridge rectifier, capacitor, switching arrangement.

   The pieces to be welded are kept at a certain distance apart (1.5 mm). Switch 'S' is put on position 1. So capacitor gets charged through the bridge rectifier.

   When S is put on position 2 the capacitor discharge takes place causing formation of arc between the two pieces. The pieces are heated up at the gap and fused together. Thus welding is obtained.

2.23 APPLICATIONS OF RESISTANCE WELDING

Type of Resistance Welding	Application
1. Butt welding	For welding rods, tubes, wires.
2. Flash butt welding	For welding, tubing, bars, rods, forgings, fittings used in automotive and aircraft products, refrigerators, welding of band saw blades.
3. Spot welding	For welding components made from plates (10 mm to 12 mm in thickness). For fabricating or joining sheet metal structure.
4. Projection welding	Used for mass production work e.g. welding of refrigerator, condensers, crossed-wire welding, refrigerator racks, granites, grill etc.
5. Seam welding	Used for welding of pipes, conduits, tanks, transformers, refrigerators, gasoline tanks, various types of containers.
6. Energy storage welding	Used for welding different metals e.g. copper to aluminium or steel, silver contact tipes to copper, cast iron to steel.

2.24 SAFETY EQUIPMENTS

2.24.1 Welding Work and Protection of Welder

- Welding work is a skilled work but at the same time it is risky and needs safety equipment to safeguard the welder.

- Particularly due care to be taken of the body organs such as eyes and hands.
- To facilitate free movements the dress of the welder shall be perfect of his size not too tight or too loose of dark colour.
- Body organs shall be protected from electric sparking while working.
- Welding machines, accessories shall be handy and easy to "work on" and be kept in a good working condition.

2.24.2 Welding Equipments Accessories for Protection and Safety

- Following are the main equipments:
  1. Gloves
  2. Helmet
  3. Jacket
  4. Glasses
  5. Cap
  6. Slag (Chipping) hammer
  7. Pliers
  8. Wire brush

1. Gloves:

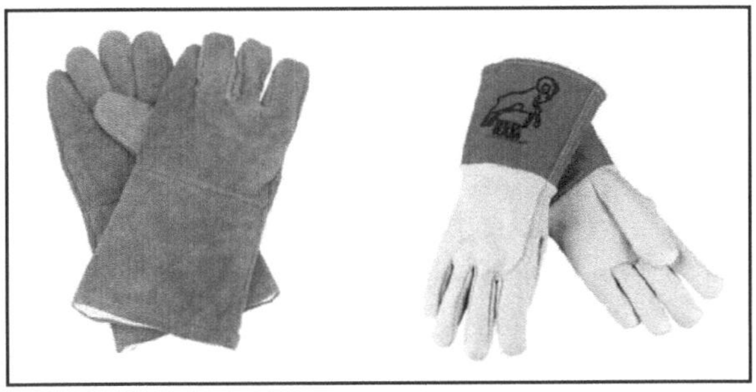

Fig. 2.53

  - Material used for gloves is a good quality thick rubber more flexible for hand movements.
  - Two types of gloves used in welding work.
  - Gauntlet type and Tig-type.
  - Gauntlet type gloves cover the forearm.
  - Tig-type gloves are high weight and designed for tig i.e. 'Healy Arc' and they give better feel for manipulating the tig-tarch and adding filler wire.
  - Use pliers to handle hot weld coupons.
  - Don't's use damp or wet gloves.
  - Replace the same if holes are developed on them.

Use:

  - Gloves are used for protecting hands and arms.

2. Helmets:

Purpose:

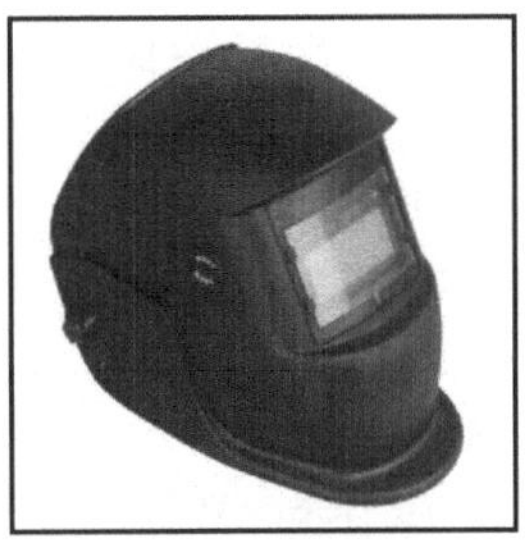

Fig. 2.54

  - To protect welder's face and head from ultraviolet and infrared rays of welding arc while allowing the welder to view the molten weld poor or puddle.
  - There are various types of helmets used for welding such as:
    (i) flip front
    (ii) auto darkening lens
    (iii) wide lens
  - The auto-darkening helmet has a lens that is clear to allow the welder to see the joint before striking the arc. When arc is struck the lens darkens to normal view of the welding puddle.

3. Jacket:

- Jackets are made of soft leather.
- It is just like a coat covering the middle portion of body from neck to wrist.
- It is used to protect upper body and neck from sparks and hot metal globules.
- Most of the jackets are having inside pocket and snap all the way upto the collar.
- A light jacket made of fire retardant material also can be used in welding shops while welding. The jacket has full sleeves.

Fig. 2.55

4. Glasses:

- These are safety glasses used for protection of eyes while welding a job.
- These glasses are called as safety glasses which protect the eyes from intense ultraviolet and infrared rays of the welding arc.
- The people working around the welders also should use these glasses for eye protection.
- These glasses also have side-shields to protect the corner of the eyes when welders are working very close.
- In case of eye injuries, it is advisable to consult a eye doctor.

5. Cap:

- The welding cap protects the head from sparks flying over the welding helmet and also protects from falling sparks while welding overhead.

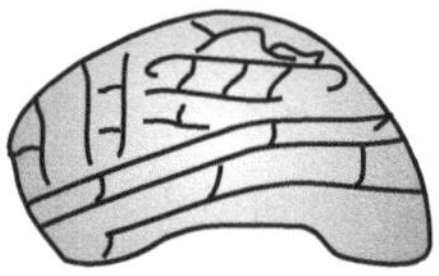

Fig. 2.56

6. Slag Hammer:

- This hammer is also called as "Chipping Hammer".
- It is very useful for removing the slag from the completed stick welds.
- The hammer is shaped on both sides.
- At one side is chisel like shape and on the other side a pointed shape.

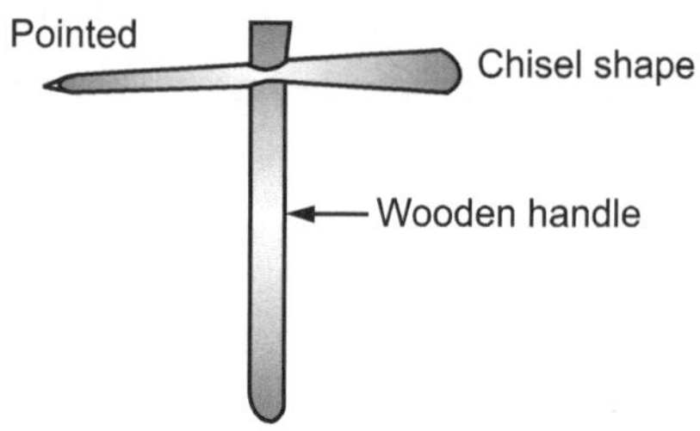

Fig. 2.57

7. Pliers:

- To handle the hot metal and hot weld coupon, a pair of pliers is needed.
- If pliers are not used then hands can burn and also the life of gloves is reduced.
- Types of pliers used are:
  - (i) Slip joint plier-pair
  - (ii) Locking plier-pair

8. Wire brush:

- Brushes are needed for
  - (i) Cleaning weld layers, and
  - (ii) Cleaning completed weld.
- Brushes are of two types:
  - (i) Typical (regular) wire brush
  - (ii) Slag-hammer and wire-brush combination.

2.24.3 Safety to be Observed in Welding Work

1. Before starting the work read the manual for that work and follow the instructions.
2. Know the details of the machine to be used.
3. Electric welding involves electric supply, so be careful.
4. Use proper clothing (jackets, helmets, caps and pliers etc.)
5. Be careful from electrical contacts.
6. Weld, clothing, pliers etc. should be dry (not damp).
7. Turn-off the electric machine when electrode holders are to be changed, ground clamps are to be changed.
8. In remote welding places, take the help of one more person.
9. Check frequently ground and electrode cables for bare spots or damage.
10. The damaged parts are to be repaired by expert technician.
11. When work is discontinued, don't leave electrodes in the holders.
12. Never drape the welding cable over your body to weld.
13. In wet conditions, water leakages, rails stop the work.
14. Don't try to repair welding machine on your own unless you are fully trained for that job.
15. If due to leakage current, a shock is sensed stop the work.
16. Don't climb on metal drums, metal tanks for welding work.
17. Falling objects may harm you. Also your co-workers or assistance to be around you and they should take care that sparks not to fall on persons passing nearby.

2.25 ELECTRIC ARC WELDING

1. Describe electric arc welding in brief. How arc is formed in electrical arc welding? State characteristics of electric arc welding.

- In this type of welding, an arc is struck between the electrode and the metal. Sufficient amount of heat is developed which melts the metal.
- The electrode is first made to touch the metal and then it is separated. Voltage is applied across the electrode and the metal. Thus, arc is struck. No mechanical pressure is required, so this type of welding is also known as non-pressure welding. The temperature obtained is upto 3000°C, at such a high temperature, mechanical pressure is not required for welding.
- Mechanical pressure is not required for metal arc welding, so it is also known as non-pressure welding. Arc is made to struck between electrode and the metal, sufficient heat is obtained which causes the metal to melt and this welding is obtained.

2.25.1 Formation of Arc

- An electric arc is the flow of electric current through gases (or ionized air) accompanied by heat and bright glow. Arc is struck by shorting the two electrodes or one electrode and job, and then separating them apart.
- When electrode is separated, the contact area of electrode decreases causing increase in resistance $\left(R \propto \frac{1}{a}\right)$. The tip of electrode becomes red hot. The electrons from electrode are accelerated due to voltage gradient developed between the electrodes and these electrons collide with atoms and molecules of air. The air gap between electrodes is now ionized i.e. acts as a conductor. Thus, arc is initiated A.C. and D.C. supply can be used in arc welding. For A.C. supply requirement is 50 to 60 volt and for D.C. it is 70 to 100 volt. Once arc is struck the voltage required to maintain the arc is only 20 to 30 volt.

2.25.2 Characteristics of Electric Arc

- The arc has negative resistance characteristics i.e. its resistance decreases with increase in current.
- Once arc is struck/established, the current increases because the ionized air gap between the electrodes acts as a conductor, the arc diameter increases, arc resistance decreases, therefore current again increases.
- In this way, cause helps the effect which in turn helps the cause. Thus, an electric circuit with constant voltage supply and an arc load is basically unstable. The arc thus possesses negative volt ampere characteristics as shown in Fig. 2.58.

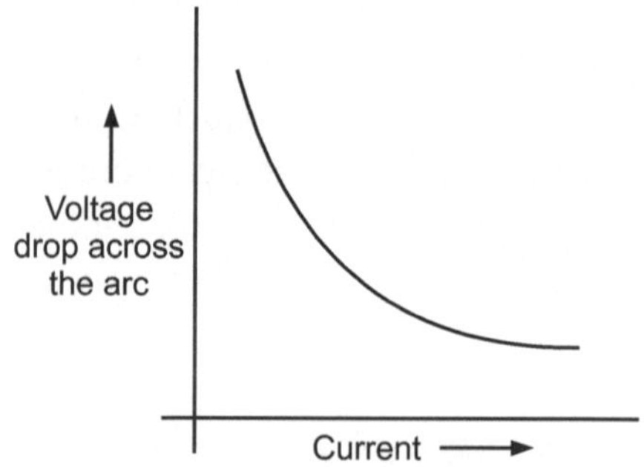

Fig. 2.58: Volt Ampere Characteristics of Electric Arc

2.25.3 Arc Blow

- Arc is a flexible conductor. It is a current carrying conductor, so it produces flux around it. Under particular condition, a force is exerted on the arc during welding. Consider Fig. 2.59.

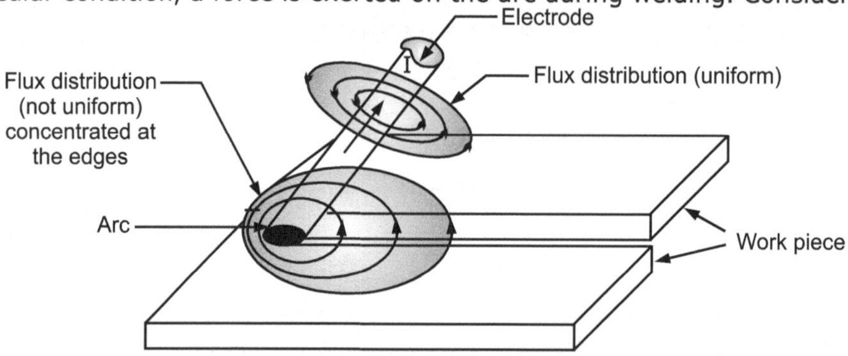

Fig. 2.59: Arc Blow

- The current direction is as shown. By applying right hand grip rule, flux direction is determined. The flux distribution in upper portion of conductor is symmetrical along its area but at the work piece the flux lines are crowded at the edge.
This is due to the fact that magnetic lines of force prefer low reluctance path so instead of taking pass through air, these are concentrated at the edge. A force is acted on the arc and hence on the electrode such that it moves away from the edge. Blow effect is also observed due to ground currents. Refer Fig. 2.60, the arc current produces magnetic field. Arc current is acted upon by a force due to the magnetic field and arc moves away from the ground. This effect is prominent in narrow sections.

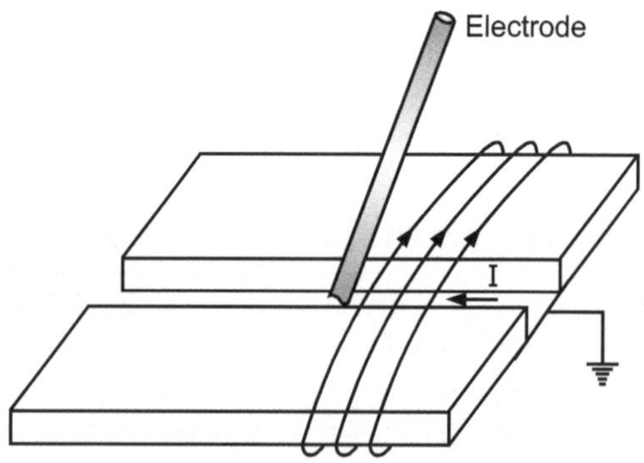

Fig. 2.60: Blow Effect

- Because of magnetic blow effect, there is improper penetration of heat and proper welding cannot be obtained.
- To minimize arc below effect, following majors are necessary:
  (i) Short arc is to be used.
  (ii) Current should be kept small.
  (iii) A.C. supply should be used instead of D.C.
  (iv) Electrode is to be held properly.
  (v) Ground connection should be made away from the joint.
- So a special supply arrangement is necessary to maintain a stable arc.

2.25.4 Factors on which Arc Length Depends (Effect of Arc Length)

- In manual arc welding, it is not possible for the operator to maintain a constant distance between the electrodes or in other words it is impossible for the operator to maintain a constant arc length.
- If the electrodes are brought closer, the arc length decreases causing decrease in arc resistance ($R \propto l$), therefore current increases causing increase in arc diameter and now resistance further decreases $\left(R \propto \frac{1}{a} \propto \frac{1}{d}\right)$. The arc will be out of control in this way.
- If the electrodes are taken away, the arc length increases and arc resistance also increases ($R \propto l$) causing decrease in arc diameter and now again resistance further increases $\left(R \propto \frac{1}{a} \propto \frac{1}{d}\right)$ and arc will be finally extinguished which is not desired.
- So in both the cases arc behaviour gradually approaches to extreme condition either it is out of control or it is extinguished.
- The stability of arc is thus important to get proper welding.

2.25.5 Power Supply for Arc Welding (Methods of Arc Stabilization)

- As described earlier arc has a negative volt-ampere characteristic. Greater voltage is required to strike the arc and once arc is struck, the voltage required to maintain it is less.
- The electrode touches the job and supply is made on, then electrode is separated from job so that arc is struck. The voltage required for this purpose is called open-circuit voltage. For D.C. supply this voltage has to be atleast 30 to 35 volt and for A.C. the range is 50 to 55 volt.
- After arc is struck, the circuit turns out to be almost a short circuit path (supply – electrode – arc – job – supply) and there is sudden in rush of current. This current should be limited. The supply voltage should be reduced so that stable current and hence stable arc can be maintained.
- The voltage source used should have a drooping characteristics. Refer Fig. 2.61.

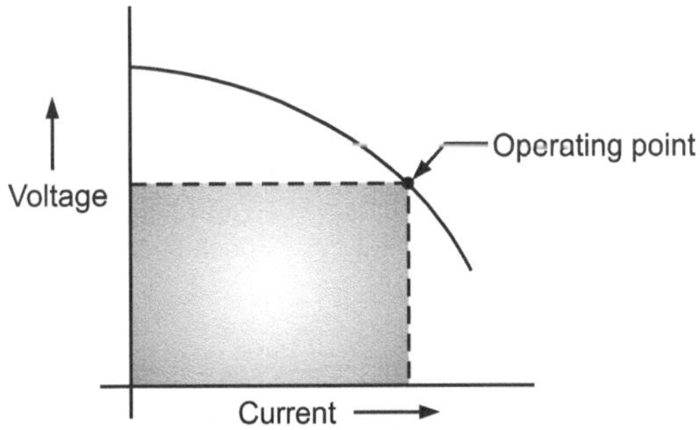

Fig. 2.61: Desired characteristics of Power Supply for Arc Welding

2.25.6 Metal Arc Welding

- Electrode used in metal arc welding is made up of same material as that of the metal to be welded. This electrode also serves the purpose of filler material. Both A.C. and D.C. supply can be used but with D.C. supply the arc blow effect is severe, hence generally A.C. supply is used (Refer Fig. 2.62).
- To get a good weld, the reaction of molten metal oxygen and nitrogen should be avoided.

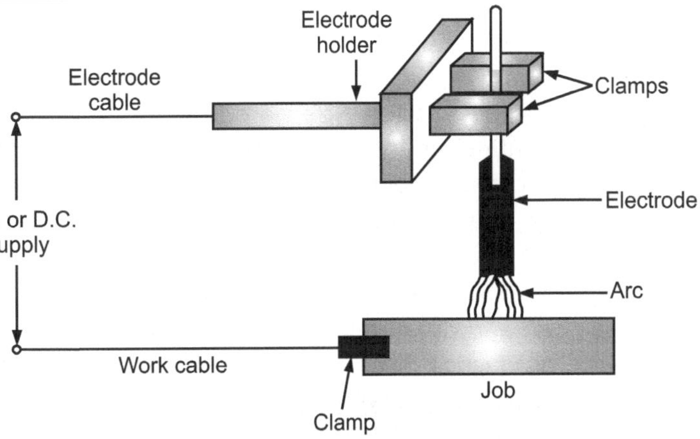

Fig. 2.62: Metal Arc Welding

- So slagging and fluxing ingredients are used. The melt is covered with slag and electrode is covered with flux.

2.25.7 Carbon Arc Welding

1. Describe carbon arc welding with neat sketch.

- The electrode used in this type of welding is made up of carbon or graphite.
- Mostly graphite electrode is preferred, it has longer life and low resistance. The arc produced between the electrode and the job and melting of job takes place, thus welding is obtained.

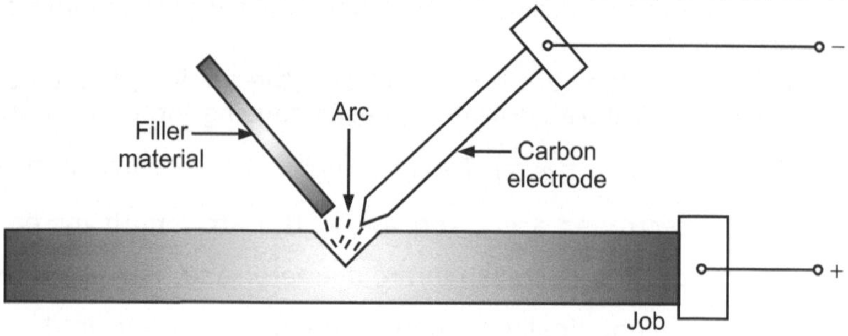

Fig. 2.63: Carbon Arc Welding

Polarity in D.C. Welding (Carbon Arc Welding)

- It should be noted that if electrode is made positive then carbon contents may flow into the weld. This will make the welding weak and brittle.
- The use of A.C. is not recommended because no fixed polarity can be maintained and above effect will cause weak and brittle weld.
- Carbon arc welding is used for welding sheet steel, copper alloys, brass, bronze and aluminium. It is not suitable for welding in vertical plane or overhead welding.

Automatic Hydrogen Welding

- In this type of arc welding, two tungsten electrodes are used. These are connected to A.C. supply and are placed in the vicinity of hydrogen (Refer Fig. 2.64). The hydrogen gas is obtained from a storage cylinder and the hydrogen gas is passed on the electrodes at 0.5 kg/cm^2 pressure by means of control valve.

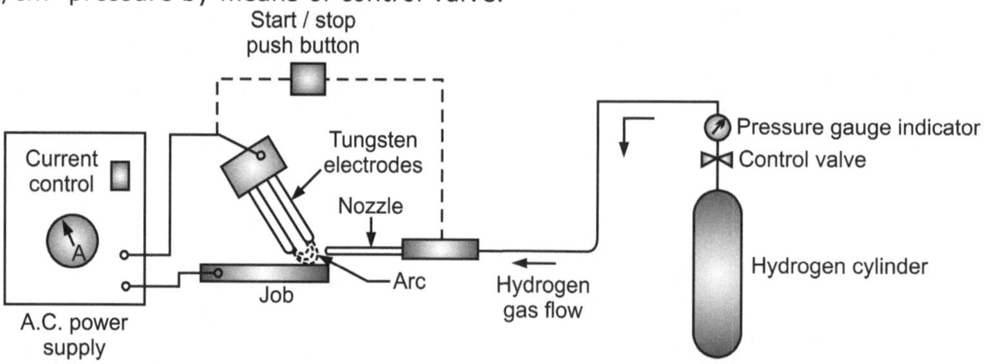

Fig. 2.64: Hydrogen Welding

- When supply is made ON, an arc is made to struck between the two tungsten electrodes this causes heating of hydrogen atom and it gets converted into atomic form. The atoms of hydrogen come in contact with base metal and atomic hydrogen is recombined and in doing so it releases a lot of heat energy. Filler material is also used and it is made up of same metal as that of the job.
- This type of welding is used for fusion welding of ferrous alloys such as chrome nickel steel, aluminium and duralumin but flux has to be used in this case. It is also used for welding thin sheets, production of tubing and repairing of expensive dies and tools.

Helium or Argon Arc Welding:

- The electrode used in this type of arc welding is made up of tungsten and it is shielded by argon or helium. A filler rod may be used if required.

- As it uses inert gas, air does not come in contact with molten metal. So flux is not required. It is possible to concentrate heat at the desired portion.

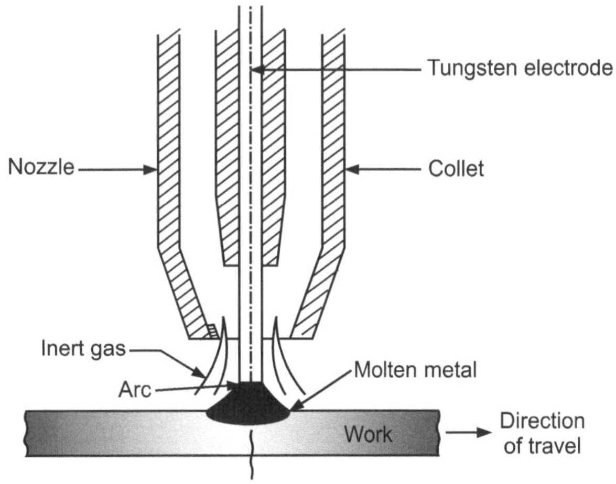

Fig. 2.65: Helium / Argon Arc Welding

2.25.8 Types of Electrodes

- Electrodes may be consumable or non-consumable type. Non-consumable electrodes are made up of carbon or graphite or tungsten. These are not consumed during welding operation. Consumable electrodes are made up of the same material as that of the job to be welded.

2.25.8.1 Advantages of Coated Electrodes

- **Bare electrodes:** These are most commonly used electrodes. During welding process, both the globules of molten electrode wire and parent metal are exposed to oxygen and nitrogen of the atmosphere. Hence, the oxidising and nitriding effect causes imbrittlement in the weld. Hence, the quality of welding obtained is not good. So these electrodes are used for unimportant welding work.
- **Light coated electrodes:** A thin coating of lime mixed soluble glass is given on the electrodes. This increases the arc stability. However, oxidation cannot be avoided for these electrodes.
- **Heavy coated electrodes:** These are also known as shielded electrodes. The heavy coating consists of chalk, aluminium, ferro-manganese, starch, kaolin, binding materials. The coating gives the following advantages:
  - (i) Slag is formed on molten metal because of kaolin. This slag floats on the surface of molten metal and prevents rapid cooling of the weld. Thus weld does not become brittle.
  - (ii) The weld is protected from oxidising and nitriding effect.
  - (iii) The arc space is kept ionized even if A.C. supply is used.

2.26 WELDING SETS (USING D.C. OR A.C. SUPPLY)

2.26.1 D.C. Welding Sets

- Drooping characteristics can be obtained by using **differential compound D.C. generator**. The generator is driven by squirrel cage induction motor.

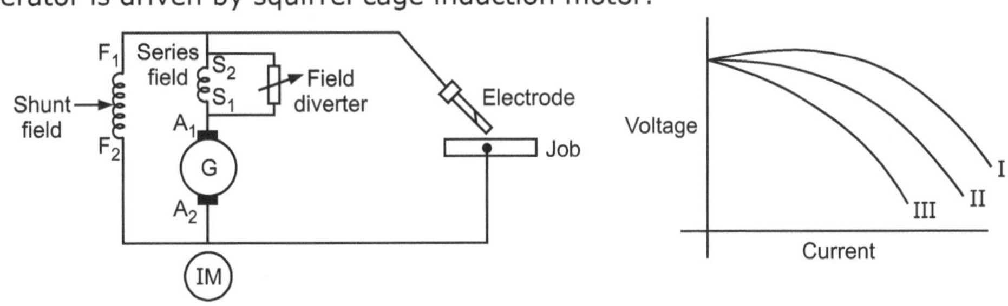

(a) (b)

Fig. 2.66: Differential Compound Generator for Arc Welding

- As the arc current increases, demagnetizing effect increases and terminal voltage decreases. Reverse happens when load current decreases. The field diverter changes the shape of the drooping characteristics (I, II, III).
- A three brush D.C. generator is also used. In that generator the drooping characteristic is obtained by demagnetizing effect of armature reaction.
- In case of A.C., a series reactor is used for arc stability. The basic property of reactor is to oppose change. The volt-ampere characteristics of reactor and arc is as shown in Fig. 2.67 (a) and (b).

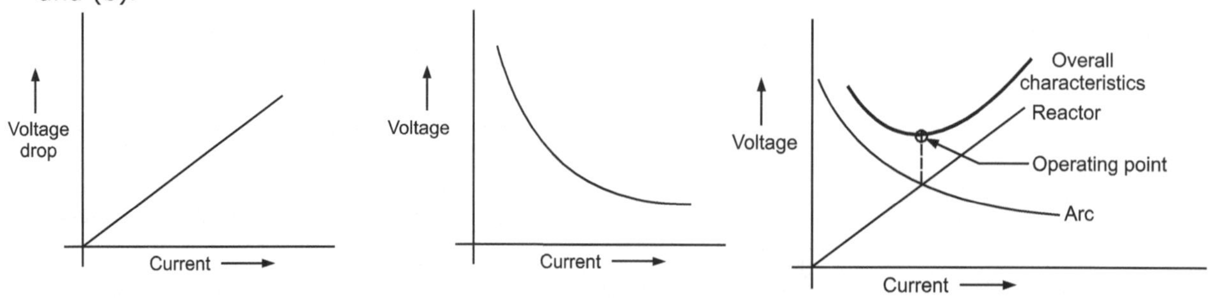

(a) V-A characteristics of reactor (b) V-A characteristics of arc (c) Overall V-A characteristics

Fig. 2.67: Arc stability using reactor

- The overall characteristics are obtained by merging the VA characteristics of arc and reactor. Thus after operating point the characteristics is linear helping to stabilize the arc and hence the current.

2.26.2 A.C. Rectifier Welding Sets

- In another scheme, a rectifier is used in conjunction with a high leakage reactance transformer and a saturable reactor.
- The reactance of the reactor is adjusted by changing the value of D.C. excitation obtained from rectifier by means of rheostat. When the current in central winding of reactor core increases, reactor approaches magnetic saturation. Reactance of reactor becomes less.
- The control action is similar to that described for series reactor method.

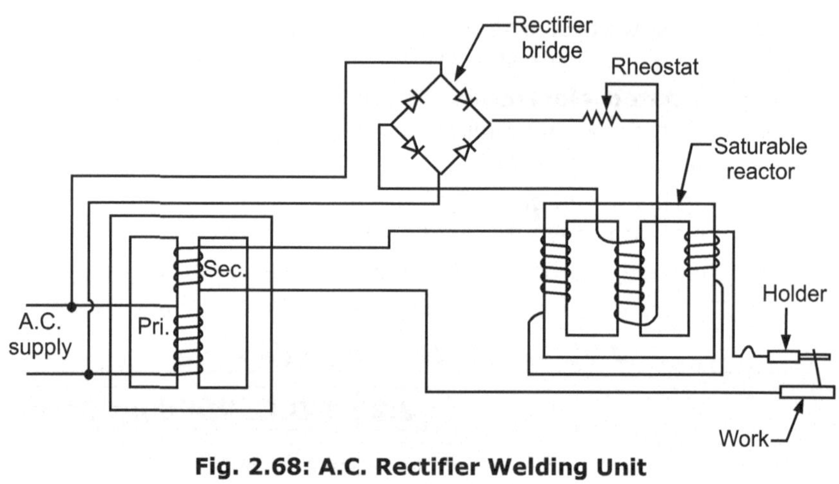

Fig. 2.68: A.C. Rectifier Welding Unit

2.26.3 A.C. Welding Sets

Welding Transformer:

- Single-phase or three-phase step-down transformers are used. The transformer is designed for high leakage reactance. The secondary voltage on open circuit is about 80-100 volt. Some control device is also used alongwith this transformer.
- The secondary winding is air cooled and madeup of copper strip. Oil cooled transformers are also used.
- The air cooled sets are cheaper in cost and lighter in weight and easier to maintain. It can be easily shifted from one location to other.
- Oil cooled sets are more compact but these are heavy, so suitable for stationary application, e.g. in workshops.

- The current control in welding transformer can be obtained by following ways:

(i) Tapped Reactor Method:

- A series reactor is used with tappings on it (Refer Fig. 2.69).
- The output current can be varied by selecting appropriate tap. Different values of currents can be obtained by this method.

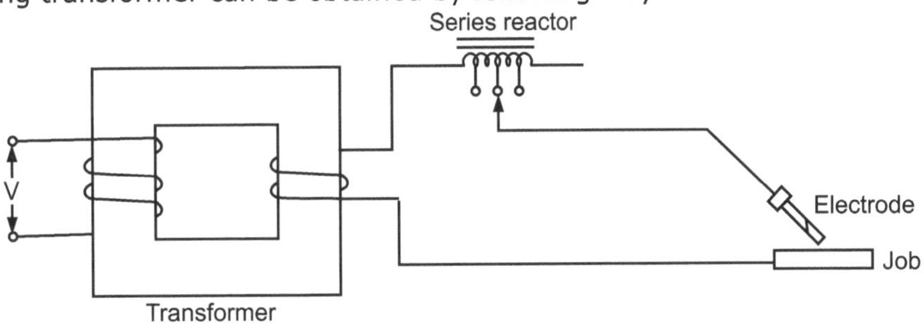

Fig. 2.69: Tapped Reactor Method

(ii) Moving Coil Method:

- The transformer used in shell type with a movable primary winding is as shown in Fig. 2.70.
- The movement of primary is possible by rotating the handle. If primary is moved near to secondary, the current increases and if it is moved away current decreases.

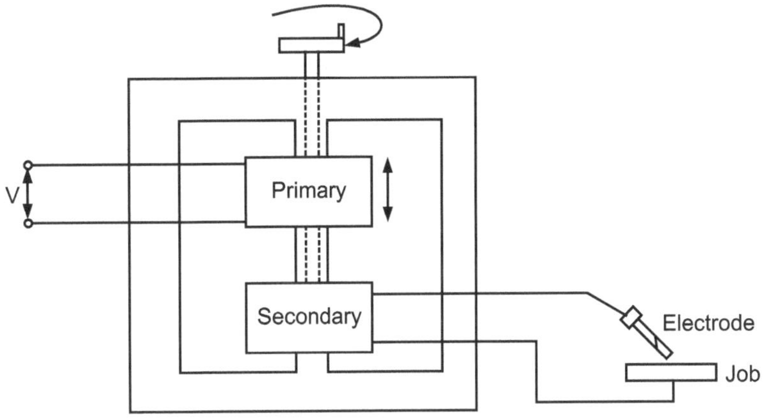

Fig. 2.70: Moving Coil Method

(iii) Magnetic Shunt Method:

- A magnetic shunt is used to divert the flux. The position of magnetic shunt can be adjusted. Hence, the value of flux in transformer core changes and thus current can be changed.

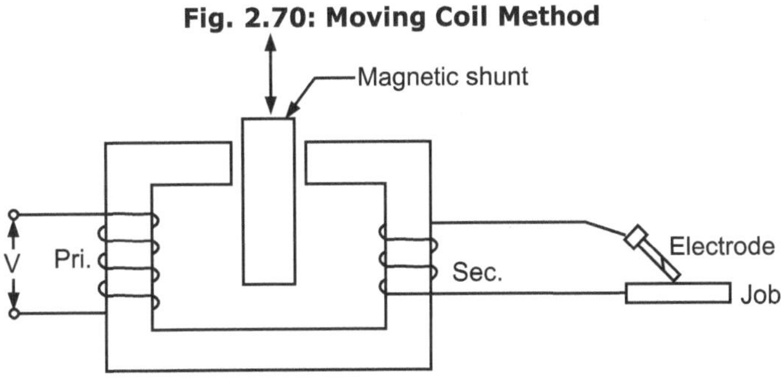

Fig. 2.71: Magnetic Shunt Method

2.26.4 Modern Welding Techniques

- **Ultrasonic welding:** In this welding technique, union between similar or dissimilar metals or non-metals is done by the application of ultrasonic vibrations generated by piezoelectric effect. The frequency of vibrations is about 20 to 60 kHz. It does not use electric current or arc or filler material etc.
- The arrangement for ultrasonic welding is shown in Fig. 2.72.
- The piezoelectric transducer produces vibrations which are transferred to the coupler and the electrode (sonotrode tip). A little bit of pressure is applied. These vibrations are thus transferred to the job. The forces developed because of vibrations produce dynamic stresses on the job or base material. Minute deformations are produced which create temperature rise. The pressure causes coalescence across the two surfaces of job causing weld.

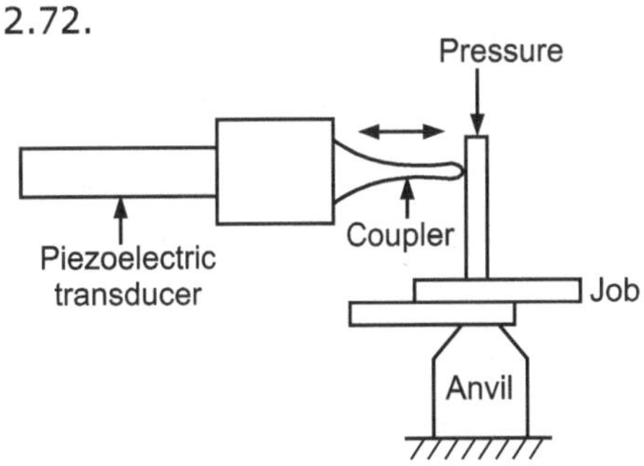

Fig. 2.72: Ultrasonic Welding

- The temperature at the weld is not raised to the melting point and there is no nugget similar to resistance welding.
- The ultrasonic welding is used in electronic industries, instrument industries, for joining plastics and for sealing.

Electron beam welding:

- An electron is made to travel in vacuum at high speed by exciting it. This electron strikes the surface of the metal to be welded. The impact gives rise to heat energy which melts the metal and thus welding is obtained.
- Since, operation is carried out in vacuum there is no need of inert gas, electrodes, flux etc.
- The arrangement is shown in Fig. 2.73.
- Tungsten filament acts as cathode and emits electrons when high voltage supply is given across it.

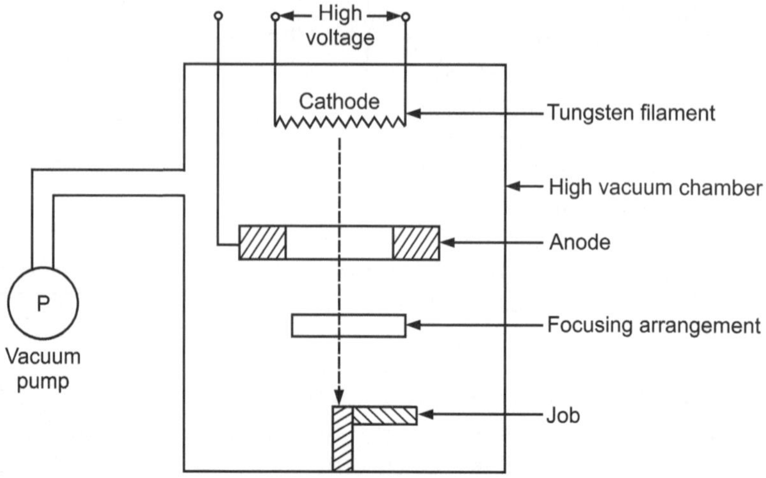

Fig. 2.73: Electron Beam Welding

- These electrons are accelerated and are focused on the job with the help of focusing arrangement. The temperature obtained is about 2500°C which is sufficient for welding the job. The major advantage of this type of welding is tremendous penetration.
- This technique is used for welding the accessories used in aerospace, automobile, aeroplane etc.

Comparison between D.C. Welding and A.C. Welding:

1. Compare D.C. welding with A.C. welding.

Sr. No.	Point of Comparison	D.C. Welding	A.C. Welding
1.	Heating	Uniform heating	Not so uniform
2.	Arc stability	Higher	Lesser
3.	Welding for non-ferrous metals	Suitable for non-ferrous metals	Not suitable for non-ferrous metals
4.	Equipment required	M.G. set or rectifier unit	A.C. supply, transformer
5.	Cost	More	Lesser
6.	Operating efficiency	Low	High
7.	Power factor	High	Low
8.	Arc blow	Pronounced	Not so
9.	Electrodes	Both bare/coated	Only coated
10.	Electrode cost	Cheaper	Expensive

Comparison between Resistance Welding and Arc Welding:

Sr. No.	Point of Comparison	Resistance Welding	Arc Welding
1.	External pressure	Required	Not required
2.	Supply	Arc	AC or DC
3.	Filler metal	Not required	Required

contd. ...

4.	Heat production	Due to contact resistance	Due to arc formation between electrode and work piece
5.	Temperature achieved	Not so high but upto 1350°	High from 3600 to 6000°C
6.	Power factor	Very poor	Poor
7.	Voltage control	Not necessary	Voltage control required
8.	Use	For mass production	Suitable for repairing works

Comparison between Spot Welding and Arc Welding:

Sr. No.	Point of Comparison	Spot Welding	Arc Welding
1.	Used	For joining or fabricating sheet metal work	Welding of Cu alloys, brass, bronze, aluminum.
2.	Pressure	Pieces to be joined are placed together by mechanical pressure exerted by electrodes	No pressure is required.
3.	Welding processes	By setting works automatically	Done manually
4.	Cost	Lower	More
5.	Work	For mass work	For repairs and small work

Comparison between Carbon Arc Welding and Metal Arc Welding:

1. Compare metal arc welding and carbon arc welding.

Sr. No.	Point of Comparison	Carbon Arc Welding	Metal Arc Welding
1.	Supply needed	Only D.C. supply	Both can be used as AC or DC.
2.	Electrodes	Carbon or Graphite	Same rod that of parent material.
3.	Applications:	(i) Welding sheet steel (ii) Copper alloys (iii) Bronze (iv) Aluminium	(i) Metal deposition in rail wagon (ii) Wheels (iii) Travelling crane wheels
4.	Filler material	Not required	Required

Important Points

Electric Heating:

- **Resistance heating:** The resistance heating process is based on heating effect of electric current i.e. I^2R effect.
- **Direct resistance heating:** In case of direct heating the current is passed through charge which produces I^2R loss in the charge and thus charge is heated up.
- **Indirect resistance heating (Resistance Ovens):** In indirect heating the current is passed through some heating elements and this heat is passed to the charge through radiation, convection.
- **Arc Furnaces:** For melting of ferrous and non-ferrous materials, a high temperature is required (3000°C to 3500°C) which is achieved by electric arc furnaces.
- **Induction Heating:** Principle of electromagnetic induction is used for heating the charge. The heat energy is transferred to the charge at extremely rapid rate much faster than other conventional methods.
- **Core Type Induction Furnace:** It consists of magnetic core having thin laminations of silicon steel. The primary is wound on one limb and a crucible type non-conducting ring is provided on the secondary.

- **Vertical Core Type Induction Furnace/Ajax-Wyatt Furnace:** It consists of a transformer with horizontal core and vertical channel for the charge.
- **Coreless Induction Furnace (High Frequency Induction Furnace:** It works on the principle of transformer. The primary winding is wound on furnace and is connected to A.C. supply. The charge inside the furnace acts as secondary. Induced currents heat up the charge.

Electric Welding:

- **Welding:** It means joining of two metals placed together by heating them to melting points.
- **Butt Welding:** Rods, pipes and wires are welded by this method. There is a transformer, across secondary of the transformer two pieces of metal to be welded are held.
- **Flash Butt Welding:** This is a modification of butt welding. The ends of the pipe to be welded are held together with a little force/pressure applied to them. Arcing at the joint is allowed to take place. This removes any unevenness at the joint.
- **Spot Welding:** In spot welding, the welding is done at certain points on metallic sheets.
- **Projection Welding:** It is a modified form of spot welding. The electrodes used are flat and not pointed as in the case of spot welding.
- **Seam Welding:** It is similar to spot welding.
- **Energy Storage Welding / Percussion Welding:** Welding is obtained by discharging the stored energy in the capacitor.
- **Gloves:** Material used for gloves is a good quality thick rubber more flexible for hand movements.
- **Helmets:** To protect welder's face and head from ultraviolet and infrared rays of welding arc while allowing the welder to view the molten weld poor or puddle.
- **Jackets:** Jackets are made of soft leather.
- **Glasses:** These are safety glasses used for protection of eyes while welding a job.
- **Cap:** The welding cap protects the head from sparks flying over the welding helmet and also protects from falling sparks while welding overhead.
- **Slag Hammer:** This hammer is also called as "Chipping Hammer".
- **Pliers:** To handle the hot metal and hot weld coupon, a pair of pliers is needed.
- **Electric Arc Welding:** In this type of welding, an arc is struck between the electrode and the metal. Sufficient amount of heat is developed which melts the metal. An electric arc is the flow of electric current through gases (or ionized air) accompanied by heat and bright glow.
- **Carbon Arc Welding:** The electrode used in this type of welding is made up of carbon or graphite.

Practice Questions

1. With a neat diagram explain salt bath furnace.
2. What precautions are necessary while operating with salt bath furnace?
3. What are the desired properties of heating element material?
4. Give different types of commercial heating elements along with their properties.
5. What are the causes of failure of heating element?
6. What are the applications of resistance heating?
7. Write the expression for Stefan's law and state meaning of each term used in the expression. Derive the mathematical expressions used for design of circular and rectangular heating elements.
8. State and explain methods of striking arc in arc furnaces.
9. With a neat diagram explain direct arc furnace.
10. With a neat diagram explain submerged furnace.
11. With a neat diagram explain indirect arc furnace.
12. Write a short note on electrodes used for arc furnace.
13. Explain the volt-ampere characteristics of arc.
14. Explain the role of series reactor in case of arc furnace.

15. Draw the circuit of arc furnace and derive the expression for maximum power output.
16. Write a short note on electrical and mechanical requirements for arc furnace.
17. What is Pinch-off effect in core type induction furnace?
18. Give applications of induction heating.
19. With a neat sketch explain coreless induction furnace.
20. Explain principle of dielectric heating.
21. On what factors the dielectric loss is dependant?
22. With a circuit diagram and phasor diagram, prove the mathematical expression for dielectric loss.
23. What is meant by dissipation factor in case of dielectric heating?
24. State the advantages and limitations of dielectric heating.
25. State the difference between graphite electrode and carbon electrode used for arc furnaces.
26. With a neat diagram explain salt bath furnace.
27. What precautions are necessary while operating with salt bath furnace?
28. What are the desired properties of heating element material?
29. Give different types of commercial heating elements along with their properties.
30. What are the causes of failure of heating element?
31. What are the applications of resistance heating?
32. Write the expression for Stefan's law and state meaning of each term used in the expression. Derive the mathematical expressions used for design of circular and rectangular heating elements.
33. State and explain methods of striking arc in arc furnaces.
34. With a neat diagram explain direct arc furnace.
35. With a neat diagram explain submerged furnace.
36. With a neat diagram explain indirect arc furnace.
37. Write a short note on electrodes used for arc furnace.
38. Explain the volt ampere characteristics of arc.
39. Explain the role of series reactor in case of arc furnace.
40. Draw the circuit of arc furnace and derive the expression for maximum power output.
41. Write a short note on electrical and mechanical requirements for arc furnace.
42. What is pinch-off effect in core type induction furnace?
43. Give applications of induction heating.
44. With a neat sketch explain coreless induction furnace.
45. Explain principle of dielectric heating.
46. On what factors the dielectric loss is dependant?
47. With a circuit diagram and phasor diagram prove the mathematical expression for dielectric loss.
48. What is meant by dissipation factor in case of dielectric heating?
49. State the advantages and limitations of dielectric heating.
50. State the difference between graphite electrode and carbon electrode used for arc furnaces.
51. Give classification of electric welding and explain each type in brief.
52. Draw a neat diagram of spot welding machine and explain its working in brief.
53. How arc is formed in electric arc welding?
54. Explain volt ampere characteristics of electric arc.
55. What is arc blow in case of arc welding? How arc blow effect is minimized?
56. Discuss the effect of arc length in electric arc welding.
57. Write short notes on automatic hydrogen welding and helium welding.
58. List the different types of electrodes used in arc welding. Write in brief about each type.
59. What is the requirement for power supply used for arc welding? Which D.C. source is suited for this purpose?
60. With a neat diagram explain A.C. rectifier welding unit.

61. Write a short note on welding transformer.
62. Explain any two methods of controlling current in arc welding.
63. Write short notes on:
    (i) Electron beam welding
    (ii) Ultrasonic welding
    (iii) Laser welding.

Problems on Resistance Heating Element Design:

1. A 15 kW, 200 volt, single phase resistance oven employs nickel-chromium for its heating element. If wire temperature is not to exceed 1000°C and temperature of charge is to be 600°C, calculate the diameter and length of wire. Assume radiating efficiency as 0.6 and emissivity as 0.9 and $\rho = 1.016 \times 10^{-6}$ Ωm. Determine also the temperature of the wire when the charge is cold.
   [Ans.: l = 24.24 m, d = 0.312 cm, temperature T_1 = 922°C]
2. If in problem (1), instead of circular wire a rectangular strip of 0.25 mm thick is used, what would be its width and length? **[Ans.:** l = 9.7 m, W = 1.228 cm]
3. A 40 kW, 3 phase, 400 volt resistance oven uses Ni-Cr strip of 0.3 mm thickness. The heating elements are star connected. If wire temperature is to be 1127°C and that of charge is to be 727°C, estimate suitable width and length of the wire required. Given: radiation efficiency = 0.6, specific resistance of Ni-Cr = 1.03×10^{-6} Ωm, emissivity = 0.9. **[Ans.:** l = 9.4 m, W = 8.075 mm]
4. A 20 kW, 1ϕ, 220 V resistance oven employs circular nichrome wire. If wire temperature not to exceed 1000°C and temperature of charge to be 400°C cal., size and length of wire required e = 0.9, K = 0.65. What would be temperature of wire when charge is cold at 20°C? Take ρ = 1.09 μΩm.
   [Ans.: l = 22.08 m, d = 1.26×10^{-5} m, T_1 = 975.32°C]
5. Determine the diameter and length of nichrome wire to be used as heating element in a 10 kW, 220 volt single-phase resistance furnace. The temperature of wire should not exceed 1000°C and the charge is to be at a temperature of 500°C. Assume radiating efficiency to be 0.7 and emissivity as 0.9. The resistivity of nichrome is 1.016 micro ohm meter. **[Ans.:** l = 17.8 m, d = 0.218 cm]
6. A 40 kW, 3 phase, 400 volt resistance oven employes Ni-Cr strip 0.25 mm thick for the 3-λ connected heating elements. If wire temperature is to be 1200°C and that of charge is to be 800°C, estimate suitable width for strip. Assume emissivity 0.9 and radiating efficiency 0.5. Also determine the temperature when the charge is cold. **[Ans.:** l = 9.13 m, W = 8.38 mm, Temperature T_1 = 1083°C]

Problems on Induction Heating / Melting Furnace:

1. Estimate the energy required to melt 500 kg of brass in single-phase Ajax-Wyatt furnace. If melt is to be carried out in $\frac{3}{4}$ hour, what must be the average power input to the furnace, given specific heat of brass = 393.6 J/kg°C, latent heat of fusion of brass = 163×10^3 J/kg, melting point 920°C.
   [Ans.: Energy input = 103 kWh, Power input = 138 kWh]
2. Find the efficiency of a high frequency induction furnace which takes 10 minutes to melt 1.8 kg of aluminium. The input to the furnace is 5 kW and initial temperature is 15°C. Given specific heat of aluminium = 880 J/kg/°C and melting point of aluminium = 660°C, latent heat of fusion = 32 kJ/kg.
   [Ans.: η = 36%]
3. An induction furnace takes 12 minutes to melt 2.5 kg of aluminium with input power of 3.75 kW. The initial temperature is 20°C. The specific heat of aluminium is 880 J/kg/°C, melting point is 650°C. Latent heat of fusion is 32 kJ/kg. Calculate the efficiency of a high frequency induction furnace. **[Ans.:** η = 54.3%]
4. 1.5 tonnes of brass is to be melted in a 3 phase induction furnace. Find the average input power if melt is to be carried out in $1\frac{1}{4}$ hours. Room temperature = 20°C, specific heat of brass is 393.6 J/kg/°C, latent heat of fusion = 163×10^3 J/kg, melting point of brass = 920°C and furnace efficiency is 70%.
   [Ans.: 246.3 kW]

5. Determine the power required to melt 500 kg of brass in 30 minutes. Assuming specific heat = 0.095 kcal/kg/^{o}C, latent heat of fusion of brass = 39 kcal/kg, melting point of brass = 920°C, furnace efficiency is 50% and temperature of cold charge is 20°C. [**Ans.:** 144.5 kW]
6. What must be the rating of a tin smelting furnace in order to smelt 50 kg of tin per hour? Smelting temperature of tin = 235°C, specific heat = 0.55 kcal/kg/^{o}C, latent heat of liquidification (fusion) = 13.37 kcal/kg. Take initial temperature to be 150°C. [**Ans.:** 1.47 kW]
7. 2000 kg of brass has to be melted in an induction furnace in 1 hour. Calculate the power of furnace for following data:

   Specific heat of brass = 0.034 kcal/kg/^{o}C
   Latent heat = 38.88 kcal/kg
   Melting point = 910°C
   Initial temperature = 20°C
   Efficiency of the furnace = 70% [**Ans.:** 407 kW]

Problems on Electric Arc Heating:

1. Calculate the time taken to melt 3 tonnes of steel in 3-phase arc furnace having following data:

   Current = 5000 Amp.
   Arc voltage = 60 volt
   Reactance of transformer = 0.005 Ω
   Reactance of transformer = 0.003 Ω
   Specific heat of steel = 0.12 kJ/kg/^{o}C
   Latent heat of fusion = 8.89 kcal/kg
   Initial temperature = 18°C
   Melting point = 1370°C

   Assume overall efficiency as 60%. Find also the power of electrical efficiency of the furnace.
   [**Ans.:** Time = 53 min., Electrical η = 88%]
2. A 3-phase electric arc furnace has following data:

   Current drawn = 5000 Amp.
   Arc voltage = 50 V
   Resistance of transformer referred to sec. = 0.002 Ω
   Reactance of transformer referred to sec. = 0.004 Ω

   Calculate:
   (i) p.f. and kW drawn from supply.
   (ii) If the overall efficiency of furnace is 65%, find the time required to melt 2 tonnes of steel when latent heat of steel = 8.99 kcal/kg, specific heat of steel = 0.12, melting point of steel = 1370°C and initial temperature of steel = 20°C. [**Ans.:** p.f. = 0.948, 900 kW, time = 40 min. 46 sec.]
3. A 3-phase arc furnace takes 7500 amp. when arc voltage is 75 volt. The circuit comprising the transformer and the electrode excluding the arc has reactance of 0.003 Ω and resistance 0.002 Ω. The steel has specific heat of 278 J/kg/^{o}C, a latent heat of fusion 37000 J/kg and melting point 1370°C. Assuming furnace overall efficiency 50% and initial temperature of steel as 15°C. Find:
   (i) the time taken to melt 5 tonnes of steel,
   (ii) the kVA taken from supply,
   (iii) electrical efficiency. [**Ans.:** Time = 2040 sec., 2088 kVA, Electrical η = 83.3%]
4. A 3-phase arc furnace is required to melt 5 tonnes of steel in one hour. Find out the average kW and kVA input to furnace, arc voltage, arc resistance and p.f.

   Given: Specific heat of steel = 0.12 kcal/kg/^{o}C
   Latent heat of steel = 8.89 kcal/kg
   Melting point of steel = 1370°C

Initial temperature of steel = 20°C
Overall efficiency of furnace = 50%
Current input = 9000 Amp.
Transformer resistance = 0.003 Ω
Transformer reactance = 0.005 Ω

[**Ans.:** kW = 2080, kVA = 2410, V_a = 50.5 volt, Arc resistance = 0.00556 Ω, p.f. = 0.865]

5. Find the current per line in a 3 phase arc furnace to melt 5 tonnes of steel in one hour at an overall efficiency of 50%. If arc voltage is 115 volt, initial temperature 30°C, melting point of steel is 1370°C, specific heat of steel 278 J/kg/°C. Latent heat of steel = 37000 J/kg. [**Ans.:** 3298 Amp.]

Problems on Dielectric Heating:

1. A piece of resin material is to be heated with dielectric heating. The dimensions of resin piece are 10 cm × 10 cm × 2 cm. The power required is 200 watt with a frequency of 30 MCPS. The material has relative permittivity of 5 and p.f. 0.05. Determine the voltage necessary and current flowing through the material. If voltage is to be limited to 600 V, what will be the value of frequency to obtain the same heating? [**Ans.:** 800 volt, 4 Amp., 53.3×10^6 cycles/sec]
2. A piece of insulating material is to be heated by dielectric heating. The size of piece is 100 cm^2 area and 2.5 cm thick. A frequency of 25 MCPS is used, the power absorbed is 350 W. Calculate the voltage necessary for heating and the current flow in the material. The material has relative permittivity of 5 and power factor of 0.05. [**Ans.:** 1588 volt, 4.42 Amp.]
3. A piece of insulating material 2 cm thick and 120 cm^2 in area is to be heated by dielectric heating. The material has permittivity of 5 and power factor 0.05. The power at 600 volt is 200 watt. Determine the frequency of supply. [**Ans.:** 65.7 MCPS]

■■■

Chapter **3...**

Elevators

Syllabus

Objectives

- Select relevant elevator machine and electric motor for the specified application with justification.
- Describe the procedure to maintain the given electric drive and elevator.

INTRODUCTION

- Elevators are electrically operated devices used for lifting up and down, moving load in suitable direction, lifts in multistoried buildings/hotels for facilitating up and down movements of occupants. So function of elevators is transporting at a fixed place.

3.1 TYPES OF ELECTRIC ELEVATORS

1. List any four types of electric elevators. Explain any one of them.
2. Write four types of elevators and explain about any one of them.

1. With respect to Service, the Elevators are Grouped as:

(a) Passenger elevators.

(b) Freight elevators: For hotels, apartments, stores etc.

(c) Special elevators: Side walk, garages.

(d) Combination of elevators: Mixed Passenger and Freight.

2. According to Capacity, the Elevators are Grouped as:

(a) Light duty elevators.

(b) Medium duty elevators.

(c) Heavy duty elevators.

(d) Extra heavy duty elevators.

3. According to Speed, the Elevators are Grouped as:

(a) Low speed elevators.

(b) Medium speed elevators.

(c) High speed elevators.

4. According to power unit, the Elevators are Divided into two Groups:

(a) Drum elevators.

(b) Traction elevators.

5. According to Location of Power unit, the Elevators are:

(a) Over mounted elevators.

(b) Under mounted elevators.

6. **According to Electric Current, these are Categorized as:**
   (a) Direct Current (D.C.) elevators.
   (b) Alternating Current (A.C.) elevators.
7. **According to Motor Drive used for Elevator, the Groups are:**
   (a) Gearless elevators.
   (b) Geared elevators.
8. **According to Rope Drive used, the Elevators are Known as:**
   (a) Half wrap elevator
   (b) Full wrap elevator.
9. **According to Control System Used in Elevators:**
   (a) Semi-magnetic type.
   (b) Variable voltage type.
10. **According to Method of Operating the Control:**
   (a) Semi-magnetic type.
   (b) Full magnetic type.
11. **According to Basic Principle of Control, these are Divided as:**
   (a) Rheostatic.
   (b) Variable voltage.
12. **According to Method of Operation of the Elevator:**
   (a) Manual
   (b) Push button
   (c) Signal
   (d) Dual.
13. **According to Velocity Ratio Between Motor and Car the Elevators are:**
   (a) Direct drive
   (b) 2 : 1 reduction
   (c) Multi-reduction.
14. **According to Method of Balancing the Load:**
   (a) Counter balanced type
   (b) Compensated type.

Thus, there are many types of elevators used according to need, situation and purpose.

3.1.1 Applications of Elevators

(i) In apartments of multistoried buildings.
(ii) In hotels.
(iii) In stores etc. for passengers and goods.
(iv) For side walks and garages.
(v) In mines to lift and transport materials etc.
(vi) In traction.

3.1.2 Ideal Requirements of Elevators

(i) Motor drive: Smooth working, easy control, reliable operation.
(ii) Size: Suitable (Explained in following sub-topics).
(iii) Speed: For different purposes of suitable speed.
(iv) Location: Suitable place.

These points are elaborated in the following topics.

3.2 SIZE AND SHAPE OF ELEVATOR CAR

1. Describe size and shape of elevator car.
2. State the factors to be considered for selection of shape and size of the car of elevator.
3. Describe about: Size and shape.
4. What are the factors that decide the size and shape of elevator car?

- In planning an elevator installation, consideration should be first given to the size and shape of the elevator car. It is mainly dependent on the design of building and also it should be such as to obtain the highest operating efficiency in the movement of the traffic.
- By experience, the best shape for an elevator car is one having a wide front and shallow depth. The wide front provides ample door opening, facilitating quick transfer of passengers at floor level.
- The size of the car is determined either from the number of passengers to be carried or by limitations in the building design. The usual practice is to allow a space of 2 sq. ft. per person (i.e. 0.2 m^2 per person) and the maximum load capacity of elevator is figured on the basis of 75 lbs per sq. ft. (340 kg/m^2) which assumes the average weight of a passenger to be 150 lbs (68 kgs).

3.3 SPEED OF ELEVATORS

1. Describe about : Speed.
2. What is meant by speed of elevator? Give typical values of elevator speed for different installations.

- The measure for the speed of a car is the number of feet the car will travel during the period of one minute.

Table 3.1: The usual practice for passenger elevators

Type	Speed	
	Feet/minute	m/minute
1. Private residence:	100 –150	30 to 54 m/min.
2. Hotels and apartment houses:		
Upto 5 floors	100 –200	30 to 60 m/min.
6 to 10 floors	200 – 350	60 to 105 m/min.
3. Department stores:		
Upto 5 floors	100 –200	30 to 60 m/min.
6 to 10 floors	200 – 300	60 to 90 m/min.
4. Office buildings:		
Upto 10 floors	250 – 450	75 to 135 m/min.
10th floor and above	400 –800	120 to 240 m/min.

3.4 LOCATION OF ELEVATOR MACHINE

1. Location of elevator machine.
2. List the types and essential features of elevator machines.

- Mostly the elevator machine is installed in a room called the pent house which is located at the top of the building. Only in some exceptional cars the machine is installed in the basement. Basement installations are objectionable because they require much longer hoisting cables and more idler sheaves than for overhead machines, they occupy valuable space in the basement.

3.5 TYPES OF ELEVATOR MACHINES AND POWER TRANSMISSION GEARS BRAKING

1. State types of elevator machines.
2. List the types and essential features of elevator machines.

- There is a variety of types of power unit or elevator machines employed to drive electric elevators. With respect to the method of transmitting the power to the car ropes, elevator machines are divided into two general classes:
  1. Drum
  2. Traction.
- An elevator machine consists essentially of a motor, drum or traction pulley, brake and motor-drum drive.
- The drive may consist of a direct shaft connection between the motor and drum or it may be trough gearing system.

3.5.1 Drum Elevator Machines

- It consists of a drum which is in the form of a spool on which the ends of ropes are attached and around which they are wound and unwound in the operation of the car.
- There are generally two types of drums used. The drum here means a winding drum. The other drum used is a traction drum which is simply a type of pulley sheave. In construction, grooves are provided on both types of drum, the grooves being spiral for drum elevator and straight for traction elevators.

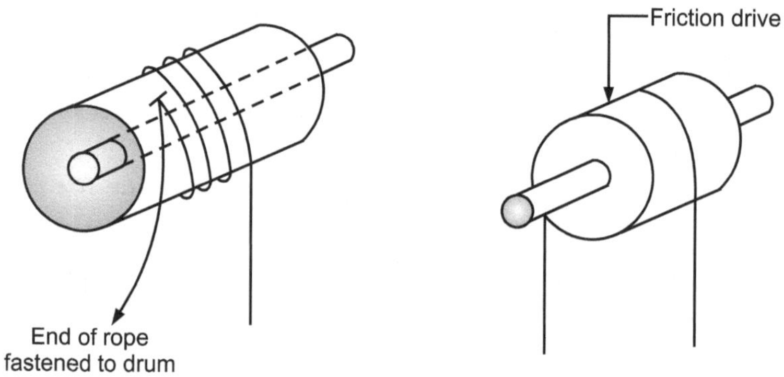

(a) Winding Drum **(b) Traction Pulley**

Fig. 3.29: Drum, Traction Elevators

- In the operation of a drum elevator machine, power when applied to the driving gear, turning it in one direction winds the ropes upon the drum and causes the car to ascend and when the power is reversed, the drum turns in the opposite direction, paying out the ropes and causing the car to descend.
- The weight of the car is balanced by a counter weight, thus reducing the energy to be expended in operating the car. Automatic devices are used to ensure proper and safe control of the movements of the car.
- With respect to location there are two types of machines known as:
  1. Over mounted
  2. Under mounted.

  According to whether the machine is placed above the hoistway or in the basement.
- The over mounted machine gives direct transmission, that is no pulleys are required between drum and car, also the drum may be so located that one face is over the centre of the car and the other face over the counter balance, thus allowing the car ropes to be fastened at one end of the drum and the counter balance ropes at the other end of the drum grooves. The car ropes occupy the entire surface of the drum when the car is up and the counter balance rope is suspended as shown. This arrangement requires minimum length of rope.

- If the drum type machine be installed overhead, it will give remarkably long rope life because there is no possible rope creepage and the bending is always in one direction. However, if the machine be installed in the basement, reverse bends are necessary in the ropes and this material reduces rope life.

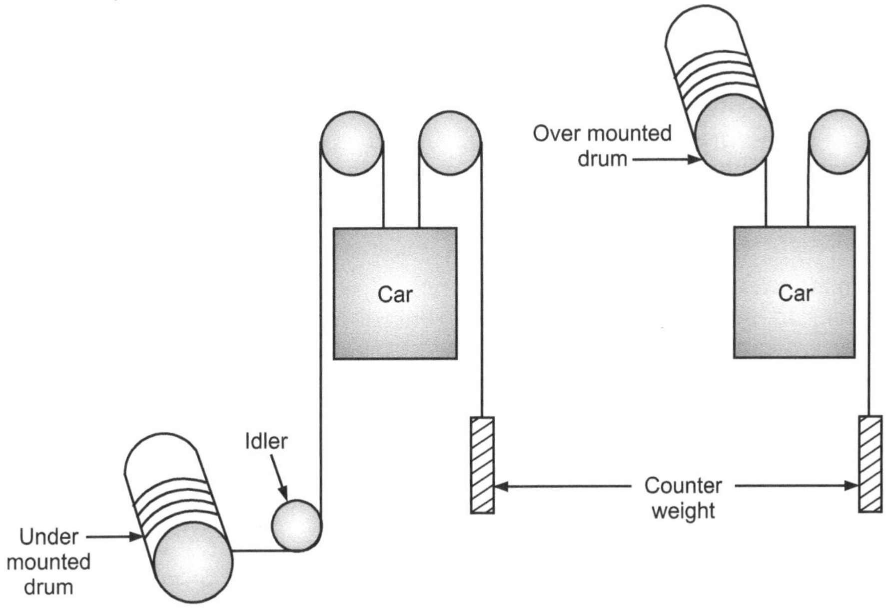

Fig. 3.30: Over Mounted and Under Mounted Machine

- One reason for the decreased use of the drum machine is because it requires a modification of the machinc for each installation, as the length of travel determines the length of the winding drum. Another bad feature is that if overtravel limit switches fail to stop the car at the terminals, the machine may continue to travel and finally pull the ropes from their socket.

3.5.2 Traction Elevators

- In this type, the motion is obtained by means of traction i.e. through the friction existing between the driving pulley and hoisting ropes.
- The ropes are not wound on a drum, but a continuous rope from car to the counter weight passes over a driving pulley.
- The traction machine carries the hoist ropes over the hoisting sheave and down to the counter weight and relies on the grip of the ropes on the sheaves to give enough traction to lift the load.

3.6 ELEVATOR MOTORS

1. State the requirement of elevator motor. State with reason best suitable motor for elevator.
2. Give suitable motors for each type of elevator machine.
3. State the requirement of elevator motor. State with reason best suitable motor for elevator.

- In order to ensure rapid acceleration and retardation, the inertia of moving parts must be kept as low as possible. For this reason motors for elevator service are designed with armatures of comparatively small diameter slow speed motor are preferable.
- Generally, the motor speed should not exceed 900 revolutions per minute.

Power Calculations:

- In order to determine the size of motor required for a given installation three factors must be considered:
  1. Load or net weight to be hoisted.
  2. Speed of the car.
  3. Efficiency.

- While calculating the load to be hoisted, it is necessary to consider that the load of the car and part of the load is counter balanced by the counter weight. So only the unbalanced load must be taken into account.
- The efficiency is the overall efficiency which considers various frictional and electrical losses. H.P. required can be found out by using the formula:

$$\text{H.P} = \frac{L \times S}{E \times 33{,}000}$$

Starting Torque Requirements:

- An elevator requires greater starting torque than is necessary to keep the same elevator running at its rated speed and load. This is mainly due to static friction. A motor therefore, to be suitable for elevator work, should be capable of developing from $2\frac{1}{2}$ to 3 times as much starting torque as is represented by its horse power rating (i.e. full-load torque).
- The motor builders do not design all their motors with the same ratio of starting torque to full-load torque. In some cases, in rush current, limitations are placed which cut down the amount of starting torque which could otherwise be obtained.

D.C. Motors:

- The type of D.C. motor which is best suited to elevator service depends to a considerable extent on the type of service, that is the speed and frequency of slopes.
- The elevator motor must not only raise and lower the load, but one of its principal duties is to accelerate and decelerate the car rapidly. This must be accomplished smoothly without jerks which might cause discomfort to the passengers. The starting torque should be at least 225% of the rated full-load torque. Motors having more than 15% variation in speed from no load to full load are not recommended for elevator service.
- Single-phase motors of self starting repulsion induction type may also be used for elevator service.
- For freight service, compound wound motors are recommended. The series winding provides a high starting torque, an essential factor for heavy duty work.
- For passenger service, either compound wound or shunt wound motors give satisfactory results.
- Most direct current motors, whether shunt or compound wound have suitable commutating pole winding so as to ensure sparkless commutation in both directions of rotation.

A.C. Motors:

- The two general types of A.C. motors suitable for elevator service are:
  (i) Squirrel cage motor
  (ii) Slip ring motor.
- The squirrel cage motor is used extensively upto about 20 hp because of its simplicity and because it requires only a relatively simple form of controller.
- The power consumption of squirrel cage motor is slightly higher than that of the shipping machine, but due to the lack of slip rings and fewer controller parts it is somewhat more reliable.
- The slip ring motor of same rating is more expensive and has a somewhat lower power factor than the squirrel cage motor. A.C. motors of low speed type are now available and are being used successfully on elevators whose car speeds are as high as 400 ft. per minute. As in D.C. service all A.C. elevator motors should specially for elevator work.

Elevator Control System

- The elevator control systems are broadly divided into two groups:
  1. Rheostatic control.
  2. Variable voltage control.
- In rheostatic control method, rheostat is included in the field and armature circuit of motor, and speed control can be achieved by varying it.
- In variable voltage control, the input voltage applied to motor is changed by some means and speed control is achieved.

3.7 SAFETY IN ELEVATORS

1. What are different safety and protective devices used in electric elevator? State function of each device.
2. Write six safety precautions to be observed in an elevator.
3. State any four safety measures used in electric lifts.
4. What are the different safety and protective devices used in elevators? Also state functions of each device.

Safety and Protective Devices:

These may be classified as electrical and mechanical.

The principal safety devices are:

1. Guide grips.
2. Overspeed governor with governor switch.
3. Car operating switch.
4. Car safety switch.
5. Terminal limit switches.
6. Over travel limit switches.
7. Slack cable switch.
8. Door switches.
9. Compensating rope sheave switch.
10. Buffers and air cushions.

1. **Guide Grips and Overspeed Governor:** The mechanism is mounted below the car with a small winding drum which is connected to the overspeed governor by a steel rope. The governor is arranged with a grip so that if normal speed of the elevator exceeds, it holds the governor rope and forces the grips against the guide rails so as to stop the car.
2. **Car Operating Switch:** The car can be automatically set to different speeds. If operator's hands are removed from lever it returns to 'off' position.
3. **Car Safety Switch:** This switch for the purpose of stopping the car in emergency in case of the failure of the car operating switch.
4. **Terminal Limit Switches:** These are for bringing the car smooth rest on landing.
5. **Over Travel Switches:** Its function is to stop the car in case of the failure of the regular terminal stop.
6. **Slack Cable Switch:** Ordinarily this is used on a drum type elevator to open the control circuit in case of slack rope caused by the car or counter weight being caught in the guides.
7. **Door Safety Switches:** These switches in combination with door locks, prevent the car operating unless all doors are closed and locked.

Protective Devices:

1. **Main line service switch and fuses:** They are mounted in an accessible location in the elevator machine, room and usually enclosed in a metal cabinet.
2. **Circuit breaker and overload relay:** Overload relays are used in order to secure protection against overloading of the elevator itself. They are normally set below fuse rating.
3. **Overspeed slowdown relay:** Overspeeding of the elevator in either direction will cause the relay to operate and thus automatically retard the speed.
4. **Phase failure protective relay:** The failure of any phase causes the relay to open the controller circuit and thus disconnect the motor from the supply lines.
5. **Phase reversal protective relay:** The reversal of phases immediately open the controller circuit and prevent the elevator motor being connected to the lines until the relation of phases is corrected.

3.8 BOMBAY ELECTRICITY ACT 1939

- Bombay Electricity Act lays down provisions for regulating construction maintenance and safe working of certain classes of lifts and all machineries, apparatus pertaining thereto and also prescribes procedure for issue of licence for working of lifts. This act also provides power to make rules about the specifications, testing and licensing of lifts etc.

Administrative Machinery:

- This Act is being enforced by the inspector of lifts and officers appointed to assist the inspector of lifts. The electrical inspector is the Inspector of lifts and assistant electrical inspectors are the officers appointed to assist the inspectors of lifts to take action against defaulters.

Penalty:

- The penalty clause is also provided in the Act, which prescribes a penalty amount for breach of any provisions of the Act or Rules made there under and for continuous offence a penalty shall be charged for every day after the first day during which such breach continues. The inspector of lifts inspects life installed in any building located except the building maintained by PWD and MES to check whether the provisions of this act and rules made there under have been compiled with and issue notices to the defaulters for rectification of defects if any and also issues licenses/NOC to those lifts which are adhering to the provisions of this Act.

The Delhi Lift Rules, 1942:

- **Enforcement:** These rules provide procedure for applying/obtaining licenses for working on lifts, specification of lift components/machinery and safety devices to be used in the lift and procedure for inspection/testing of lifts. These rules are basically safety regulations which are being enforced for safe and proper construction, working, maintenance and inspection of lift for ensuring safety and users and also smooth working of the same. As per the provisions of these rules whoever desires to install passenger lift, shall have to submit an application on a prescribed performa to the inspector of lifts for obtaining permission to erect the lifts. On receipt of the application the designated field staff visits the site and if the same is found suitable for erection of lift then permission is granted to applicant to install the lift. On completion of lift, the owner of the lift shall inform the inspector of lifts on a prescribed form along with the license fees varying depending upon the speed of the lift. On receipt of such application, the inspecting staff inspects the electrical installation of lift, its layout, safety mechanisms and safety gears etc. to see whether the provisions of lift Act/Rules and the guidelines laid down in relevant Bureau of Indian Standards have been compiled with. If upon inspection, the same is found in accordance with relevant provisions of lift act and rules, the license for working of lift is issued. The lifts are also periodically inspected almost once in a year to check whether the same are being maintained properly and also whether the safety provisions are being followed or not.

Important Points

- Elevators are now-a-days needed for many places like (i) Multistoried buildings, (ii) Mines, (iii) Hotels.
- These are electrically operated drives to move load up and down.
- It's function is transporting at a fixed place.

Practice Questions

1. Give types of electric elevators.
2. Write a short note on size and shape of elevator car.
3. What is meant by speed of elevator? Give typical values of elevator speed for different installations.
4. Comment on location of elevator machine.
5. Describe working of drum elevators with a neat sketch.
6. Describe working of traction elevators.
7. Explain overmounted and undermounted machine in context with elevators.
8. Write a short note on motors used for elevators.
9. What are the requirements of torque for elevator motor?
10. Write a short note on safety devices used for elevator.
11. State Bombay Lift Act.

■■■

Chapter **4...**

Electric Drives

Syllabus

4.1 Introduction:
- What is Drive?
- Drives: Mechanical Drive and Electric Drive.

4.2 Advantages and Disadvantages of Electric Drive.

4.3 Factors Governing Selection of Electric Motors.

4.4 Nature of Electric Supply: Three-phase & Single-phase A.C and D.C.

4.5 Type of Drive: Group Drive and Individual Drive.

4.6 Nature of Load: Nature of the Mechanical Load, Matching of the Speed - Torque Characteristics of the Motor with that of the Load, and Starting Conditions of the Load.

4.7 Electrical Characteristics: (Only DC Series, Three-phase and Single-phase Induction Motors are to be dealt)
- **Running Characteristics:** Three Typical Speed - Torque Characteristics: Inverse, Constant Speed and Drooping.
- **Starting Characteristics:** Starting Torque only (No Starters).
- **Speed Control:** Suitability to Economic and Efficient Speed Control Methods (Above and Below Normal Speed).
- **Braking Characteristics:** Plugging, Rheostatic Braking and Regenerative Braking, as applied to DC Series and Three-phase Induction Motor. Only characteristics and applications of following motors: D.C. Series, Single-phase A.C. Series, Single-phase I.M., Three-phase I.M., Universal Motor, Stepper motor. Requirements of Motors used in following applications: Stone Crushing, Textile Industry, Paper Manufacturing Industry, Rolling Mill, Chemical Industry.

4.8 Mechanical Features
- Types of Enclosures as per IS
- Types of Bearings
- Types of Transmission for Drive
- Noise Level.

4.9 Size and Rating of Motor
- Load Cycles for – Continuous Loads, Short Time Loads, Intermittent Loads, Continuous Operation with Short Time Loads and Continuous Operation with Intermittent Loads.
- Duty Cycles.
- Standard Ratings for Motors as per ISS.
- Estimation of Rating of a Motor. (Simple numericals on Estimating Size of Continuously Rated Motor)
- Load Equalization. (No calculations)

4.10 Cost:
- Capital Cost
- Running Cost (Losses, P.F., Maintenance).

Objectives

- Differentiate salient features between the given types of electric drives.
- Recommend relevant motor for the given application with justification.
- Select the relevant enclosure of motor for the given atmospheric condition with justification.
- Select the power transmission drive of the electric motor for the given application with justification.
- Estimate relevant size and rating of electric motor for the specified load cycles.
- Explain with sketches relevant braking system for the given electric motor.

4.1 INTRODUCTION

1. What is a drive? Define electric drive.

- **Drive:** In simple words, drive is a device that has to do required motion for useful work. Drive can be mechanical and electrical type.
- Examples of mechanical drive are: steam turbines, water turbines, wind mills, IC engines.
- Example of electrical drive is electric motor.
- A variety of different types is available for this drive with effective control mechanism.
- Large number of equipments for domestic and industrial applications make use of electric motors.
- Industrial applications include – rolling mills, textile mills, machine tools, pumps, fans, hoists, crains, paper machines, transportation systems, robot and many more.
- Domestic applications include mixer, washing machine, air conditioners, coolers, small pumps, fans etc.
- These applications involve motion control.
- The combination of an electric motor, energy transmitting shaft and control devices, for controlling the performance of motor (motion control) is known as **electric drive.**
- In the simplest form, the electric drive can be represented with block diagram as per Fig. 4.1.

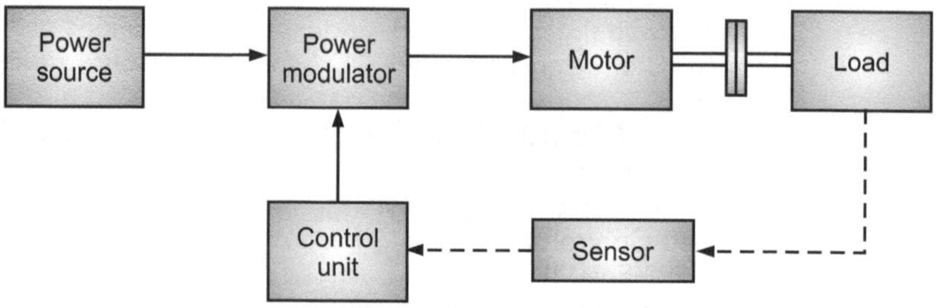

Fig. 4.1: Electric Drive

- The load may be pump, fan, robot etc. The motor is supposed to drive the load. The power modulator has to perform following functions:
  (i) To provide input power to motor in a proper form (e.g. for D.C. motor, A.C. is converted to D.C. and then fed to motor).
  (ii) To start the motor in a proper way (i.e. performing function of starter).
  (iii) To feed modulated power to motor such that it gives desired speed-torque characteristics.
- The control of power modulator is done through control unit. The control unit takes feedback from sensor about the performance of motor. If performance is not as per the requirement, the control unit generates control signals and send it to power modulator, the power modulator then controls the performance of motor till desired output is obtained.

4.2 ADVANTAGES AND DISADVANTAGES OF ELECTRIC DRIVE

1. Give advantages of electric drive.
2. State four advantages of electric drive.
3. State any six advantages and any two disadvantages of electric drives.
4. State advantages associated with electric drives.
5. State two advantages and two disadvantages of electric drives.

- Electric drive has become popular because of the following advantages:
  (i) **Electric drive is more flexible:** This talks about the control part. The performance can be controlled effectively by using solid state devices such as SCRs, IGBTs and microcontrollers.
  (ii) It is more economical.
  (iii) It is more clean; no fumes, ash etc. is produced, hence do not pollute the environment.
  (iv) This is available in wide range of torque, speed and power.
  (v) It is compact in size, hence occupies less space.
  (vi) Unlike other prime movers, it does not require warm-up time or refueling. It can be started immediately.

(vii) It can be remote controlled.

(viii) It has easy starting method.

(ix) It is adaptable to different operating conditions such as submerged in liquid, explosive chemical or mining environment, radioactive environment.

(x) It has long life.

(xi) It is reliable source of drive.

(xii) It is powered by electrical energy which has number of advantages over other sources of energy.

- Because of the above advantages, the mechanical energy already available from a non-electrical prime mover is first converted to electrical energy by means of a generator and then back to mechanical energy by means of electric motor.

Disadvantages:

(i) Requirement of additional control circuitary.

(ii) Cost is more.

Comparison of Electric drive and Mechanical drive:

Sr. No.	Electric Drive	Mechanical Drive
1.	Drive is suitable only on electrifide tracks.	As no electricity is required, so suitable on any track.
2.	Very clean, free from smoke, dust, flue gases.	Due to burning of fuels, it is not clean, due to fumes, ash, smoke.
3.	Chance of shut down on failure of electricity.	No such chance.
4.	Compact unit requiring less space.	Large space is required due to so many mechanical components.
5.	Easy to modify electrical characteristics.	Mechanical characteristics not easy to be modified.
6.	Drive is flexible and easy control.	Complicated mechanical control.
7.	Maintenance is very easy and less costly.	Maintenance is frequent and costly, time consuming.
8.	Braking system is reliable, prompt and easy.	Braking involves wear and tear.
9.	Starting/stopping is very easy by electric control.	Starting/stopping is not that easy.

4.3 FACTORS GOVERNING SELECTION OF ELECTRIC MOTORS

1. State the factors governing selection of electric motors.
2. Explain any four factors governing the selection of motors.

- The success of electric drive for any purpose depends on the correct choice of driving motor for particular application.
- While selecting motor for a particular application, following points should be considered:
  1. Nature of electric supply.
  2. Nature of drives:
     (i) Group drive
     (ii) Individual drive
     (iii) Multimotor drive.
  3. Nature of load.
  4. Electrical characteristics: This includes:
     (i) Starting characteristics.
     (ii) Operating or running characteristics.
     (iii) Speed control.
     (iv) Braking.
  5. Size and rating of motor: This includes:
     (i) Continuous rating.
     (ii) Intermittant or variable rating.
     (iii) Load capacity and pull-out torque.

6. Mechanical factors: These include:
   (i) Types of enclosures
   (ii) Types of bearings
   (iii) Transmission of drive
   (iv) Noise level.
7. Cost:
   (i) Capital cost
   (ii) Variable cost

These factors will be discussed in brief.

4.3.1 Nature of Electric Supply

- Types - D.C. and A.D. supply.
  - D.C. Supply - 110 volts, 220 V
  - Single-phase A.C. - 230 volts
  - Three-phase A.C. - 415 V, 440 V

4.4 TYPES OF ELECTRIC DRIVES

1. State the types and applications of different drives.
2. State any four difference between group drive and individual drive.
3. Compare group drive and individual drive based on following factors:
   (1) Definition, (2) Cost, (3) Total HP, (4) Appearance, (5) Safety, (6) Flexibility, (7) Performance, (8) Any one application of each.
4. Compare group drive with individual drive.

Electric drives can be classified into following categories:
(i) Group drive (ii) Individual drive (iii) Multimotor drive.

4.4.1 Group Drive

1. Define group drive.

- This is also known as line shaft drive. A big size motor (main motor) drives a common shaft. To this shaft, number of small machines are connected. Thus a common shaft drives small machines.

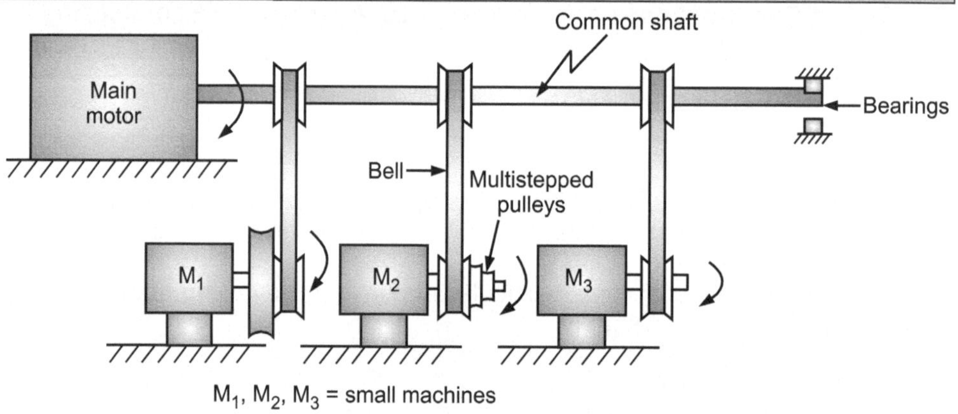

Fig. 4.2: Group Drive

- The small machines are connected to the common shaft by means of multistepped pulleys as shown in Fig. 4.2.

1. State advantages of group drive.

Advantages of group drive are:

(i) Single motor of big capacity is used, so there is considerable saving in cost.
(ii) Maintenance is less.
(iii) Efficiency and power factor is better if the motor operates near rated capacity.
(iv) If machines are overloaded, the group drive is advantageous because the overload occurring for main motor is very less.

> **1.** State disadvantages of group drive.

Disadvantages of group drive are:

(i) Any fault on main motors causes stopping of all small machines connected to the common shaft.

(ii) Machines have to be installed at places which is convenient for shafting.

(iii) There is no flexibility of layout i.e. shifting of a small machine from one location to other is hardly possible.

(iv) If only few number of machines are needed to be operated, the main motor operates with a very less capacity, hence its efficiency and power factor is poor.

(v) The appearance due to shaft, pulleys etc. is not good.

(vi) Speed control of individual small machine is cumbersome.

(vii) Considerable amount of power is wasted in power transmission.

(viii) Noise level is high.

(ix) Future expansion is hardly possible.

Applications:

- Group drive is adopted when existing factories are changed over from engine drive to electric motor drive. It is used in grain processing industries, food grinding mills, paper industries.

4.4.2 Individual Drive

> **1.** Define individual drive.
> **2.** State advantages of individual drive.

- Individual machine is fitted with its own motor. e.g. drilling machine, lathe machine etc. Each operator has complete control on his machines.

Advantages:

(i) Flexibility in layout.

(ii) Each machine operates near to rated capacity, so efficiency and power factor is better.

(iii) Fault on any one machine does not create problem for other machines.

(iv) The appearance of layout is good.

(v) It is economical on long-term basis.

(vi) Speed control of individual machine is easily possible.

(vii) Future expansion is possible.

(viii) Noise level is comparatively less.

Disadvantages:

(i) High initial cost.

(ii) Power loss in transmission (gears, pulleys, etc).

4.4.3 Multimotor Drive

- In multimotor drive, separate motors are used for operating different parts of same mechanism. e.g. in case of overhead crane, different motors are used for hoisting (up and down motion), long travel motion and cross travel motion.

Advantages:

(i) Better control of process.

(ii) Complicated process can be automated.

Disadvantages:

(i) Cost is more.

(ii) Maintenance of one motor causes disturbance in the entire process.

Applications:

(i) Craines

(ii) Lifts

(iii) Hoist

Comparison of Group Drive and Individual Drive:

Sr. No.	Group Drive	Individual Drive
1.	One primeover motor is required to drive common shaft.	Separate motor drive is required for each individual machine.
2.	Total power in H.P. required for complete section is lesser.	As each machine requires separate motor, total sum of HPs is more.
3.	Cost involved is much less.	Cost is much more.
4.	As overload capacity is concerned, this system has high overload capacity.	This system has less overload capacity.
5.	System has clumpsy appearance.	Clean appeance.
6.	Speed control of individual machine is not possible.	Speed can be controlled separately for each motor.
7.	Operation is not that safe.	Safe operation.
8.	Layout is fixed and cannot be changed.	Machines can be installed as per requirement and space available.
9.	Future extensions are not possible.	Can add more separate machines.
10.	Fault in main motor drive stops operation of all machines and work is held up.	Fault in drive does not disturb other drives.
11.	Power losses are more.	Less power losses.
12.	Uses - in textile mills, workshops.	Lifts, cranes, shapers, lathe machines.

4.5 MECHANICAL FEATURES OF DRIVE

4.5.1 Purpose

- Enclosure is necessary for motors for the following purposes:
  (i) To protect the operating personal from live parts of motor like winding terminals, slip rings, brushes etc.
  (ii) To protect the motor against foreign body and severe conditions. In many cases, motor has to operate in an atmosphere covered with dust, dirt, presence of water or moisture, chemical fumes, radiations etc. So inner parts of motor should be protected from such things. Also some solid parts like spanners, nut bolts etc. may enter the motor when operator is doing some work on the machine.
- Indian Standards (IS) 4691 have described codes for different types of enclosures. It consists of the letter IP followed by two numbers. The first number designates the extent of protection of persons against contact and of the machines against ingress of solid foreign bodies. The second number designates the extent of protection to the machine against harmful ingress of water.

4.5.2 Types and Applications of Various Types of Enclosures

- Fig. 4.3 shows different types of enclosures used for motors. The details about the application are shown in Table 4.1.

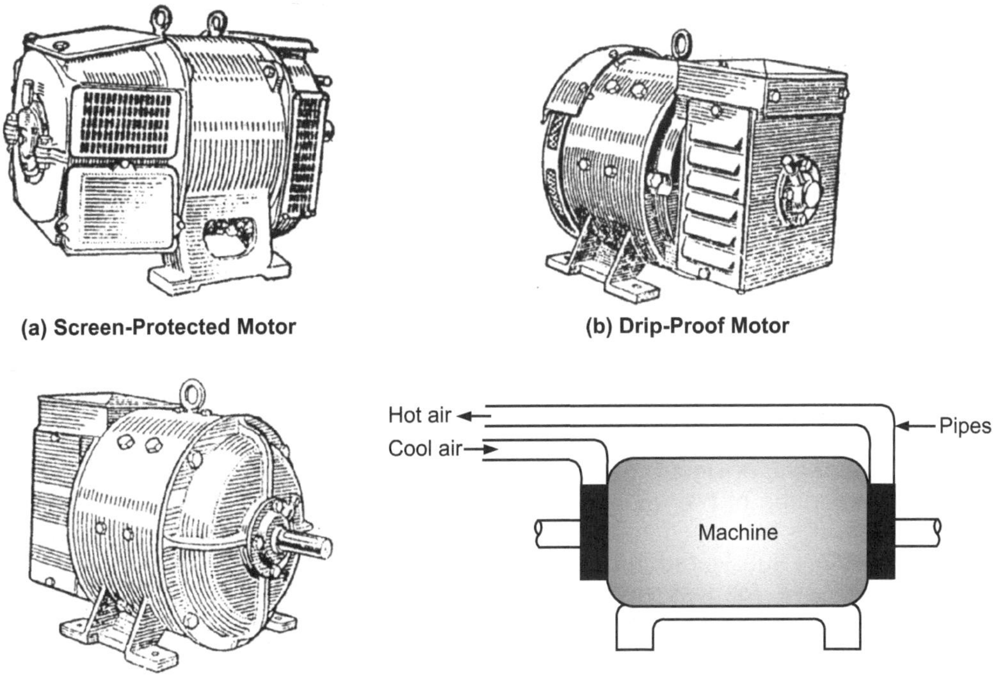

(a) Screen-Protected Motor (b) Drip-Proof Motor

(c) Totally-Enclosed Motor (d) Pipe Ventilated Motor

Fig. 4.3: Types of Enclosures for Motor

Table 4.1: Details of enclosures used for motors

Industrial location	Type of contamination	Type of enclosures recommended
General industrial installations (non-explosive atmosphere).	(i) No contamination. (ii) Clean and dry air. High humidity. Water jets splashing. Dust abrasive or conducting or oily textile dust.	Open general purpose or open drip proof, screen protected. Open drip proof type and splash proof. Splash proof / Drip proof. Totally enclosed, pipe ventilated.
Chemical plant (Non-explosive atmosphere)	Dust, corrosive acid vapours or gas, mist.	Totally enclosed.
Metal working machinery (Non-explosive)	Dust metallic oily	Totally enclosed.
Mines or other hazardous locations	Explosive gas or vapours or dust.	Explosive proof.
Outdoor installation	Rain and snow, Duty storm, Nearly clean.	Weather proof. Totally enclosed. Open drip proof.

4.6 BEARINGS

- The bearings provide support to rotating part (shaft) and to permit free rotation with minimum amount of friction.

4.6.1 Types of Bearings

- There are two types of bearings:
  1. Ball bearing or roller bearings.
  2. Sleve or bush bearings.

1. **Ball bearing or Roller bearings:**

   The shaft is supported by balls or rollers which are also movable. The action of ball or rollers is not rubbing but of rolling. The balls/rollers are placed between shaft and housing as shown in Fig. 4.4. Proper greasing is done to reduce friction. Motors upto 75 kW use these type of bearings.

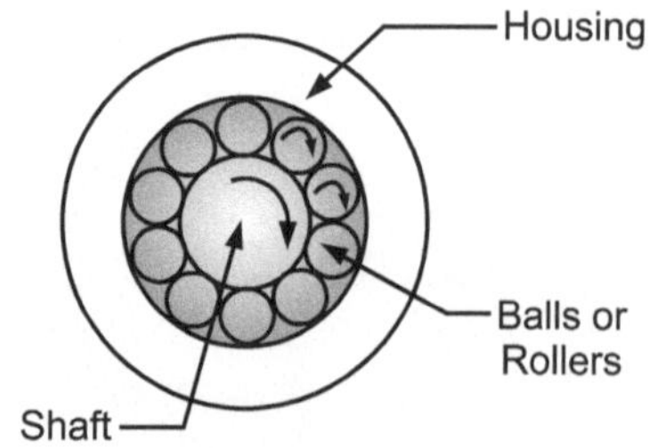

Fig. 4.4: Roller Bearing

Advantages:

- These bearings require less space and less maintenance. The life of bearings is more.

Disadvantages:

- These are costly. Operation is noisy.

2. **Sleve or Bush bearings:**

   The bearing component is rigidly fixed to the frame of machine. These bearings support the shaft. These are madeup of bronze but for universal and fan-cooled motors, sintered powder or porous metal bearings are used because of their self-lubricating property. Sleve bearings on very small motors are lubricated with an oil wick which is pressed lightly against the shaft by spring. These are used for less noise level. These are always used with shaft in horizontal position.

4.7 TRANSMISSION OF MECHANICAL POWER

- The motor power output available at the shaft is then transmitted to respective load through transmission media like chain, belt, gear, rope etc.
- The choice is governed by most economical speed of motor. The dimensions and hence the cost of motor of given output is inversely proportional to the speed.

4.8 DIRECT DRIVE

- The motor shaft (driving member) is directly connected to the driven member. e.g. Motor-generator set in electrical lab.
- A flexible coupling is used for such purpose. This can be used only where speed requirement of driven member matches with speed of the driving member.

4.9 INDIRECT DRIVE

- As described earlier, the mechanical power available at source is transmitted through some medium like gear, chain etc. The power loss in case of indirect drives is more and hence transmission efficiency is less.

4.9.1 Belt Drive

- This is a simple and cheap drive. Following types of belts are used: Leather belt, nylon belt, terylene belt, rubberized cotton ply belt, hair belt.

  The belt drives are of two types:

  (i) Falt belt drive

  (ii) V-belt drive.

(i) Flat belt drive:

It is a long distance drive. The belt is generally made up of leather. Other material listed in above can also be used for belt. It is preferred in horizontal position. The slack side of the belt is kept at top so that natural sag increases the arc of contact of the two pulleys.

Advantages:

1. Simple.
2. Greater flexibility.
3. Any desired speed ratio can be obtained.
4. When sudden load change occurs there is a tendency of belt to slip on pulley. This is better for both driving and driven member.

Disadvantages:

1. More space is required.
2. Maximum power it can transmit is 150 to 200 kW. So transmission capacity is limited.
3. It exerts side pull on bearing causing wear and tear of bearings.

(ii) V-belt drive:

1. It is suitable for motors with rating upto 450 kW. Because of wedge effect, there is increased friction between the belt and groove. This gives a better grip of about 3 times as that of the flat belt.

Advantages:

1. Less wear and tear of bearings.
2. Short distances can be maintained.
3. Less noise.

4.9.2 Rope Drive

- It is used for power ranges beyond the limit of V-belt drive. It is a long centered drive. Main advantages of rope drive are negligible slip and ability to take sudden loads.

4.9.3 Chain Drive

- It has no slip and no initial tension is present, hence bending stresses are eliminated. It can be used for high speed ratio (7: 1). It is particularly suitable for damp and dirty conditions.
- The chain must run concentrically on the sprocket and at right angles to the shaft.

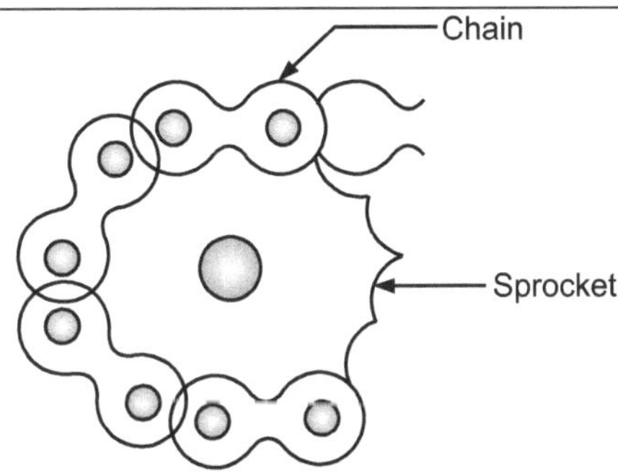

Fig. 4.5: Chain Drive

4.9.4 Gear Drive

- It is a short centered positive drive. Proper alignment is very essential, otherwise, motor shaft may bend.
- Modern straight cut gears and particular variation found in worm reduction drive are common features in many machines such as cloth calendar.

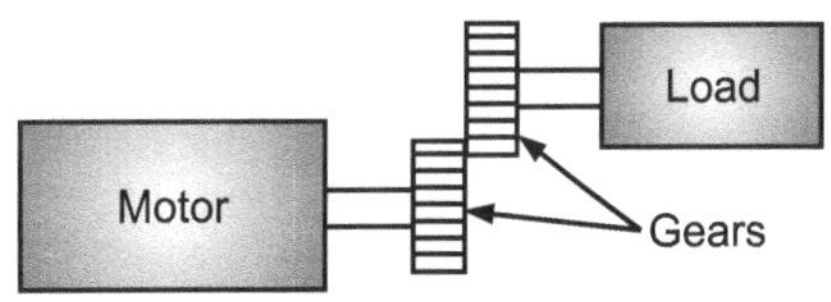

Fig. 4.6: Gear Drive

4.9.5 Vertical Drive and its Applications

- In case of vertical drive, the shaft is vertical. The axis of rotation is thus vertical. The examples of vertical drive are water turbine used in power plants, hard disc-motor mechanism in computers, washing machine drive.

4.10 NOISE

- Noiseless operation of motor is very much important in domestic theatre, hospital applications. The motor noise may cause vibrations in the surrounding structure and may cause resonance.

4.10.1 Reasons for Production of Noise

- The reasons for production of noise are as follows:
  (i) Vibrations of lamination due to improper clamping.
  (ii) Improper foundation work.

(iii) Improper alignment.

(iv) Improper coupling.

(v) Magnetic pulsations because of improper magnetic material section.

4.10.2 Method of Noise Reduction

- If motor is installed near radio aerials, it causes radio interference i.e. when motor current interrupts, the magnetic field of it collapses and since radio aerials are in vicinity of motor, this field also links with it. This causes distortion in radio reception. The problem is more if commutator motor is present.
- The avoid this, condensers are connected to motor circuit as shown in Fig. 4.7.
- The noise level of motor can be reduced by resilient or spring mounting of motor.
- Resilient mounting uses cushioning material such as felt mat, cork and rubber slabs. These can absorb audible frequency range noise.
- Spring mounting can absorb any range of vibration frequency.

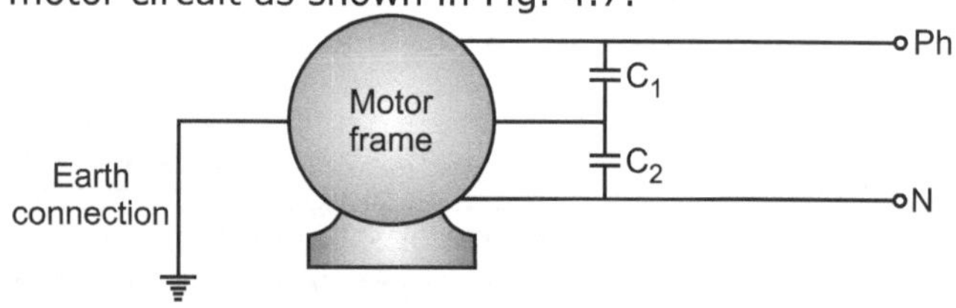

C_1 and C_2 = 0.1 to 0.025 μF

Fig. 4.7

4.11 SIZE AND RATING OF A MOTOR

1. Define the terms continuous rating and short time rating as per IS. Also state with sketches various types of load conditions on which motor work.
2. Define: (i) Continuous rating and (ii) Continuous maximum rating of the machine.

- The size and rating of motor mainly depends on following factors:
  (i) Nature of load
  (ii) Starting conditions of load
  (iii) Matching of speed torque characteristics.

Nature of load:

- The load may be having different nature as:
  (i) Hoist, crains: Torque is constant irrespective of speed.
  (ii) Fluid friction: Torque is proportional to speed.
  (iii) Blowers, fans: Torque is proportional to $(\text{speed})^2$.
  (iv) Load torque is proportional to $\frac{1}{\text{speed}}$.

Starting conditions of load:

- Some applications like traction require heavy torque at start with low speed.
- Applications like: drilling machine, lathe machine need medium torque at normal speed. The motor rating is to be selected accordingly.
- Matching of speed torque characteristics. The load dynamics should be properly known. The motor characteristics should match with load characteristics.

4.11.1 Standard Ratings of Motors as per IS

- As per IS specifications, following are the ratings:

(a) Continuous rating: A motor is capable of giving output continuously without exceeding the temperature rise. It can also give 2.5% overload for 2 hours.

(b) Continuous Maximum Rating: Similar to the above without any overload capacity. It is used for motors of size larger than 2.5 HP per rpm.

(c) Short time rating: It is the output that a motor can give for a specified short time, say $\frac{1}{2}$ hour or $\frac{1}{4}$ hour without exceeding the specified temperature rise.

4.12 LOAD CYCLES / DUTY CYCLES

1. Describe concept of load cycle and explain: (i) continuous loading, (ii) short time loading, (iii) long time loading, (iv) continuous operation with short time loading.

- The motor has to supply power to variety types of load. They are as discussed in the subarticles.

4.12.1 Continuous Loading

- The load is constant for long duration of time. So motor has to supply a constant and continuous load without excessive temperature rise e.g. pump load. Fig. 4.8 illustrates continuous load.

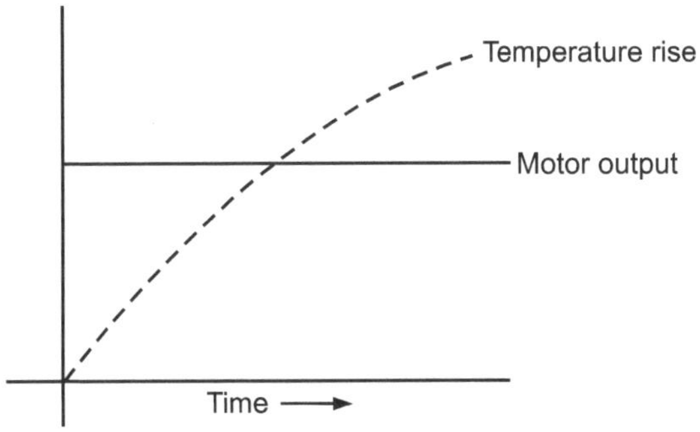

Fig. 4.8: Continuous Load

4.12.2 Short Time Loading

- The load is present at regular intervals. In between the intervals motor is OFF. Fig. 4.9 illustrates short time load. The motor is loaded during ON period and its temperature rise takes place. The OFF period is sufficient to cool the motor to ambient temperature.

  e.g. motors used for lock gates, bridges.

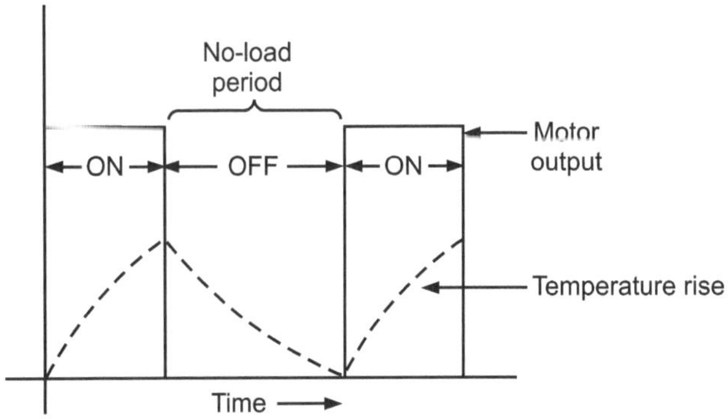

Fig. 4.9: Short Time Load

4.12.3 Long Time (Intermittent) Loading

- It is that way similar to previous type but frequency of occurrence of load is more in intermittent load. Fig. 4.10 illustrates intermittent load. The OFF period of motor is shorter than previous case. Hence, off period is not sufficient to cool the motor to ambient temperature. So temperature continuously rises with time.

  e.g. motors used for cranes, lifts, rolling mills.

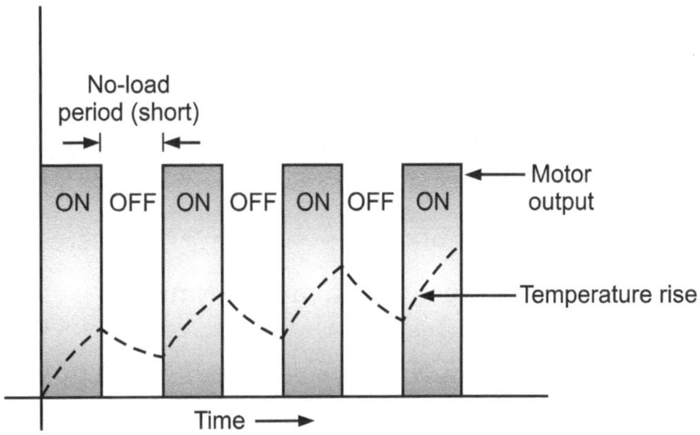

Fig. 4.10: Intermittent Load

4.12.4 Continuous Operation with Short Time Load

- In continuous operation with short time load, there is no OFF time. But during short time load, motor runs at no-load condition. So motor runs continuously. The heat produced by motor is carried away by proper ventilation arrangement. e.g. motor used to drive generator in Ward-Leonard system.

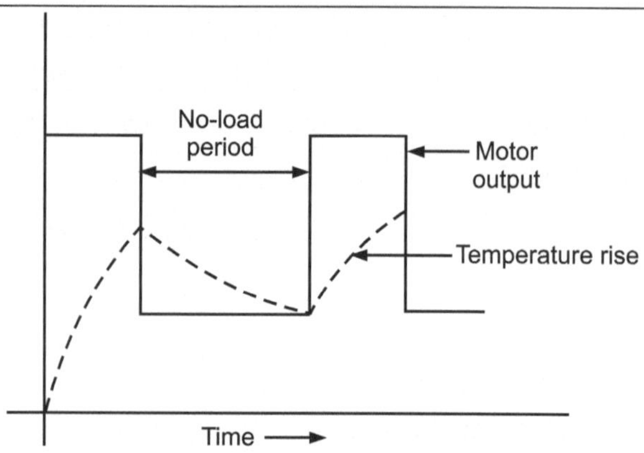

Fig. 4.11: Continuous Operation with Short Time Load

4.12.5 Continuous Rating with Long Time Loading

- It is similar to intermittent operation except that OFF period is replaced by no-load operation of motor.
- Since, no-load period is small and frequency of occurrence of load is more, the motor does not cool to ambient temperature but its temperature goes on increasing.

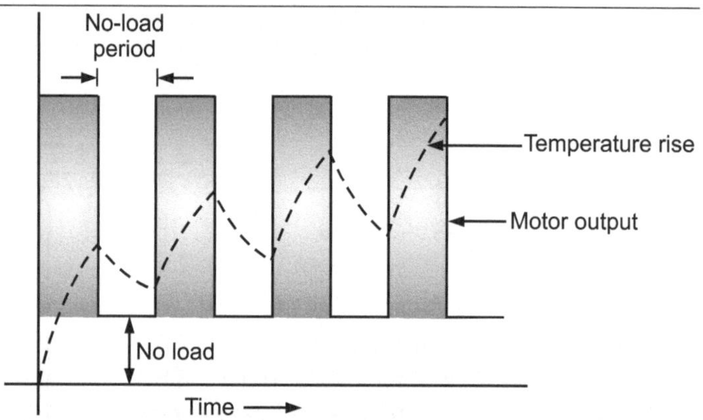

Fig. 4.12: Continuous Rating with Intermittent Load

4.13 EXPRESSION OF RATINGS OF MOTOR

- The applications such as air compressors, blowers, pumps, fans comes under constant load category. Choice of motor for these types of load is simple.
- The motor choosen should not be too small otherwise it will be overloaded. The motor should not be big sized otherwise it will operate much below its rated capacity giving poor performance i.e. poor efficiency and power factor. So optimum size of motor has to be selected.
- For the above applications there is no possibility of overloading because load pattern is constant, so a motor with continuous maximum rating can be selected. Such a motor will be smaller in size and cheaper in cost and can operate at higher temperature (55°C).
- The size / rating of motor can be determined as follows:
- Output power of motor = $\frac{2\pi NT}{60}$ ... where N is speed in rpm, T is torque in Nm.

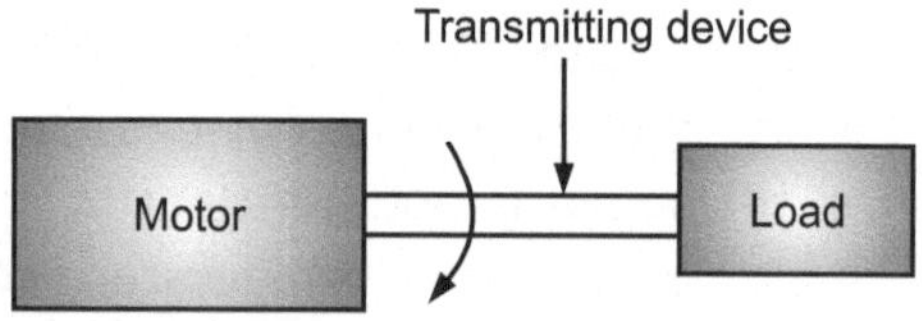

Fig. 4.13: Motor Supplying Power to Load

Considering the efficiency of driven load and transmitting device by means of factor η.

$$\eta = (\text{Efficiency of driven load}) \times (\text{Efficiency of transmitting device})$$

$$\text{Power in kW} = \frac{2\pi NT}{60} \times \frac{1}{\eta} \times \frac{1}{1000}$$

- The value of η is obtained by experience. The above expression is for rotary motion of motor.

- For linear motion of motors following expression is used:

$$P = \frac{F \times 9.81 \times v}{\eta} \times \frac{1}{100} \text{ kW}$$

where, F = force caused by load in kg

v = linear velocity of load in m/sec

- The example of linear load is hoist or lift. So above expression can be used with some modification for hoist load. The expression becomes

$$P = \frac{Fv}{2 \times 102\,\eta} \text{ kW} \quad \ldots v = 0.5 \text{ to } 1.5 \text{ m/s for passenger lifts.}$$

- In case of pumps, the rating of water is given by,

$$P = \frac{Q \times \rho \times h}{102\,\eta} \text{ kW}$$

where Q = delivery of pump in m^3/sec

ρ = density of liquid in kg/m^3

h = gross head in m

- Rating of fan motor is given by,

$$P = \frac{Qh}{102\,\eta} \text{ kW}$$

where Q = volume of air or gas in m^3

h = pressure in mm of water or in kg/m^2

- Other methods of finding the rating of motor for continuous duty are as follows:

1. Equivalent Current Method Expression:

In this method, an assumption is made that actual variable current is equal to the equivalent current I_e. This I_e produces some losses as produced by the actual variable current.

$$I_e = \sqrt{\frac{I_1^2 t_1 + I_2^2 t_2 + \ldots + I_n^2 t_n}{t_1 + t_2 + \ldots + t_n}}$$

2. Root Mean Torque Method / Equivalent Torque Method Expression:

The motor current $I \propto$ Torque

Heating $\propto I^2$

$\therefore$ Heating $\propto (\text{Torque})^2$

In case of short time and intermittent loads (Refer Fig. 4.14) the load or torque does not remain same, so equivalent torque is considered.

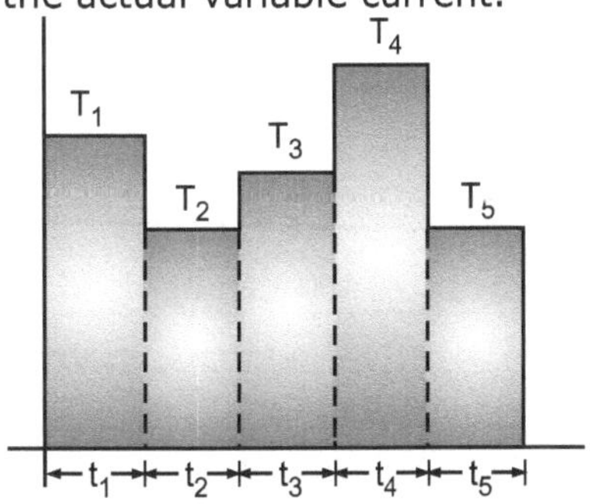

Fig. 4.14

The equivalent torque T_{eq} produces same heating as produced by $T_1 + T_2 + T_3 + \ldots + T_n$.

$$T_e = \sqrt{\frac{T_1^2 t_1 + T_2^2 t_2 + \ldots + T_n^2 t_n}{t_1 + t_2 + t_3 + \ldots + t_n}}$$

3. Heating Time Curve Method Expression:

The final temperature attended by a motor is proportional to loss or I^2 or $(\text{load})^2$.

Let, θ_f = Permissible temperature rise with continuous load of P watt

θ_f' = Maximum temperature rise that takes place with short time load of P' watt

We can write, Losses at load of P watt $\propto$ P

Losses at load of P' watt $\propto$ P'

Since, temperature rise $\propto (\text{load})^2$

$$\theta_f \propto P^2$$

$$\theta_f' = P'^2$$

$$\therefore \quad \left(\frac{P'}{P}\right)^2 = \frac{\theta_f'}{\theta_f}$$

θ_f' can be calculated from heating time curve equation. Then θ_f' value is put in above equation to get the value of P'.

SOLVED EXAMPLES

Example 4.1: *A certain motor has to perform following duty cycle:*

100 kW for 10 min.

No load for 5 min.

52 kW for 8 min.

No load for 4 min.

The above duty cycle is repeated indefinitely. Assume that losses are proportional to square of the load. Determine the size of continuously rated motor.

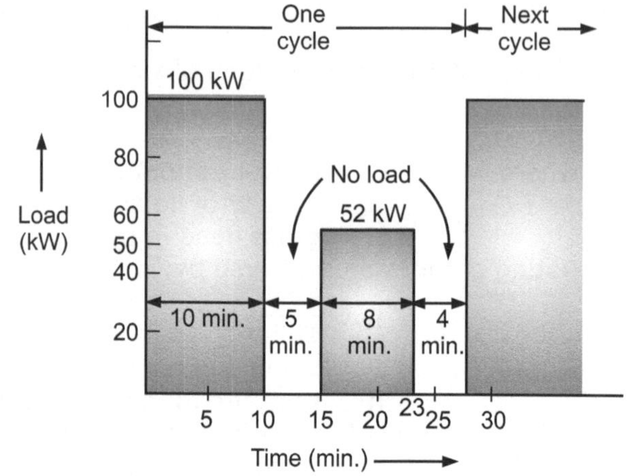

Fig. 4.15

Solution: The duty cycle is shown in Fig. 4.15.

We know that, $I_e = \sqrt{\dfrac{I_1^2 t_1 + I_2^2 t_2 + \ldots + I_n^2 t_n}{t_1 + t_2 + \ldots + t_n}}$

and power $\propto$ I, so current can be replaced by power and equivalent power rating can be found. In the similar way if power is expressed in HP, then HP $\propto$ I and current will be replaced by HP. The generalized expression will be

$$P_e = \sqrt{\frac{P_1^2 t_1 + P_2^2 t_2 + \ldots + P_n^2 t_n}{t_1 + t_2 + \ldots + t_n}}$$

Putting the respective values,

$$P_e = \sqrt{\frac{[(100)^2 \times 10] + [(0)^2 \times 5] + [(52)^2 \times 8] + [(0)^2 \times 4]}{10 + 5 + 8 + 4}} = \sqrt{\frac{121632}{27}}$$

$$= \mathbf{67.12\ kW}$$

So a motor of 70 kW can be used.

Example 4.2: *An electric motor has load variations as given below:*

(i) Torque 250 Nm for 20 min.

(ii) Torque 150 Nm for 10 min.

(iii) Torque 300 Nm for 10 min.

(iv) Torque 200 Nm for 20 min.

If speed of motor is 720 rpm find the power rating of motor.

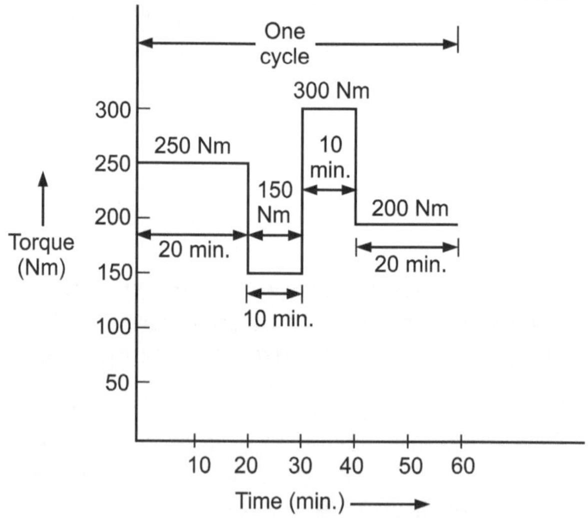

Fig. 4.16

Solution: The torque variation is as shown in Fig. 4.16.

The equivalent torque,

$$T_e = \sqrt{\frac{T_1^2 t_1 + T_2^2 t_2 + \ldots + T_n^2 t_n}{t_1 + t_2 + t_3 + \ldots + t_n}}$$

Putting the respective values, we get

$$T_e = \sqrt{\frac{[(250)^2 \times 20] + [(150)^2 \times 10] + [(300)^2 \times 10] + [(200)^2 \times 20]}{20 + 10 + 10 + 20}}$$

$$= \sqrt{\frac{3175 \times 10^3}{60}} = 230.036 \text{ Nm}$$

$$\text{Power} = \frac{2\pi NT_e}{60} = \frac{2 \times \pi \times 720 \times 230.036}{60}$$

$$= \mathbf{17344.306\ watt = 17.34\ kW}$$

17.5 kW rating can be taken.

4.14 LOAD EQUALIZATION AND MEANING OF LOAD EQUALIZATION

1. What is load equalization? Explain it with the help of diagram.
2. State the necessity and principle of load equalization.
3. What is load equalization?

- The process of smoothing out fluctuations in load is known as load equalization.
- In industrial drives, load varies between very wide limits in a very short intervals of times e.g. steel rolling mill in which the load is very large during the rolling operation and very little when charges pass by. A typical example of load cycle is shown in Fig. 4.17.

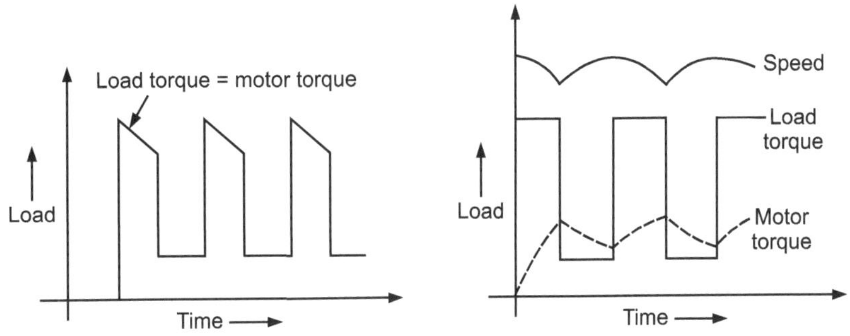

Fig. 4.17: Rolling Mill Load Cycle

- It should require a motor of large output through an average, the load on motor is not large. The load cycle is quite regular, so it is possible to look for some way by which large torque is met for a short while and if it is done there is considerable economy in motor size.

4.14.1 Method of Load Equalization

1. Explain load equalization.

- The best way to do this is to increase the moment of inertia of motor by flywheel.
- During the period when load is small flywheel accelerates and stores kinetic energy. Thus demand on supply side remains constant. When load increases during peak period, flywheel gives a part of its kinetic energy, thus helping in decreasing the demand on supply side.
- It is thus, practical to have a motor, whose torque is not as large as the required load torque, the gap being adjusted by the flywheel.

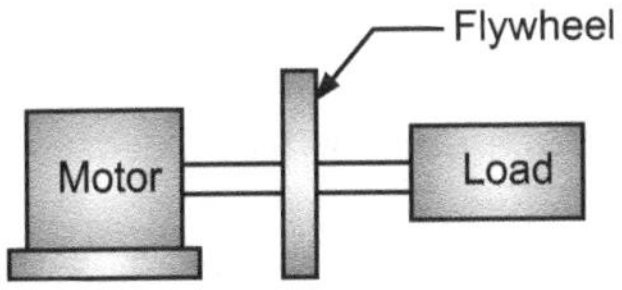

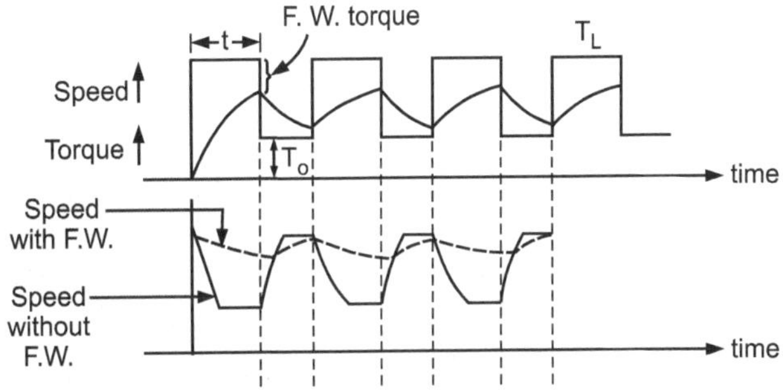

Fig. 4.18: Load Equalization

4.14.2 Conditions of Load Equalization

- In load equalization process, the fluctuations in load are to be smoothened. Flywheel is used for this purpose.
- The fluctuations may occur due to two conditions:
  - (i) Load increasing
  - (ii) Load decreasing.

Load increasing condition:

- When load on the motor increases, the flywheel deaccelerates and gives up some part of kinetic energy to the load.

  Thus, $T_m = T_L - T_F$

  where T_m = Motor torque

  T_L = Load torque

  T_F = Flywheel torque

Load decreasing condition:

- When load on the motor decreases, the flywheel accelerates and stores the kinetic energy.

  Thus, $T_m = T_o + T_F$

  where T_m = Motor torque

  T_o = No-load torque

  T_F = Flywheel torque

4.15 BRAKING DEFINITION

- Braking is a process of stopping motion by some means.
- For electrical motor, breaking means stopping the rotation of motor.

4.15.1 Requirements of Ideal Braking System

- Following are the requirements of ideal braking system:
  (i) It should be simple.
  (ii) Wear and tear to parts should be less.
  (iii) Excessive heating should not take place.
  (iv) Braking should be comfortable i.e. it should not produce jerk.
  (v) It should be convenient.
  (vi) It should be quick.
  (vii) There should not be time lag in braking system.
  (viii) It should be cost effective.

4.15.2 Advantages and Disadvantages of Electrical Braking over Mechanical Braking

1. State advantages and disadvantages of electrical braking over mechanical braking.
2. State advantages of electrical braking over mechanical braking.
3. Compare electrical braking with mechanical braking.

Advantages:

1. Operation smooth.
2. Reduced wear and tear on mechanical brakes.
3. Heat can be developed at a convenient place.
4. No dust.
5. Cost of replacement is very less.
6. No time lag.
7. Some part of the energy can be returned to supply, therefore saving in energy.

Disadvantages:

1. Requires additional equipment such that switches, resistance, therefore cost is more.
2. Complicated electric connections.

4.15.3 Types of Electrical Braking System

1. Write three types of electrical braking.

- When the running motor is to be stopped either mechanical breaks are applied, otherwise, the motor can be stopped electrically first and at the last mechanical breaks are applied.
- Electrical braking is the braking where kinetic energy of motor and tool is converted into electric energy and finally it is dissipated into heat energy or part of the energy is returned to the supply (Regenerative). Electric braking does not bring the motor to standstill but bring the motor to low speed and mechanical brakes are applied.

- The methods of electric braking are:
  (i) Plugging (Reversing armature connection) Reverse Current Braking
  (ii) Rheostatic Braking
  (iii) Regenerative Braking.

4.15.4 Plugging (Reverse Current Braking)

1. What is plugging? Explain plugging of three-phase induction motor with simple circuit diagram.
2. With neat diagram, explain the plugging methods applied to d.c. series motor.
3. Describe how plugging is done with series and shunt motor.

- **Plugging applied to DC shunt and DC series motor:** In this type, the connections of the armature are reversed. Thus torque produced is reversed. Thus the braking effect is produced. Due to reverse connection of armature, the direction of e.m.f. is reversed. Thus the voltage becomes $E_b + V$. The machine must be stopped at proper instant, otherwise, it will rotate in reverse direction. During plugging $I_a = \frac{V + E_b}{R_a}$, thus current will be very high. A high resistance is therefore required to limit current. Method is simple but requires power from supply.

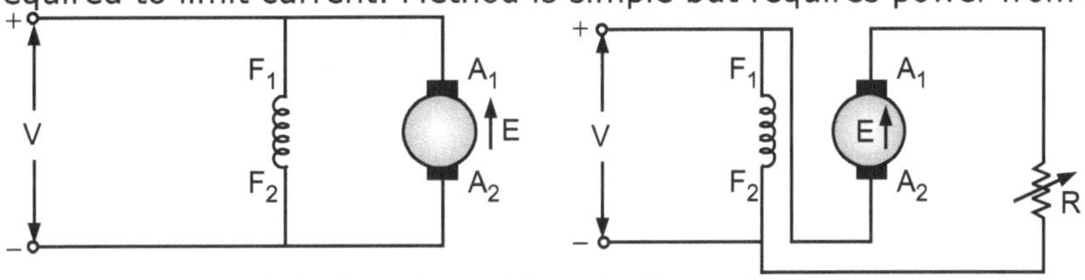

(a) Plugging with D.C. Shunt Motor

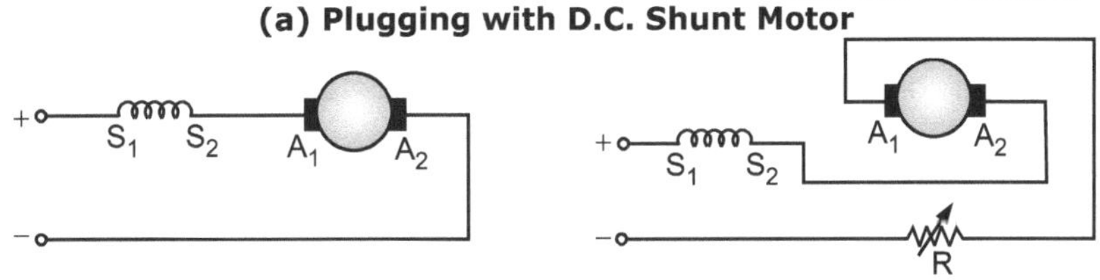

(b) Plugging with D.C. Series Motor

Fig. 4.19

Plugging with induction motor:

- The connections of any two supply phases are interchanged so the direction of rotating magnetic field is reversed. Thus, motor tends to develop torque in opposite direction causing braking. (Refer Fig. 4.20).

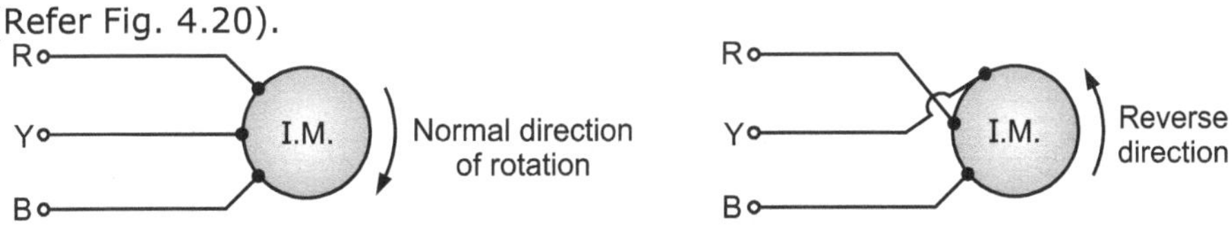

(a) (b)

Fig. 4.20: Plugging with Induction Motor

- The supply must be cut off at a proper instant otherwise motor continues to accelerate in reverse direction. The stator and rotor current magnitude is high during plugging, so high resistances are put in rotor or stator circuit for limiting current.

4.15.5 Rheostatic Braking

1. Explain with neat sketch rheostatic braking as applied to d.c. series motor.
2. With neat diagram, explain rheostatic braking.

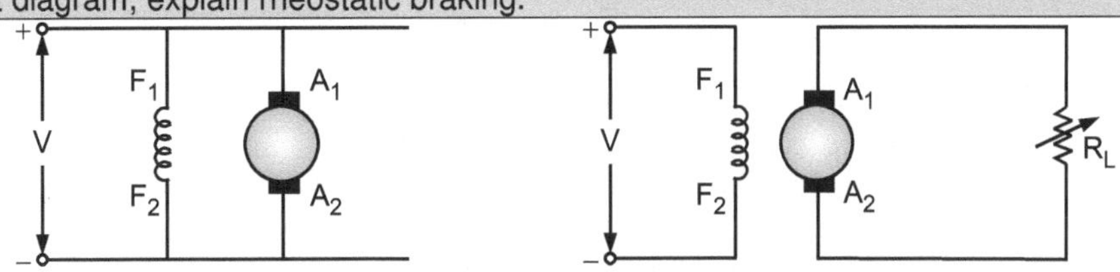

Fig. 4.21: Rheostatic Braking – Shunt Motor

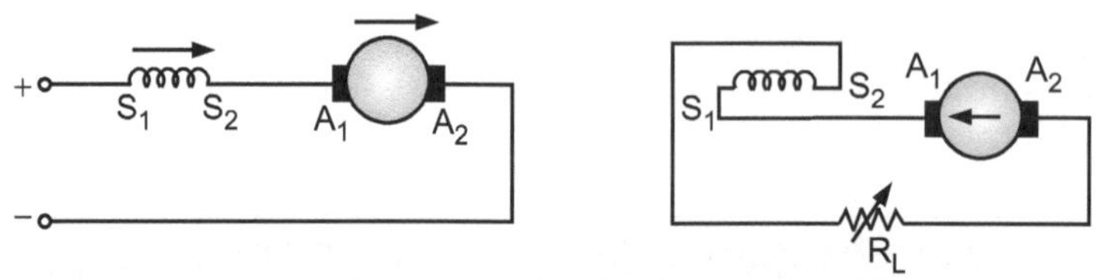

Fig. 4.22: Rheostatic Braking – Series Motor

- The arm is disconnected from the supply and load resistance (R_L) is connected across it. Machine works as a separately excited generator and induced e.m.f. circulates current in R_L. The kinetic energy of the rotor is converted into electrical energy and finally heat in the resistance. For series machine field connection must also be reversed to build e.m.f. Also total resistance should be smaller than critical resistance.

Induction motor

- For induction motor, the stator is disconnected from the supply and direct steady current is passed through stator winding. A flux is produced. The rotor still rotating in normal direction, cuts this flux and e.m.f. is induced in it. This e.m.f. now opposes the cause of producing it i.e. it opposes rotation. Thus braking is achieved.

4.15.6 Regenerative Braking

- The necessary conditions:
  In case of regenerative braking, the motor is made to run as a generator and it remains connected to the line. Thus, the kinetic energy is converted into electrical energy which is feedback to the line.

Conditions for regenerative braking:

1. The necessary condition for regenerative braking to take place (in case of D.C. motors) is that the back e.m.f. or generated voltage must exceed the supply voltage.
2. The braking system should have mechanical stability. i.e. braking system should be in a position to apply more and more braking torque in case of powerful overhauling (step down gradient).
3. Braking system should have electrical stability i.e. braking torque applied should be independent of supply voltage fluctuations.
4. If the energy feedback to the line is surplus i.e. supply is greater than demand then the generating station should be able to waste the energy, otherwise, regenerative braking is ineffective.

- The regenerative method is obviously preferable to other two methods because the braking energy is not wasted but it is fed back to the line. This fed back energy can further be utilized by some other apparatus connected to the line.

4.15.7 Regenerative Braking for D.C. Shunt Motor

1. What is regenerative braking? State the conditions for regenerative braking.
2. Explain regenerative braking for D.C. series motor.
3. Explain regenerative braking as applied to D.C. shunt motor.

- Suppose decrease in load on motor causes increase in speed above normal with field current remaining the same. Thus, the back e.m.f. will increase and it may exceed supply voltage. When this happens, the power is fed back to the line and K.E. decreases gradually thereby decreasing the speed and thus braking is achieved (Refer Fig. 4.23).
- Alternatively the field current can be increased and same effect can be achieved. Regenerative breaking can thus easily be applied with shunt motor.

  Note: Direction of field current should remain same.

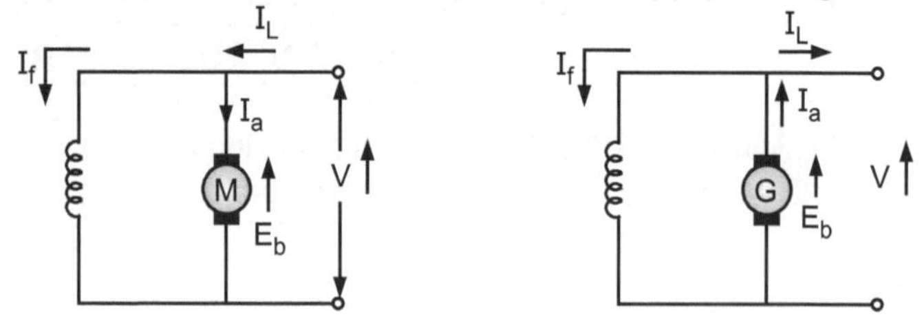

Motoring action $V > E_b$ (a)

Generating action (Regenerative braking) $E_b > V$ (b)

Fig. 4.23: Regenerative Braking – Shunt Motor

4.15.8 Regenerative Braking for D.C. Series Motor

1. Explain regenerative braking for D.C. series motor.

- The conditions with a series motor for regenerative braking are not so simple, since the armature current necessary to produce regeneration also flows through field winding. Thus, the direction of field current also reverses. This causes short-circuit (Refer Fig. 4.24) condition.
- Other possibility is the field flux will be neutralized during the transition.
- Thus, the requirement is to reverse the field winding connection at the exact instant but the method is still useless because at the instant of reversal, the e.m.f. generated will be less i.e. supply voltage will be greater than generated e.m.f., hence regenerative braking is not possible.
- Thus, the main trouble associated with regenerative braking by series motors is seen to be due to the lack of control of the field. There are various methods of overcoming this difficulty, either by modification of the windings or by supplying the machine with separate excitation.

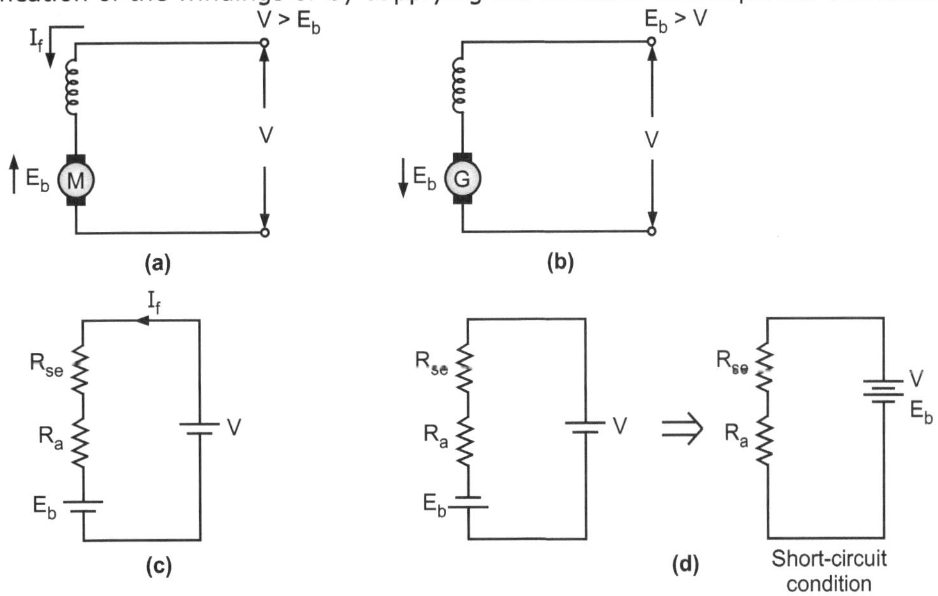

Fig. 4.24: Regenerative Braking – Series Motor

- These methods described below are mainly used in electric traction.
- In Fig. 4.25 a device called pantograph is shown, the pantograph acts as a current collector. It makes a continuous sliding contact with overhead line and sends the current to motor installed in locomotive.

1. French Method:

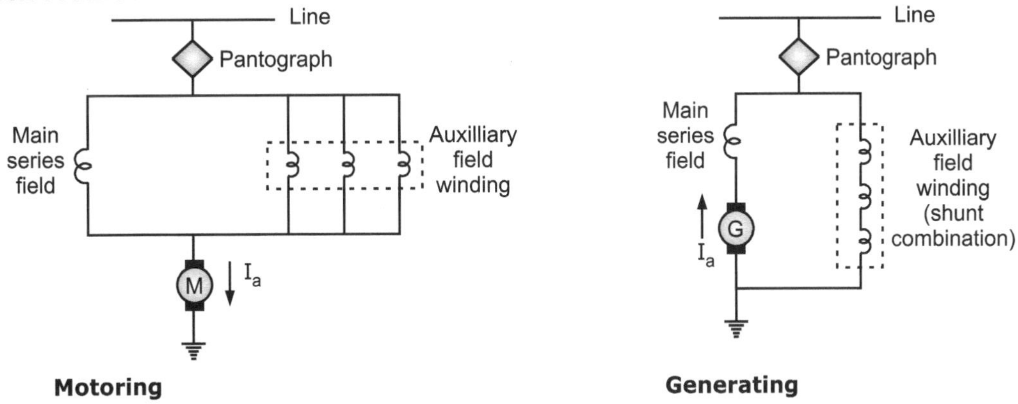

Fig. 4.25: French Method

- During motoring action, the main series field winding and auxiliary field windings are connected in parallel and this whole combination is in series with the armature (Refer Fig. 4.25).
- In generating mode, the auxiliary field winding group is connected in series and this combination is connected to the motor circuit as shown. Thus the machine now behaves as a compound generator with a slightly differential action.
- The shunt combination is sensitive to voltage fluctuations. Note that slight changes in supply voltage frequently occur in the traction system. The shunt combination provides necessary balancing action and thus maintains potential difference between V and voltage generated (with $E_b > V$).
- Suppose during regenerative braking the line voltage increase beyond the generated voltage. Same increased voltage is present across the shunt combination, thus it will cause increase in exciting current through it. Hence, as the voltage developed the generator also increases and again becomes greater than supply voltage. Reverse will happen when line voltage falls. The method is applicable for trolley bus and trams.
- The idea of French method can be implemented for locomotives also. In locomotives 4 to 6 motors are used at a time. In that case the main field winding of motor is retained and others are made to function like auxiliary winding.

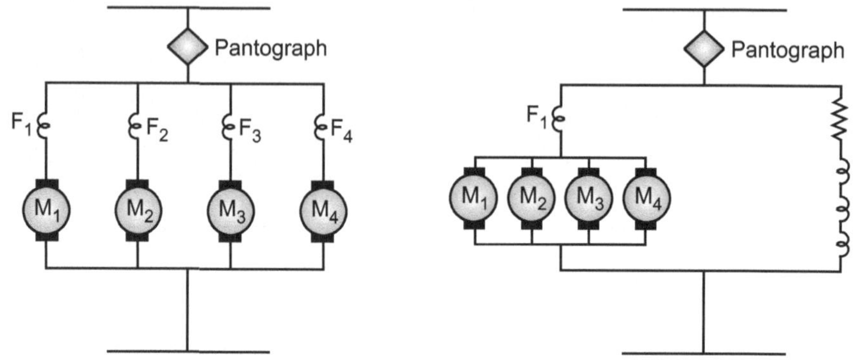

Fig. 4.26

2. Regenerative braking using exciter:

- The series field windings of motor are connected in series and these are excited by the exciter. The exciter unit has got its own field windings F_1 and F_2. F_1 is again connected to a separate D.C. source. The field current of F_1 is to be controlled by the operator. F_1 and F_2 are connected in such a manner that their fluxes act in opposite direction.
- Suppose during regenerative braking, the line voltage decreases, hence (E_b – V) increases, hence the regenerative current increases and same current is flowing through F_2, therefore excitation or flux of F_2 increases. Since, F_1 and F_2 oppose each other, the e.m.f. E falls on account of reduced excitation. (Action similar to differential computer generative). As soon as the e.m.f. of the exciter falls, the current in the field windings of M_1 and M_2 decreases and therefore regenerative e.m.f. also decreases. Thus compensation for decrease line is provided.
- In another arrangement Fig. 4.27 (b) armature of exciter is connected in the circuit of field windings and stabilizing resistance. The current in stabilizing resistance is the summation of regenerated current and exciter current.
- Suppose during regenerative braking, the line voltage falls, then (E_b – V) increases, therefore, regenerative current increases. Therefore, voltage drop across stabilizing resistance increases. Therefore, the balance of voltage or net voltage available in the exciter armature circuit reduces. This causes reduction in exciting currents of field windings of motor, hence regenerated e.m.f. decreases. Thus compensation is provided.
- The function of stabilizing resistance is to prevent the current surges when the vehicle crosses one section of supply to another and to compensate for variation in line voltage.

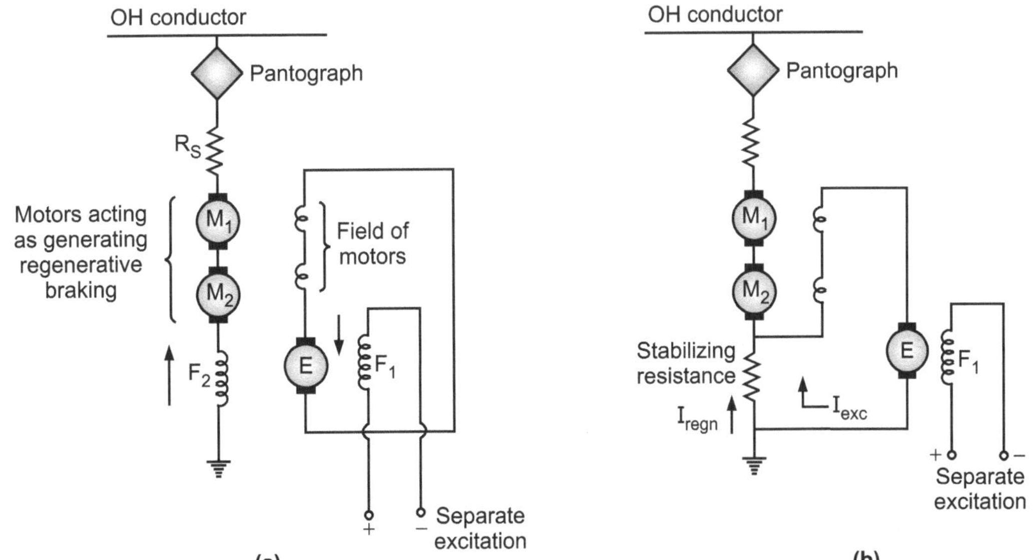

Fig. 4.27: Regenerative Braking using Exciter

4.15.9 Regenerative Braking with Three-phase Induction Motor

- It is achieved automatically when the motor run at a speed slightly above the synchronous speed. The motor starts working as an induction generator, the induction generator however is not self exciting and must be connected to a system supplied from synchronous generator i.e. its field winding/rotor coinciding will be supplied with D.C. by an exciter of some other synchronous generator system.

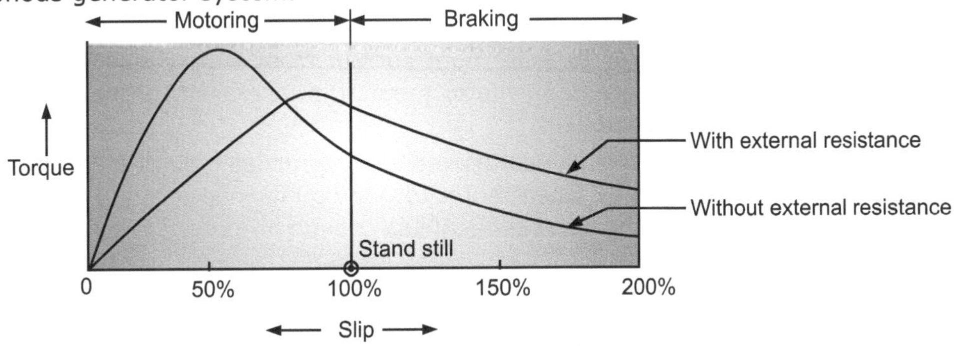

Fig. 4.28: Regenerative Braking with Induction Motor

Comparison of Rheostatic braking and Regenerative braking:

Sr. No.	Rheostatic braking	Regenerative braking
1.	Kinetic energy of moving masses is converted into electric energy as in this system, motor is made to work as generator.	Motor is not disconnected from supply and it works as a generator.
2.	Electric energy is dissipated in resistance and heat is produced.	Generated energy is fed back to the supply.
3.	Not economical due to loss of energy.	Most economical as energy is fed back.
4.	Braking torque: Due to generation, action energy is dissipated, produce braking torque.	Braking torque: Due to magnetic drag produced on account of generating action.
5.	Adjustment in braking torque: (i) by D.C. excitation control (ii) by varying rotor resistance	Adjustment by gradual increase in field excitation of separately excited motor.

4.16 SIZE AND RATING OF A MOTOR

- While selecting motor for a particular application, the motor and load characteristics should be known.
- Following factors are important while deciding size and rating of motor for a particular application:
  (i) Starting conditions of load.
  (ii) Matching of speed-torque characteristics.
  (iii) Nature of load.

4.16.1 Starting Conditions of Load

- The torque requirement is different for various applications e.g. For electric traction, conveyor belts, hoists etc. the starting torque required is large and it should be provided at low speed. So a motor with high starting torque is to be selected. The motor starting current should not be very high.
- For applications like water pump, lathe machine, drilling machine etc. starting torque requirement is medium.
- The motor rating is selected considering above factors.

4.16.2 Matching of Speed-Torque Characteristics

- The basic problem in choosing proper motor for a particular application is in matching the speed-torque characteristics to the load requirement. The load mechanics should be fully known i.e. the requirements of load should be known properly.

Steady state stability:

- For stable operation of a motor, if load torque demand increases, the motor speed decreases. The decreased speed causes motor to take more current from supply side. Thus input power increases and motor torque increases and motor meets the increased torque demand. Reverse happens if load torque demand decreases.
- Consider a motor which supplies mechanical power to a particular type of load. This motor load system operates in steady state only when motor torque equals load torque (equilibrium point). Any disturbance in load (increase or decrease) causes change in motor speed.
- Refer Fig. 4.29 in which motor load characteristics is shown.
- The equilibrium point is said to be stable when the operation is restored to equilibrium point even if there is some disturbance on load or motor.
- The equilibrium point in Fig. 4.29 is point A. At this point, motor and load characteristics are intersecting and motor torque is equal to load torque. i.e. $T_M = T_L$. The speed of motor is N r.p.m. at this condition.
- Suppose a small disturbance causes reduction in motor speed from N to N_1. At this new speed (N_1), the motor torque is greater than load torque ($T_{M_1} > T_{L_1}$). The motor is now accelerated and operation is restored to point A.

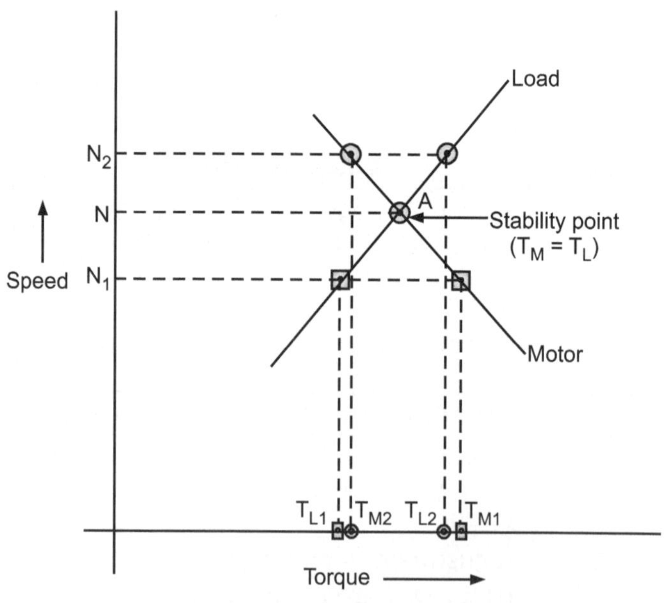

Fig. 4.29: Motor load characteristics

- Suppose a small disturbance causes increment in speed from N to N_2 (point C). At this new speed (N_2), the motor torque is less than load torque ($T_{M_2} > T_{L_2}$). The motor is now de-accelerated and operation is restored to point A.
- The overall operation of motor is thus steady state stable at point A.
- Now consider the same motor, this time supplying mechanical power to a different kind of load. (Refer Fig. 4.30).
- The operation is stable at point A ($T_M = T_L$). Suppose a disturbance causes reduction in motor speed from N to N_1. At this new speed (N_1), the motor torque is less than load torque ($T_{M_1} > T_{L_1}$). The motor further decelerates and operating point now moves away from A.
- Similarly, when a small disturbance causes increase in motor speed from N to N_2, the motor torque becomes greater than load torque ($T_{M_2} > T_{L_2}$) and motor further accelerates and operating point now moves away from A.
- Thus, we can conclude that point A is unstable point of equilibrium.

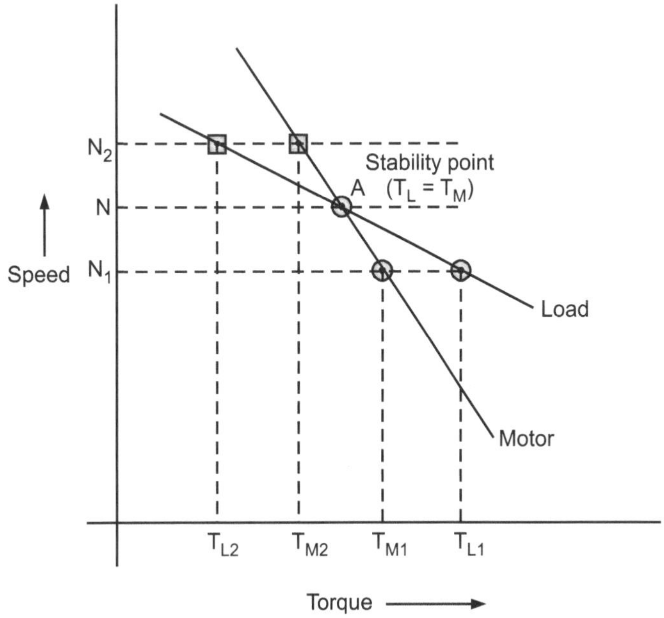

Fig. 4.30: Motor-load characteristics

- From above two cases we can say that for stable operation, the load and motor characteristics should intersect at 90°. (Fig. 4.29). If these are not intersecting at 90° (Fig. 4.30), the operation is unstable.
- After the point of intersection of characteristics, the motor torque should decrease with increase in speed and the load torque should increase with increase in speed. We can again trace it in Fig. 4.31. This corresponds to stable operation.
- Let us take a practical example now. Suppose we want a motor for driving a fan. The speciality of fan load is that its torque varies a square of the speed.
- We can choose induction motor or D.C. shunt motor for this purpose.

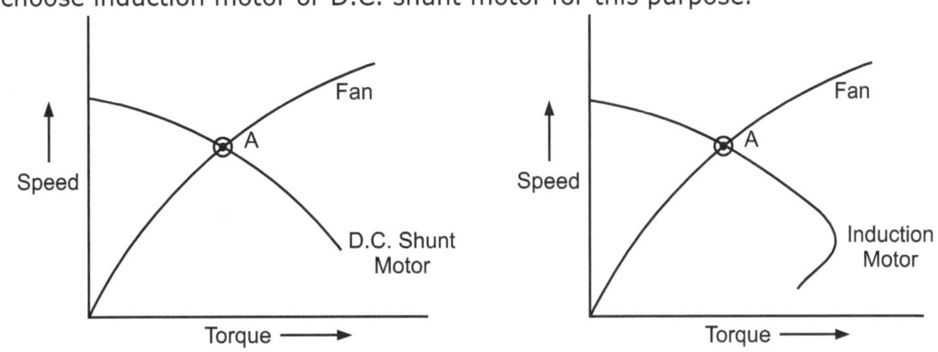

Fig. 4.31: Matching of speed-torque characteristics

- The characteristic of fan load along with two motors is shown in Fig. 4.31. We can trace the stability after point of intersection (point A). After point A, the motor torque decreases with increase in speed and load torque increases with increase in speed.
- The operation is thus stable. Thus, these motors are suitable for fan load.

4.16.3 Nature of Load

- There are four types of load:
  - (i) Load torque remaining constant irrespective of speed e.g. hoist, crane etc.
  - (ii) Load torque $\propto$ speed. e.g. fluid friction where lubricant is used.
  - (iii) Load torque $\propto$ (speed)^2. e.g. air and fluid friction such as blowers, fans, water wheels.

    Load torque $\propto \frac{1}{\text{speed}}$ i.e. load torque decreases with increase in speed.

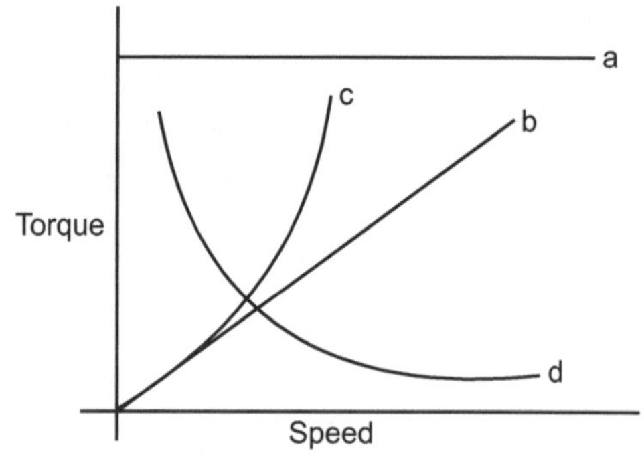

Fig. 4.32: Types of load

- The characteristics of above types are shown in Fig. 4.32.
- In practical world, a load may have combined characteristics of above basic types.

4.16.4 Standard Ratings as per IS

1. Define: (i) Continuous rating, (ii) Continuous maximum rating, (iii) Short-time rating.

(a) Continuous Rating: A motor is capable of giving output continuously without exceeding the temperature rise. It can also give 2.5% overload for 2 hours.

(b) Continuous Maximum Rating: Similar to the above without any overload capacity, it is used for motors of size larger than 2.5 HP per rpm.

(c) Short Time Rating: It is the output that a motor can give for a specified short time, say $\frac{1}{2}$ hour or $\frac{1}{4}$ hour without exceeding the specified temperature rise.

4.17 MOTORS AND THEIR APPLICATIONS

Sr. No.	Types of Motor	Applications
1.	Three-phase slip ring induction motor	Lifts cranes generators, line shaft, haulage, pumps.
2.	3-phase squirrel cage motor	Small scale - college industries, pumps, tube wells etc.
3.	Single-phase induction motors	Household domestic appliances like washing machine, refrigerator record players, vacuum cleaners, air conditioners, air compressors.
4.	Repulsion motor	Industrial machines, gauge-equipment pumps, small compressors.
5.	Shrage motor	Lifts, pumps, belt-conveyors, printing presses, cranes, textile mills, machine tools etc.
6.	Single-phase AC series motor	Household appliances, refrigerators, washing machines, vacuum cleaner, record players, air conditioner.
7.	Three-phase AC series motor	Haulage etc.
8.	Synchronous motor	Rolling mills, compressors, line shaft, M-G set, fans, frequency changers, electric clock (small motor).

4.17.1 Typical Applications of Motors/Electric Drives

1. Suggest suitable electric drive for following applications:
   (i) Paper mills, (ii) Stone crusher, (iii) Textile mills and (iv) Electric traction.

Rolling mills:

- In blooming and slabbing mill it is required to produce heavy torque at low r.p.m. Also the motor should have robust construction, it should be capable of having wide speed variation, it should be able to develop a short time torque of 2 to 2.5 times rated torque, it should maintain its speed nearly constant from no load to full load.
- Cascaded induction motors are used where speed variation required is not more than 30%. D.C. motor is most suitable. By varying the armature voltage and keeping field excitation constant, it is possible to get higher torque at low speeds.
- Secondly the rolling operation in final pass requires low torque and higher speed. This is achieved by weakening the excitation to the field after the rated speed of motor is attained (i.e. first the motor is made to run at rated speed by varying applied voltage and then field excitation is weakened).

Textile mills:

- Loom motors should operate in atmosphere which is laden with wool lint and moisture, motor should also have high starting torque (2-2.5 times full-load torque), loom motors have to start and stop several times in a day. This increases the motor heating during starting.
- So motors having low temperature rise under full-load condition are selected because they ensure frequent starting and stopping without exceeding permissible temperature rise.
- Totally enclosed, fan cooled, high torque squirrel cage induction motor is selected. If only wool lint is present and no moist atmosphere, green protected motors are used. Splash proof motors are used in wet locations such as dye house.

Cranes:

- Basic requirement is high starting torque and should withstand great number of switching operations.
- D.C. is used in cranes in steel mills. D.C. series motor is used. D.C. series motor adjusts its speed to load. This characteristic prevents excessive power demand from supply.
  (a) **A.C.:** Slip-ring induction motor is used, speed control is obtained by rotor resistance variation. Speed range of slip ring induction motor is less than that of cables motor.
  (b) For outdoor applications, totally enclosed motors of above types are suitable.

Mines:

- Motors should have flame proof enclosure. Slip ring motor with D.C. dynamic breaking ward Leonard.

Requirement:

- Speed control, breaking capacity.
- **Paper mills:** Requirements:
  (i) **Speed variation:** To vary speed of rolls for different papers.
  (ii) Relative speed between rolls should be constant otherwise tearing of paper results.
  (iii) Speed of any roll group relative to other should be adjusted for drawing on paper. Schrage motor is suitable.

Marine drive:

- Electrical drive for propeller of ship is widely used. The power to electrical drive is supplied by diesel generators.
- Electrical drive in ship application has following advantages:
  1. The diesel engine runs at very high speed (3000 rpm) and requirement of propeller of ship is only 400-800 rpm. So mechanical coupling has to be done by gears arrangement. Instead of that if electrical drive is used then any speed can be obtained for most efficient operation.
  2. Reversal of direction of rotation is easier in electrical drive.
  3. Variation of speed is easier in D.C. motors and I.M.S.
  4. Control is simpler and easier.
  5. Less space.

For big ships and battle ships:

Turbo alternators feeding 3ϕ syn. motor Turbo alternators feeding 3ϕ I.M.	Speed control is done partly by voltage variation and partly by frequency control.

For cargo ships:

D.C. motors if D.C. generation is available or synchronous motor, I.M.	Ward Leonard speed control is used

In workshop:

Punches, shears : D.C. compound motor.
Lathe grinding machine : Constant speed, D.C. shunt motor.

Pumps:

- Totally enclosed surface cooled motors are normally employed. If pump speed does not come in range of motor speed then V belt is employed.
- Starting torque necessary for centrifugal pumps is 40 to 50% of full-load torque, squirrel cage induction motor is used.
- Starting torque necessary for reciprocating pump is 100 to 200% of full-load torque: slip ring induction motor is used.

Refrigeration and air conditioning:

- In refrigeration during vapour compression cycle, the motor is used to drive the compressor. The motor is controlled by thermostat. Motor should be above to give starting torque of 200 to 250% of full-load torque. Single-phase induction motor is used.
- For small unit air conditioners 230 V, single-phase, capacitor type motor with D.O.L. starter is used. For large installations, high torque squirrel cage induction motor or slip ring induction motor are used. For very large plant size synchronous motor driving turbo compressor is used because p.f. is also one of the factors there.
    - **Excavators:** Totally enclosed motors of rugged construction.
    - **Bleat conveyors:** High starting torque squirrel cage or slip ring induction motor.
    - **Cement work:** Due to dusty atmosphere, pipe ventilated or totally enclosed motors are used.
    - **Printing Press:** Constant speed motors (synchronous motors).
    - **Steel mills:** Induction motors.
    - **Lifts:** D.C. series motor, slip ring induction motors.
    - **Floor mill:** Totally enclosed.

Sugar Factory

- The duty cycle involved in a sugar factory is as shown in Fig. 4.33. The operations performed are charging, intermediate spin, spinning, super synchronous, breaking, reverse current breaking and plugging.
- All these operations involve very wide variation of speed. The speed variation may be as high as 1:30.
- The motor used is four speed pole changing motor. It has 2 sets of stator winding. Depending upon the requirement of speed for the operation, pole configurations are changed to get desired speed. Automatic control can also be obtained by two methods. In first method, time relays are used. But this method is not used due to some difficulties.

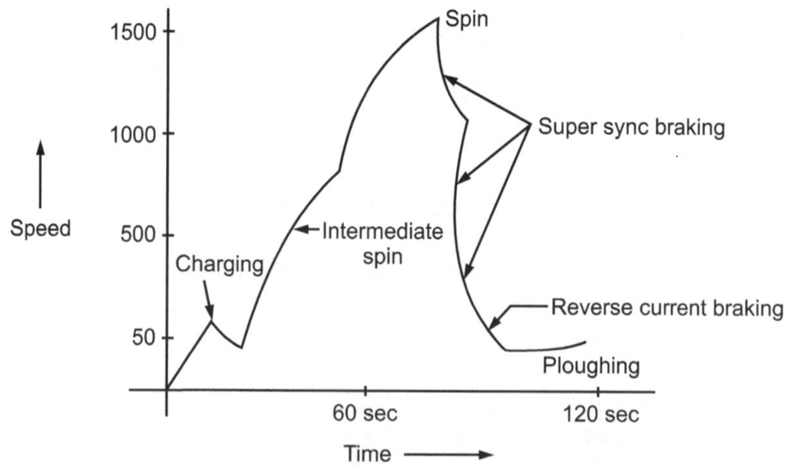

Fig. 4.33: Duty cycle of centrifuge

- In second method automatic control is made speed sensitive by using a tacho generator and this tacho generator is coupled to centrifuge motor. The selection of number of poles is made appropriately.
- There are wide fluctuations in current drawn by motor, over current trips are not employed. Because they may trip frequently, so thermal elements or sensors are embedded in stator winding. Operations of these elements will directly trip the motor supply.

4.17.2 Specifications of Motors

Motor Specifications/Ratings etc.:

- To define the basic information/parameters, mounting details the National Electrical Manufactures Association (i.e. NEMA), defines some basic design and dimensional parameters in NEMA standard MG-1.
- These parameters are then coded into the motor name plate to give a basic definition of what we have received.
- Manufacturers also include additional information to define some key motor features.
- See name plate of a motor manufactured in some country by manufacturer which gives the complete specifications/ ratings / information.

Table 4.2

Manufacturer				LOBO			
Identification No. 08RS0001006G			001 VU	365T			
HP	75	VOLTS 230/460	PHASE 3	DESIGN	B	TYPE	P
RPM	1780	AMPS 175/876	HZ 60	AMB	40°C	SF	1.15
Code no. of bearing	65BC03J30X			DUTY	CONT	INC Class	F
Code of bearing	65BC03J30X			ENCL	TEFC	CODE	G
			Maximum power factor	84.9	Maximum Efficiency		94.1
			Maximum kVAR	17.0	Guaranteed Efficiency	93.6	
					Motor weight		795 LBS

4.18 SUITABILITY OF DIFFERENT MOTORS FOR TRACTION WORK

- The desirable characteristics of traction motor are discussed in previous section. In following section, different motors used for traction motor along with suitable features for traction are discussed.

4.18.1 D.C. Series Motor

1. Describe any four points that prove the suitability of D.C. series motor in traction system.
2. "DC series motor is used for traction purpose". Justify your answer with any six characteristics.
3. Give any four justifying features which make DC series motor suitable for traction work.

1. Suitable speed-torque characteristics:

- The speed-torque characteristics of DC series motor are steep as shown in Fig. 4.34.
- So for high torque (T_1), the speed is low (N_1) and for low torque (T_2) the speed is high (N_2). This is the exact requirement of traction system. At start, heavy mass of train is to be accelerated, so torque required is high and speed should be low (T_1N_1).

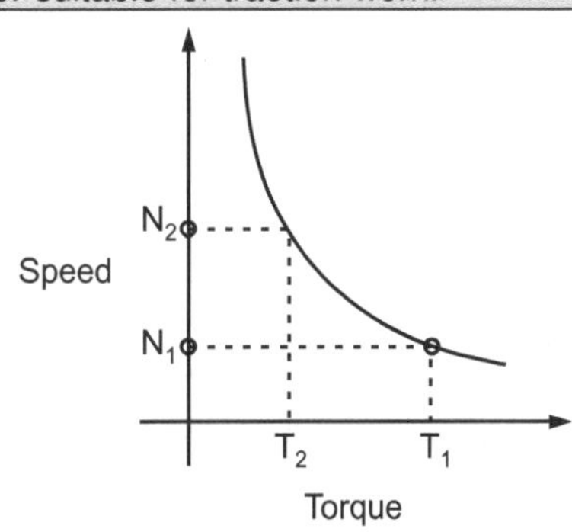

Fig. 4.34: Speed-Torque Characteristics of D.C. Series Motor

- As train gathers speed, momentum is developed, speed increases and torque requirement decreases (T_2N_2).

2. Parallel running:

- Due to steep speed - torque characteristics, the parallel running is effective. So even if there is difference in mechanical load on motors, there is no drastic difference in the currents taken by motors. So DC series motors give good performance for parallel running.

3. Voltage fluctuations:

- The torque produced is not much affected due to supply voltage fluctuations.

4. Temporary interruption of supply:

- The supply interruption takes place at tunnels, bridges, etc. and usually down gradient is present, so the motor can easily cope up with temporary interruption of supply.

5. Overload capacity:

- It can sustain overload effectively. There is not much increase in current even if load increases.

6. Self-relieving property:

- The output of motor is given in HP.

$$\text{Output} = \frac{2\pi NT}{60}$$

$$\therefore \quad \text{Output} \propto T \cdot N$$

or $$\text{HP} \propto T \cdot N$$

For series motor, $$N \propto \frac{V}{I_a}$$

$$\therefore \quad \text{HP} \propto T\frac{V}{I_a}$$

Also, $$T \propto I_a^2$$

$$\therefore \quad I_a \propto \sqrt{T}$$

$$\therefore \quad \text{HP} \propto T\frac{V}{\sqrt{T}}$$

$$\therefore \quad \text{HP} \propto \sqrt{T}\,V$$

From above equation we can say that for given increase in torque, there is not much increase in HP rating. This is called self-relieving property.

7. Speed Control:

- The speed of DC series motor can be controlled effectively by various methods like - series-parallel control, bridge transition method, solid state control etc.

8. Electric braking:

- There are some issues while carrying out regenerative braking for DC series motor. Additional components and changes in circuit connections are needed for regenerative braking.

9. Weight and Size:

- The DC series motor has moderate size. Due to commutator, the axial length is somewhat more.

10. Robustness:

- It is robust.

11. Efficiency:

- Efficiency is good at different loading conditions.

12. In case of D.C. series motors:

$$T \propto \phi \cdot I_a$$

$$T \propto I_a \cdot I_a \quad \text{... since } \phi \text{ is produced by } I_a$$

$$\therefore \quad \phi \propto I_a$$

$$\therefore \quad T \propto I_a^2 \quad \text{... upto magnetic saturation.}$$

For shunt motor, $T \propto \phi \cdot I_a$

$T \propto I_a$ only since ϕ is constant

Thus, for series motor, $T \propto I_a^2$

and for shunt motor $T \propto I_a$

If torque requirement increases, the series motor will draw less current as compared to shunt motor i.e. drain on electric current is less in case of series motor.

4.18.2 A.C. Series Motor

- The traction application requires steep speed-torque characteristics. The D.C. series motor fulfils this requirement and hence it can be used for traction work.
- The A.C. series motor characteristic is similar to D.C. series motor.

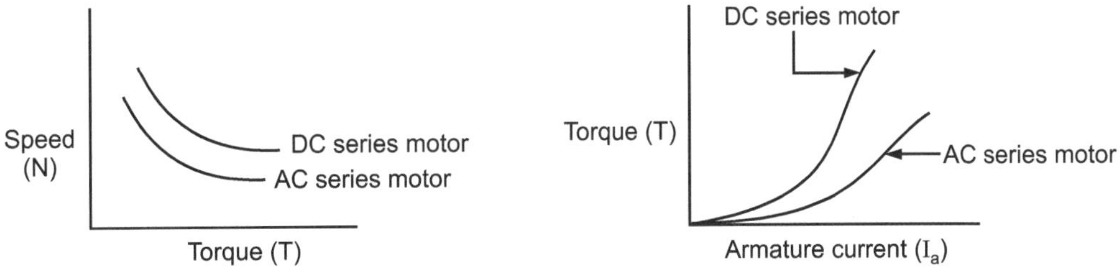

Fig. 4.35: Characteristics of A.C. series motor and D.C. series motor

- The construction of A.C. series motor is almost similar to D.C. series motor but some modifications are done as follows:
  (i) Compensating winding is used.
  (ii) The motor is operated at lower flux density to reduce iron losses.
  (iii) Air gap is very small to reduce magnetizing current.
  (iv) Field system (core) is laminated to reduce eddy current loss.
  (v) The motor is operated at lower frequency (25 Hz, 16.66 Hz). Due to this commutation is improved, also efficiency increases.
  (vi) Shunted compoles are used.
    - The size of A.C. series motor is almost twice of D.C. series motor of same capacity. The power factor of A.C. series motor is poor during starting, so starting torque is somewhat less. The maximum operating voltage is 400 volt.
    - In case of traction A.C. series motor is used for main line services where moderate acceleration is required.
    - It is not used for urban and suburban service where high acceleration is required.

4.18.3 Three-Phase Induction Motor

1. Explain the suitability of three-phase induction motor for traction service.

- Three-phase induction motor have the following features:
  1. Robustness.
  2. Less maintenance.
  3. Trouble free operation.
  4. Three-phase induction motor can be used for high voltage. The current requirement is less. Because P = V·I, hence for same power, voltage is high, current required is less.
  5. Automatic regenerative braking is possible.
  6. It can sustain overload.

Following are the drawbacks w.r.t. traction work:

1. Speed-torque characteristics is flat (Fig. 4.36).
2. The overhead conductor system is complicated because three conductors are needed.
3. If single-phase supply is tapped from OHE then single-phase to three-phase converter is needed in the locomotive.
4. The speed control method is complicated than D.C. motors.

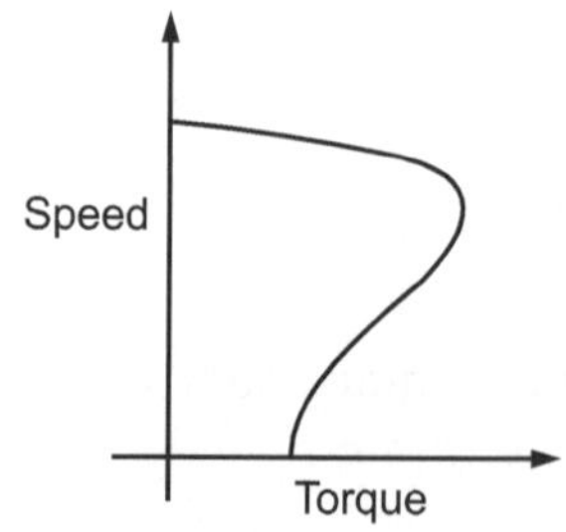

Fig. 4.36: Speed-torque characteristics of three-phase induction motor

Important Points

- **Drive:** In simple words, drive is a device that has to do required motion for useful work. Drive can be mechanical and electrical type.
- Examples of mechanical drive are: steam turbines, water turbines, wind mills, IC engines.
- **Electric drive:** The combination of an electric motor, energy transmitting shaft and control devices, for controlling the performance of motor (motion control) is known as electric drive.
- **Individual drive:** Individual machine is fitted with its own motor. e.g. drilling machine, lathe machine etc. Each operator has complete control on his machines.
- **Multimotor drive:** In multimotor drive, separate motors are used for operating different parts of same mechanism. e.g. in case of overhead crane, different motors are used for hoisting (up and down motion), long travel motion and cross travel motion.
- **Direct drive:** The motor shaft (driving member) is directly connected to the driven member. e.g. Motor-generator set in electrical lab.
- **Indirect drive:** The mechanical power available at source is transmitted through some medium like gear, chain etc.
- **Belt drive:** This is a simple and cheap drive. Following types of belts are used: Leather belt, nylon belt, terylene belt, rubberized cotton ply belt, hair belt.
- **Flat belt drive:** It is a long distance drive. The belt is generally made up of leather.
- **V-belt drive:** It is suitable for motors with rating upto 450 kW.
- **Gear drive:** It is a short centered positive drive. Proper alignment is very essential, otherwise, motor shaft may bend.
- **Noise:** Noiseless operation of motor is very much important in domestic theatre, hospital applications. The motor noise may cause vibrations in the surrounding structure and may cause resonance.
- **Continuous loading:** The load is constant for long duration of time.
- **Short time loading:** The load is present at regular intervals. In between the intervals motor is OFF.
- **Long time (Intermittant) loading:** It is that way similar to previous type but frequency of occurrence of load is more in intermittent load.
- **Continuous Operation with Short Time Load:** In continuous operation with short time load, there is no OFF time. But during short time load, motor runs at no-load condition.
- **Continuous Rating with Long Time Loading:** It is similar to intermittent operation except that OFF period is replaced by no-load operation of motor.
- **Load Equalization:** The process of smoothing out fluctuations in load is known as load equalization.
- **Braking:** Braking is a process of stopping motion by some means. For electrical motor, braking means stopping the rotation of motor.

Practice Questions

1. What are the factors governing selection of motors?
2. What are meant by plugging, rheostatic braking, regenerative braking?
3. Explain rheostatic braking (dynamic braking) for induction motor.
4. Explain the difficulties in regenerative braking of D.C. series motor.
5. Explain regenerative braking with:
   (i) Induction motor.
   (ii) D.C. shunt motor.

6. What is load equalization? How it is achieved?
7. State various systems of power transmission.
8. State the reasons where fly-wheel for load equalization is mounted on the system.
9. What are the control methods used in controlling D.C. motors? Explain with diagram rheostatic control method.
10. What is power transmission system? Write any two requirements of mechanical load.
11. Give types of transmission of drive.
12. What is electric drive? Explain in brief with a suitable sketch.
13. Give advantages of electric drive.
14. What is group drive?
15. What are the advantages and disadvantages of group drive?
16. What is individual drive? Give its advantages and disadvantages.
17. Write a short note on matching of speed torque characteristics of motor and load.
18. Explain plugging for induction motor.
19. Explain French method of regenerative braking for D.C. series motor.
20. Write short notes on:
    (i) Types of enclosures.
    (ii) Types of bearings.
    (iii) Noise level.
21. What are meant by continuous load, intermittent load?
22. Give standard ratings of motor as per IS.
23. What are the different methods of finding rating of motor for continuous duty?
24. Give the three factors which are considered while deciding the size and rating of the motor.
25. Write a full note on matching of speed torque characteristics.
26. Give the four types offloads.
27. What are the standard ratings as per I.S. of the motors.
28. Give the applications of any five motors used as drives.
29. Suggest the suitable motors for following applications:
    (i) Rolling mills
    (ii) Textile mills
    (iii) Mines
    (iv) Marine drive
    (v) Ships
    (vi) Refrigeration/Air-conditioning
30. Give the specifications of any one motor.
31. Drawing speed/torque characteristics explain D.C. series motor drive.
32. Drawing speed/torque characteristics explain A.C. series motor drive.
33. Drawing speed/torque characteristics explain three-phase I.M. series motor drive.

Problems with Answers

Numericals on Estimation of Continuous Rating:

1. A certain motor has to perform the following duty cycle:
    (i) 100 kW for 10 min.
    (ii) No load for 5 min.
    (iii) 50 kW for 8 min.
    (iv) No load for 4 min.

   The above duty cycle is repeated continuously. Assume heating is proportional to square of the current, calculate the suitable size of continuously rated motor. **[Ans.:** 16994 watt ≃ 17 kW]

2. A motor has to perform the following duty cycle:
   (i) 100 hp for 10 min.
   (ii) No load for 5 min.
   (iii) 60 HP for 8 min.
   (iv) No load for 4 min.

   The above duty cycle is repeated continuously. Determine a suitable size of a continuously rated motor.

   [**Note:** Problem is similar to example 2, the rating is mentioned in HP instead of kW]

   [**Ans.:** 21.45 HP, so a motor of 25 HP will be adequate]

3. A motor performs the following duty cycle:
   (i) 50 HP for 10 min.
   (ii) No load for 5 min.
   (iii) 60 HP for 5 min.
   (iv) No load for 10 min.

   The above cycle is repeated continuously. Determine the continuous rating of motor.

   [**Ans.:** 32.87 HP so a motor of 40 HP can be used]

4. The load cycle of a motor for 15 min. in driving some equipment is as follows:
   (i) 0 – 5 min. 30 HP.
   (ii) 5 – 9 min. No load.
   (iii) 9 – 12 min. 45 HP.
   (iv) 12 – 15 min. No load.

   The above load cycle is repeated indefinitely. Suggest a suitable size of continuously rated motor for these conditions.

   [**Note:** Time duration is to be considered carefully

   i.e. t_1 = 5 min. for 30 HP (5 – 0 = 0 min.)
   t_2 = 5 min. for No load (9 – 5 = 4 min.)
   t_3 = 3 min. for 45 HP (12 – 9 = 3 min.)
   t_4 = 3 min. for No load (15 – 12 = 3 min.)

   [**Ans.:** 26.55 HP so a 30 HP motor is suitable]

■■■

Chapter **5...**

Electric Traction

Syllabus

5.1 Requirements of an Ideal Traction System

5.2 Traction Systems

- Non-electric Traction Systems
- Electric Traction Systems: Straight Electric Traction, Its advantages and disadvantages. Diesel Electric Traction, Its advantages and disadvantages.

5.3 Systems of Track Electrification: DC System, Composite System – Single-phase to Three-phase System and Single-phase AC to DC System (Kando System). Advantages and Disadvantages of Single-phase 25 kV AC System over DC System.

5.4 Traction Mechanics

- Units Used in Traction Mechanics
- Types of Services
- Speed-Time Curve
- Simplified Speed-Time Curve (No Derivation)
- Average Speed and Schedule Speed
- Factors affecting the Schedule Speed
- Tractive Effort [No Derivation]
- Specific Energy Consumption [No Derivation]
- Factors affecting Specific Energy Consumption
- Coefficient of Adhesion

(Simple Numericals on Simplified Speed - Time Curves and Specific Energy Consumption)

5.5 Traction Motors

- Desirable Characteristics of Traction Motors, Special Features of Traction Motor, Only advantages, disadvantages and applications of following motors:
- Suitability of D.C. Series Motor for Traction
- Suitability of Three Phase Induction Motor for Traction, LIM, Single-phase A.C. Series Motor

5.6 Traction Motor Control:

- Requirements
- Traction Control of D.C. Locomotives and EMUs: Series Parallel Control Combined with Rheostatic Control, Transition from Series to Parallel Combination (Open Circuit Transition, Shunt Transition and Bridge Transition), Energy Efficiency and Limitations of Series Parallel cum Rheostatic Control, Chopper Control of Motors in D.C. Traction Systems
- Traction Control System of A.C. Locomotives: Tap Changer, Stepless Voltage Control through the use of Thyristors, PWM

5.7 Braking:

- Requirements of a Braking System.
- Mechanical Braking: Vacuum Braking, Compressed Air Braking, Hand Brake for Parking.
- Electric Braking: Rheostatic Braking and Regenerative Braking. (No Derivation and No Numericals)
- Sequence of Braking
- Dead Man's Handle

Objectives

- Recommend relevant traction system for the given application with justification.
- Select the relevant track electrification system for the specified traction services with justification.
- Differentiate salient features between the given types of track electrification system.
- Draw the speed-time curve for the specified electric traction application.
- Differentiate between the given types of traction services based on the given criteria.
- Determine average and schedule speed for the given traction services.
- Describe the procedure to maintain the given electric traction system.

INTRODUCTION

- Railways in public service, carrying passengers and goods, have existed for over 170 years and for longer than that in mining and industrial works.
- In early days, railways were laid down for horse drawn trains of wagons in coal mines. The track (hauling way) was made-up of stone slabs or timber wood. As the industrial revolution progressed, these were replaced with cast iron/wrought iron plates. In early nineteenth century steel rails were developed which enabled use of flanged wheels. Steam locomotive was in use for many years during 18th century and early decades of 19th century.
- Today there are two main sources of power for locomotive – oil and electricity. The steam locomotive is not in use due to several drawbacks.
- **Traction:** The act of drawing or pulling by power is known as traction.
- There are number of systems of producing this power like steam engine drive, diesel engine drive, electric drive, battery drive, hybrid drive etc.

5.1 REQUIREMENTS OF AN IDEAL TRACTION SYSTEM

1. What are the requirements of traction system?

(i) The starting torque should be high because initially heavy mass is to be accelerated rapidly.

(ii) It should be a self-contained unit which can run on any route.

(iii) It should be compact.

(iv) It should be capable to withstand temporary overloads.

(v) It should cause minimum wear and tear of track.

(vi) Its speed control should be easy and comfortable.

(vii) Initial cost and running cost should be low or reasonable.

(viii) Efficiency should be high.

(ix) Braking system should be effective. It should be smooth. It should cause minimum wear and tear of brake shoes. If possible regenerative breaking should be employed.

(x) It should not pollute the environment.

(xi) It should not produce interference with neighbouring communication lines.

(xii) Speed should be high.

5.2 TRACTION SYSTEMS USED IN INDIA

- The traction system can be classified into two categories:

  (i) Non-electric traction system: This traction system does not use electricity. e.g. steam engine drive, direct internal combustion engine drive (direct IC engine drive).

  (ii) Electric traction system: This traction system uses electricity partly or completely. e.g. straight electric drive, battery electric drive, diesel electric drive.

5.2.1 Non-electric Traction System

1. Steam Engine Drive:

Reciprocating steam engine is used to get necessary motive power.

Advantages:

(i) Simplicity of design.
(ii) Operational dependability.
(iii) Simplified maintenance.
(iv) Easy speed control.
(v) Not tied to any particular work.
(vi) No interference with telecommunication lines.

Disadvantages:

(i) Low thermal efficiency (η).
(ii) Need of adequate supplies of feed water at regular intervals.
(iii) Periodic stops are necessary to clean ash pan.
(iv) Cannot be put into service/operation immediately.
(v) Coefficient of adhesion is less so difficulties while negotiating curves.
(vi) Power to weight ratio is low.
(vii) Centre of gravity is high, this limits the speed while negotiating curves.
(viii) Power to weight ratio is low.
(ix) Centre of gravity is high this limits the speed while negotiating curves.
(x) Considerable wear and tear of track.
(xi) Not clean due to coat dust, smoke.
(xII) Not sultable for underground rallway system due to smoke.
(xiii) Heavy repair cost.
(xiv) Low acceleration and retardation rate 0.6 to 0.8 kmphps.
(xv) Low starting torque.

2. Direct Internal Combustion Engine Drive or Diesel Engine Drive:

Advantages:

(i) Self-contained unit.
(ii) Low initial cost.
(iii) Simple speed control.
(iv) Simple braking system.
(v) It can run on different road/track conditions.
(vi) Higher efficiency than steam locomotive.

Disadvantages:

(i) For our country oil is to be imported, so it is not good for economy.
(ii) It cannot be overloaded.
(iii) Maintenance and running cost is high.
(iv) Life is comparatively shorter.

5.2.2 Electric Traction System

1. Enlist eight advantages of electric traction system.
2. State advantages of A.C. traction system.
3. State any four advantages and disadvantages of electric traction.

5.2.2.1 Straight Electric Traction System

- This system uses D.C. series motor or single phase A.C. series motor or three phase induction motor. The electrical input is directly fed to the motor and torque is generated.

Advantages:

(i) Cleanliness: Absence of smoke, ash etc. So suitable for underground railway also.

(ii) High starting torque, high acceleration and retardation 1.5 to 2.5 kmphps.

(iii) High traffic holding capacity suitable for urban area.

(iv) Flexibility of operation. This gives better traffic handling.

(v) Maintenance cost is 50% of steam locomotive.

(vi) Starting time is very very less.

(vii) Regenerative braking can be applied about 80% of electric energy is fedback to supply lines. Also due to this, less requirement of mechanical braking. So less maintenance, wear and tear of brake shoes, wheels, rails.

(viii) Centre of gravity is lower than steam locomotive, so curvatures can be negotiated at higher speeds.

(ix) Power to weight ratio is high.

(x) Absence of unbalanced forces, so coefficient of adhesion is more.

(xi) Saving is high grade coal used for metallurgical purpose.

(xii) Railway electrification encourages rural electrification.

(xiii) Operational convenience on severe gradients.

Disadvantages:

(i) High capital cost due to overhead equipments, so electric traction becomes uneconomical unless heavy traffic is to be handled.

(ii) Power failure of few minutes causes distortion in traffic.

(iii) Traction is tied to electric routes.

(iv) Provision of negative booster is necessary, otherwise it causes corrosion of underground pipes.

(v) Causes disturbance in neighbouring communication lines.

5.2.2.2 Diesel Electric (Internal Combustion Engine) Drive

- A diesel engine is used which is coupled to a D.C. generator. The speed of diesel engine is maintained constant. The D.C. generator supplies power to traction motors which are connected to driving axles.

Advantages:

(i) High acceleration and braking retardation.

(ii) It is a self-contained unit and therefore it is not tied to any particular route.

(iii) No modification in existing track is required.

(iv) Power loss in speed control is low.

(v) Starting time is very less, hence can be put into service immediately.

(vi) Passenger capacity is more than steam locomotive.

(vii) Initial cost is low as compared to electric traction because it does not require overhead equipments.

(viii) It can operate at higher sustained speeds with less damage to the track.

Disadvantages:

(i) The diesel engine cannot be overloaded.

(ii) Life of diesel engine is comparative shorter.

(iii) Separate cooling arrangement is necessary for diesel engine.

(iv) The dead weight of this type of locomotive is more.

(v) Running cost and maintenance cost is high.

(vi) Regenerative braking is not possible.

(vii) For our country, the oil is to be imported which is not good for our economy.

1. Battery Drive:

- These came into existence after 1830 A.D. The vehicles carry secondary batteries which supply power to driving motors (D.C. series motor). These are suitable for daily run of 40 to 60 km and are generally used in industrial works for carrying goods. These are also very popular in gassy mines. The motors and batteries used in that case are flame proof.
- The speed control is very easy and effective. Maintenance is easy and cheap.
- However the batteries need charging at regular intervals.

2. Hybrid Drive:

- This uses power partly from overhead lines and partly from batteries. The system is generally used in mines where long tunnels are to be excavated. The locomotive takes power from overhead lines for most of the journey and at the end, Batteries are switched ON so now locomotive is powered by batteries. The charging of batteries is done through overhead lines.

3. Tramways:

- It is also known as tram. This vehicle runs on a track laid on street. This is operated in single unit and it is driven by electric motors (D.C. series motors).
- 600 volt D.C. supply is tapped from overhead conductor with a sliding contact arrangement and it is fed to motors in the locomotive. The track laid on street forms return path for current. Earlier, the tramways were powered by batteries but it was expensive and inefficient way. The tramways cannot manoeuvre in conjusted areas.

4. Trolley Bus:

- It is also known as trackless trolley. This vehicle runs with rubber wheels or tyres and the motors are powered by electricity drawn from two overhead lines or wires. It can run on normal roads. The supply used in 600 volt D.C. The trolley bus can effectively manoeuvre in conjusted areas.
- The trolley bus is also provided with secondary batteries. These batteries are used in emergency i.e. if overhead supply fails.

5.2.3 Factors Governing Selection of Electric Drive

1. State the factors governing selection of electric drive.

- The success of electric drive for any purpose depends on the correct choice of driving motor for particular application.
- While selecting motor for a particular application, following points should be considered:
  1. Nature of electric supply.
  2. Nature of drives:
     (i) Group drive
     (ii) Individual drive
     (iii) Multimotor drive.
  3. Nature of load.
  4. Electrical characteristics: This includes:
     (i) Starting characteristics.
     (ii) Operating or running characteristics.
     (iii) Speed control.
     (iv) Braking.

5. Size and rating of motor: This includes:
   (i) Continuous rating.
   (ii) Intermittant or variable rating.
   (iii) Load capacity and pull-out torque.
6. Mechanical factors: These include:
   (i) Types of enclosures
   (ii) Types of bearings
   (iii) Transmission of drive
   (iv) Noise level.
7. Cost:
   (i) Capital cost
   (ii) Variable cost

- These factors will be discussed in brief.

5.2.3.1 Comparison between various Traction Systems

Sr. No.	Point of comparison	Steam engine drive	Diesel engine drive	Electric drive
1.	Starting time	High	Less	Least
2.	Design	Complicated	Less complicated	Simple
3.	Electrified route	Not required	Not required	Required
4.	Control	Complicated	Less complicated	Easy
5.	Acceleration	Less	More	More
6.	Maintenance required	More	Less	Least
7.	Cleanliness	Less	More	Most clean
8.	Centre of gravity	High	Low	Low
9.	Power to weight ratio	Low	Moderate	High
10.	Speed control	Complex	Easy	Easy
11.	Braking system	Complex	Less complex	Easy
12.	Efficiency	Poor	Better	Good
13.	Self-contained unit	No	Yes	No
14.	Interference with communication lines	No	No	Yes
15.	Power obtained from	Steam	Diesel	Electricity

5.3 SYSTEMS OF TRACK ELECTRIFICATION

1. Write the different systems of track electrification.

1. D.C. track electrification.
2. Single-phase low frequency system (25 kV system).
3. 3ϕ A.C. system.
4. Composite system.

5.3.1 D.C. Track Electrification

1. Explain which system is preferred for traction work in India. Why?
2. State different systems of track electrification. State advantages of single-phase 25 kV A.C. system over D.C. system.
3. Explain following systems of track electrification:
   1. Single-phase AC
   2. Three-phase AC

- It uses 600 V to 750 V D.C. for urban services and 1500 V to 3000 V D.C. for main line service. Current collection system may be of third rail type or overhead conductor type. The substation consists of a transformer and mercury arc rectifier or rotary converters. These substations are usually automatic and remote controlled.
- Main advantage of using D.C. is due to D.C. series motors. These motors have the exact characteristics which are required by traction system. But initial cost of D.C. substation is very high, also they are to be placed at shorter spans. The provision of negative booster is a must.

5.3.2 Single-Phase Low Frequency A.C. (25 kV System)

- If low frequency 1ϕ A.C. supply is used for A.C. series motor, it improves its commutation properties, p.f. and η. High supply voltage (25 kV) reduces the current and low frequency reduces the impedance drop, so substations can be spaced at longer spans. The substation consists of trans-former and frequency converter equipment. The frequency converter equipments consist of a synchronous motor (3ϕ) and alternator set. The locomotive carries a stepdown transformer. The speed of A.C. series motor is varied by the following arrangements.

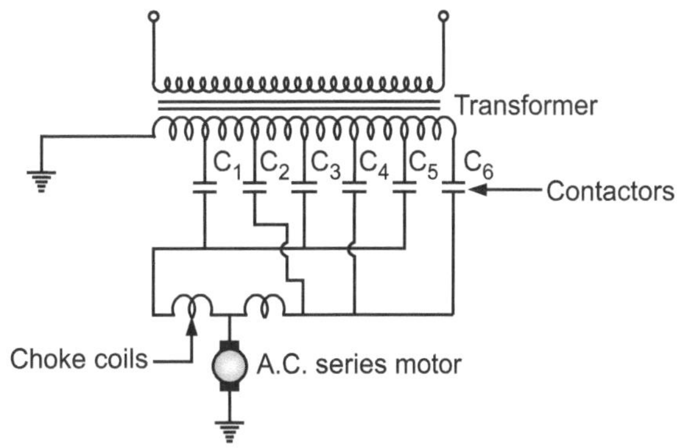

Fig. 5.1: Speed variation system

- Contactors C_1 and C_2 are closed so current flows corresponding to the tappings short circuited. The magnitude of this current is limited by choke coils which produce flux in opposite direction to that of short circuited coil. This reduces reactance effective. Then C_1 is opened and C_3 is closed. So voltage given to motor is corresponding to tapping C_2 and C_3. The voltage given to motor is mean of the voltage of corresponding two tappings. Thus by varying applied voltage, speed can be controlled.
- Advantage is - less number of substations are required.
- Disadvantage is - frequency conversion equipment is required.

5.3.3 Three-Phase Low Frequency System

- Two overhead conductors and one rail form 3ϕ supply for I.M. No need of transformer in locomotive. High supply voltage reduces the amount of current needed. Also I.M. has property of automatic regenerative braking. But two overhead conductor system is complicated while crossing the junctions. Also characteristic of I.M. is not suited for traction purpose.

5.3.4 Composite System

(i) Kando system.

(ii) Single phase A.C. to D.C. system.

(i) Kando System:

1ϕ 50 Hz supply is taken from overhead equipment and is converted into 3ϕ supply by means of phase converter equipment carried on locomotive. This 3ϕ supply is fed to 3ϕ I.M. So complication in overhead equipment is avoided as only single conductor is used. If I.M. is fed with low frequency supply with the help of inverter, its performance is improved. At low frequency it develops high starting torque without excessive current. Speed control is easily obtained by varying frequency.

(ii) Single Phase A.C. to D.C. System:

This uses 25 kV industrial frequency A.C. supply collected from overhead equipment and stepped down by a transformer in the locomotive. This supply is then converted in D.C. supply and used for traction purpose.

Advantages:

The advantages of this 25 kV A.C. over D.C. system are:

(i) Due to high voltage, current is less, therefore, cross-sectional area of conductor is small, so weight of overhead equipment reduces. Therefore cost of foundations supporting structures also reduces.

(ii) Erection and maintenance of overhead equipment is easier.

(iii) Saving in substations.

(iv) Large spacing between substations.

(v) Substations used are simple and cheap.

(vi) Capacity of A.C. substations is higher than D.C. substations.

(vii) Coefficient of adhesion is more in this system.

(viii) Starting 'η' of A.C. locomotive is higher than that of D.C. locomotive.

(ix) The kW demand during starting is less in case of A.C. locomotive than D.C. locomotive.

Disadvantages:

(i) Interference with communication lines.

(ii) Current and voltage imbalance is produced.

5.3.5 Comparison between Single-Phase 25 kV A.C. and D.C. Track Electrification

1. Compare A.C. and D.C. track electrification.

Sr. No.	Point of comparison	1-ph 25 kV A.C. track electrification	D.C. track electrification
1.	Current required	Less	More
2.	Conductor size	Small	More than A.C. system
3.	Cost of foundation and structure	Less	More
4.	Substations required	Less	More
5.	Spacing between substations	More	Less
6.	Capacity of substation	High	Low
7.	Starting efficiency	Better	Comparatively less
8.	Interference with communication lines	More	Very less

Sr. No.	Point of comparison	Pure A.C. system	Pure D.C. system
1.	Torque	Less starting and running torque	More starting and running torque
2.	A.C. series motors cost	Cost more	Cost lesser
3.	Weight of motor	Less in weight	More in weight
4.	Efficiency	Lesser	More
5.	Maintenance	More	Lesser
6.	Regenerative braking	Less efficient	More efficient
7.	Interference with communication lines	More	Not at all
8.	Number of sub-stations	Lesser	More
9.	Distribution system	Lighter and less costly	Heavier and more costly

5.4 TRACTION MOTORS

5.4.1 Desirable Characteristics of Traction Motors

1. Write six desirable characteristics of traction motor.

1. **Suitable speed-torque characteristics:** The torque required at start or tractive effort required at start is high or maximum. During the constant speed coasting, the tractive effort requirement is not high because kinetic energy is developed and the tractive effort required is for overcoming the track resistance and gravity component.

   Thus the requirement is that motor should develop very high starting torque which should fall off at high speeds.
2. **Parallel running:** In traction work, usually more than one motor operate at a time. Hence, these motors should be capable of operating in parallel. There occurs a small difference in rotational speed of various motors because of uneven wear and tear of wheels. This should not produce wide variations in torques developed and current drawn by various motors.
3. **Voltage fluctuations:** In traction work, on account of heavy current in rush at starting, considerable voltage fluctuation of supply line is a normal feature. Traction motor should be capable of withstanding voltage fluctuations without adverse effect on their performance.
4. **Temporary interruption of supply:** This occurs when section insulators and cross overs are crossed with the controller ON. Motor should withstand these fluctuations without heavy in rush of current.
5. **Overload capacity:** The traction motors are subjected to heavy loads which cause large rush of current. This may produce large armature reaction and bad commutation. The arcing produced on commutator surface may exceed over the whole periphery and flashover may cocur.

   This is to be avoided at all costs. Thus, motors should be capable of taking heavy loads without flashover.
6. **Self-relieving property:**

   $T \propto \frac{1}{N}$ and T.N. $\rightarrow$ constant

   H.P. or O/P $\propto$ T.N. $\rightarrow$ constant

   This gives motor a self-protective property against excessive overload.
7. **Amenability to speed control:** Motor should be amenable to simple speed control methods.
8. **Amenability to electric braking:** Motor should be amenable to easy and simple methods of rheostatic and/or regenerative braking.
9. **Limitation of weight and size:** The traction motors should have high power to weight ratio. The physical size of motor depends on type of insulation employed. Traction motors are wound with class H insulation.
10. **Robustness:** Motors are subjected to severe conditions. These motors should be robust enough to withstand such severe conditions. Motors should be further provided with mechanical protection so as to prevent dirt, water, mud etc.
11. **High efficiency:** Motor should have high efficiency.

5.4.2 Various Types of Traction Motors

5.4.2.1 D.C. Series Motor

1. D.C. series motor is used for traction purpose. Justify the answer giving its characteristics.

- The D.C. series motor has features that make it suitable for traction work. These are as follows:

1. D.C. series motors develop high starting torque which is required for traction purpose.
2. The high torque is produced at low speeds and low torque is produced at high speeds.
3. D.C. series motor is ideally suitable for parallel running.

   Suppose two motors are connected to different driving axles and these are operating simultaneously. Ideally when more than one motors are operating, they should rotate at same speed with equal load on them i.e. torque should be same.

   But in actual practice there is uneven wear and tear of individual wheels, the wheels are connected to axles and axles are driven by motors. Hence, motors also run at different speeds which is not desired.

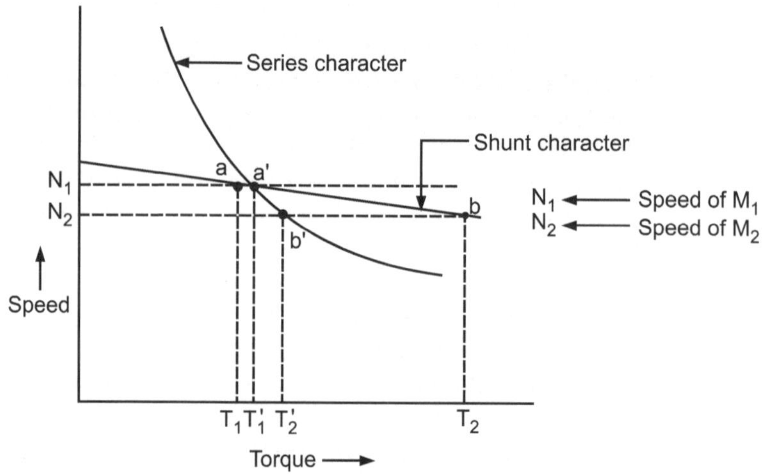

Fig. 5.2

   Now consider that the motors M_1 and M_2 are operating simultaneously and their speeds are N_1 and N_2 i.e. there is difference in their speeds.

   If motors are having shunt characteristics, with motor M_1 (with worn wheels) running at a higher speed N_1 and motor M_2 running at a lower speed of N_2. From the characteristics it is seen that motor M_1 exerts a torque T_1 which is approximately half of T_2 developed by M_2.

   But it is desired that all motors of train should share load equally i.e. torque developed by them should be approximately same even if small speed variation is there. Thus, with shunt characteristics desired performance requirements are not met.

   But with series characteristics, however, it can be seen that small difference in the speeds has a very little effect on relative torques (T_1' and T_2'). Hence, these motors can meet the performance requirement effectively.

   Induction motors have characteristics similar to shunt characteristics. These motors are also used for traction purpose. But to meet above requirement (i.e. variation in speed should not cause drastic difference in torque), the driving axles are coupled by a connecting rod so that there is no possibility of difference in speed.

4. In case of D.C. series motors:

$$T \propto \phi \cdot I_a$$

$$T \propto I_a \cdot I_a \text{ ...since, } \phi \text{ is produced by } I_a$$

$\therefore$ $$\phi \propto I_a$$

$$T \propto I_a^2 \text{ ... upto magnetic saturation.}$$

   For shunt motor,

$$T \propto \phi \cdot I_a$$

$$T \propto I_a \text{ only since } \phi \text{ is constant}$$

   Thus, for series motor, $T \propto I_a^2$

   and for shunt motor, $T \propto I_a$

   If torque requirement increases, the series motor will draw less current compared to shunt motor i.e. drain in electric current is less in case of series motor.

5. Commutating properties of series motor are good. At heavy loads, the armature speed decreases due to which commutation is better and sparking is less.

6. Speed of series motor can be controlled easily and effectively.
7. The torque is unaffected by variations in supply voltage.
8. The series motor speed-torque characteristics gives stability in control.
9. The output of motor is given in HP.

$$O/P = \frac{2\pi NT}{60}$$

$\therefore \quad O/P \propto T \cdot N$

or $\quad HP \propto T \cdot N$

For series motor, $\quad N \propto \frac{V}{I_a}$

$\therefore \quad HP \propto T \frac{V}{I_a}$

Also, $\quad T \propto I_a^2$

$\therefore \quad I_a \propto \sqrt{T}$

$\therefore \quad HP \propto T \frac{V}{\sqrt{T}}$

$\therefore \quad \mathbf{HP \propto \sqrt{T}\, V}$

- From above equation we can say that for given increase in torque there is not much increase in HP rating. This is called self-relieving property.

  Following are negative points about series motor w.r.t. traction work:

  (i) Regenerative braking is difficult to achieve. Special arrangements are essential for it.

  (ii) Due to temporary interruption of supply, the heavy in rush current is more.

5.4.3 Single-Phase A.C. Series Motor

- The characteristics are almost similar to D.C. series motor. These can operate with 400 volt A.C. The power factor of this motor is poor, therefore, the starting torque is poor. High acceleration is difficult to obtain. The performance of this motor can be improved by using low frequency supply $\left(16\frac{2}{3}\text{ Hz, 25 Hz}\right)$. If motors have to be used for 50 Hz, then their magnetic design is to be improved.

5.4.4 Block Diagram of A.C. Locomotive and Function of Each Part

1. Draw a neat labelled block diagram of A.C. locomotive and state function of each part.

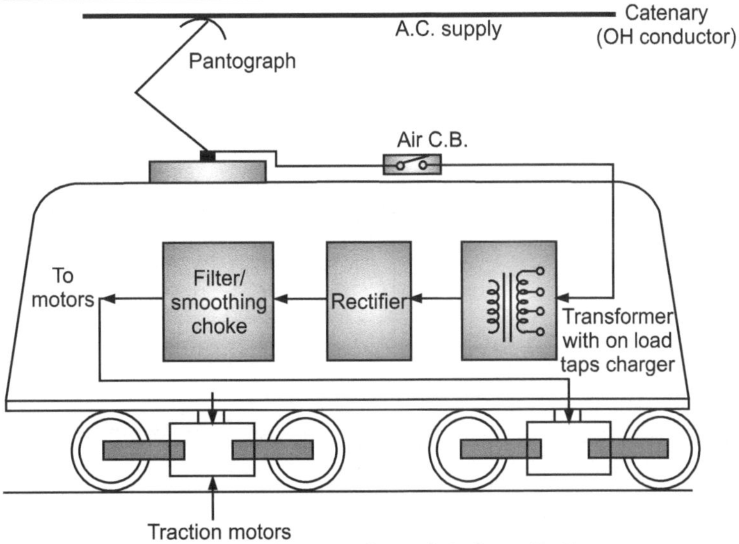

Fig. 5.3: Conversion of A.C. to D.C.

- The block diagram of electric locomotive is as shown in Fig. 5.3. The pantograph collects the current through sliding contact with the overhead conductor or catenary. Air circuit breaker is provided at the top of the locomotive which disconnects the supply in case of over voltage. Step-down transformer with tap charger on HV side is provided in the locomotive. The taps can be charged as per the speed requirement of the traction motor. The traction motor is D.C. so the D.C. supply is provided through rectifier unit. A filter or smoothing choke is also provided so that pure D.C. supply can be fed to the motors. The filter reduces the undesirable effects such as bad commutation and heating (which are produced because of alternating component of rectified A.C.).

5.4.5 Classification of Traction Services and their Comparison

<table>
<tr><th>Sr. No.</th><th>Parameters of comparison</th><th>Urban service</th><th>Suburban service</th><th>Main line service</th></tr>
<tr><td>1.</td><td>Distance between the stations</td><td>1 km</td><td>1 to 8 km</td><td>More than 10 km</td></tr>
<tr><td>2.</td><td>Acceleration in kmphps</td><td>1.5 to 4 km</td><td>1.5 to 4 km</td><td>0.6 to 0.8 km</td></tr>
<tr><td>3.</td><td>Retardation in kmphps</td><td>3 to 4 km</td><td>3 to 4 km</td><td>1.5 km</td></tr>
<tr><td>4.</td><td>Speed-time curve</td><td colspan="3">

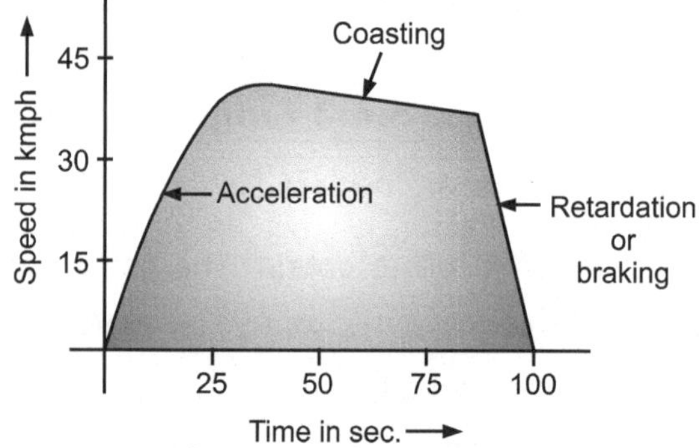

Fig. 5.4: Urban Service

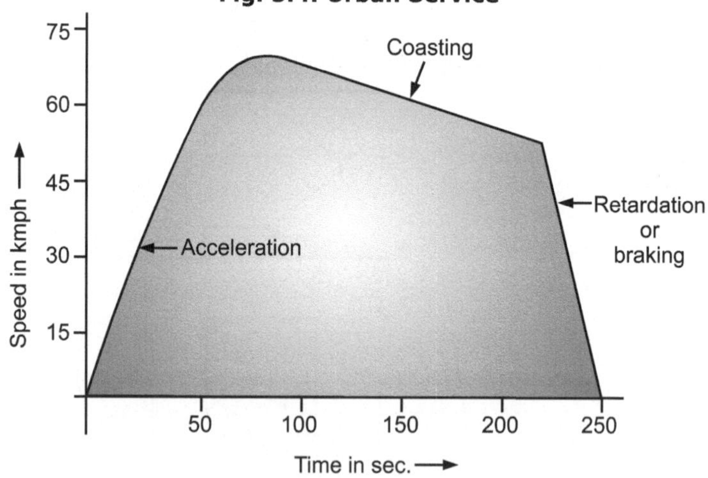

Fig. 5.5: Suburban Service

</td></tr>
</table>

contd. ...

		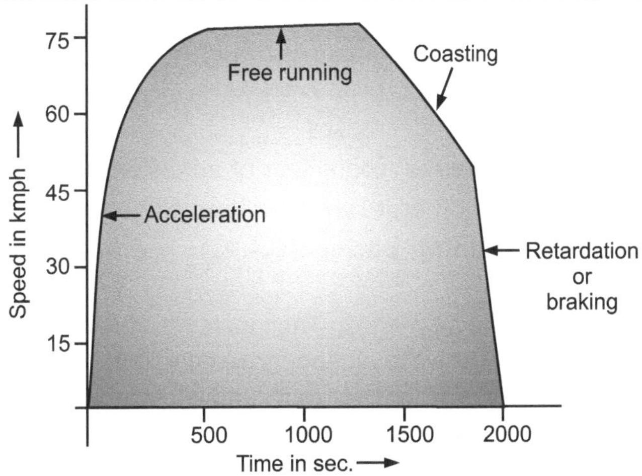 **Fig. 5.6: Main Line Service**		
5.	Comment on speed-time curve	No free running period, coasting period is small.	No free running period, coasting period is long.	Long free running and coasting period and, acceleration and braking periods are compara-tively large

5.4.6 Speed-Time Curve

1. Draw speed-time curve. Show and list various time periods associated with it.

- The movement of trains and their energy consumption is most conveniently studied by means of speed-time and speed-distance curves. However, speed-time curve is generally more useful because the slope of the curve represents the acceleration and area under the curve represents the distance travelled. A typical speed-time curve of train running on a main line is shown in Fig. 5.7.

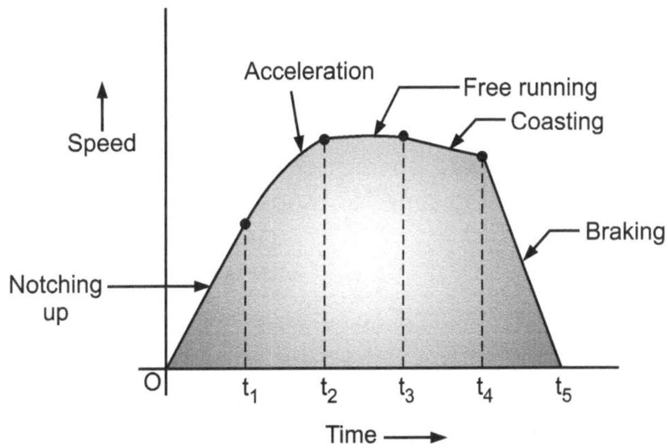

Fig. 5.7: Speed-time curve

(i) Notching up period ($0 - t_1$):

Starting resistance is being cut out during this period, so that the current and therefore the tractive effort are approximately constant. Because of notching action of starting resistance the current fluctuates between certain maximum and minimum limits, therefore, the motor and the tractive effort also fluctuate. However average tractive effort remains same and hence acceleration remains constant. The speed-time curve therefore is a straight line ($0 - t_1$).

(ii) Acceleration of speed curve (t_1 to t_2):

After the starting resistance has been cut out (at point t_1), the train continues to accelerate. The torque gradually falls as the speed increases in accordance with the torque-speed characteristics of the motors.

It is observed that acceleration is constant during ($0 - t_1$) and it decreases during ($t_1 - t_2$).

This is because of the following reasons:

(i) Speed-torque characteristics of motor (as N increases, torque decreases).

(ii) As N increases, train resistance increases very much. Hence, the tractive effort utilized for acceleration decreases.

(iii) As N increases, coefficient of adhesion decreases.

Therefore, motor torque utilized to generate tractive effort decreases.

(iv) **Free running period ($t_2 - t_3$):** The train continues to run at the constant speed reached at point t_2.

(v) **Coasting (t_3 – t_4):** The power is cut off and the train is allowed to run under its own momentum, the speed gradually falls due to train resistance, friction windage etc. So during coasting, the train runs without any input power.

(vi) **Braking ($t_4 - t_5$):** At the end of coasting brakes are applied and the train is brought to rest at the end of run.

- The acceleration and free running periods are governed by the speed torque characteristics of motors. So these parameters should be available for calculations related to curve portion ($t_1 - t_3$).
- The coasting and braking are governed by the train resistance and allowable braking retardation respectively.
- The actual shape of speed-time curve will differ from the above speed-time curve depending upon various factors such as type of service, gradients of different length etc.

5.4.7 Important Definitions

1. State difference between actual speed and schedule speed.

1. **Maximum or Crest speed:** It is the maximum speed attended by a train during a run.
2. **Average speed:** It is the mean speed maintained by a train during a run. It is given by the following expression:

$$\text{Average speed} = \frac{\text{Distance between stops or stations}}{\text{Actual time of run}}$$

3. **Schedule speed:** It is the ratio of the distance between the stops and the total time taken including time for stops to cover the distance.

$$\text{Schedule speed} = \frac{\text{Distance between the stops}}{(\text{Actual time of run}) + (\text{Stop time})}$$

1. State factors affecting schedule speed.

Factors affecting schedule speed:

- The following factors affect the schedule speed of a train:

(i) Crest speed: High crest speed results into increase in schedule speed. This is because if the acceleration and retardation are fixed, then for a constant distance, the running time will decrease if crest speed is high. And if actual run time decreases, then schedule speed increases.

(ii) Acceleration: If distance between the two stations and crest speed are constant, then increase in acceleration will reduce the running time of train. Thus increase in acceleration will increase the schedule speed. This is considerable if distance between the stations is small.

(iii) Braking retardation: Similar to acceleration, the effect of braking retardation is to increase the schedule speed. This is considerable if distance between the stations is small.

(iv) Duration of stop: If duration of stop is more, total running time will be more and schedule speed will be low.

5.4.8 Simplified Speed-Time Curve

- The speed-time curves for different services are having different nature. For making calculations easy, these speed-time curves are drawn in a simple way. Some errors will be caused because of change in shape but it is negligibly small.
- For main line service trapezoidal speed-time curve is closer approximation. Refer Fig. 5.8.

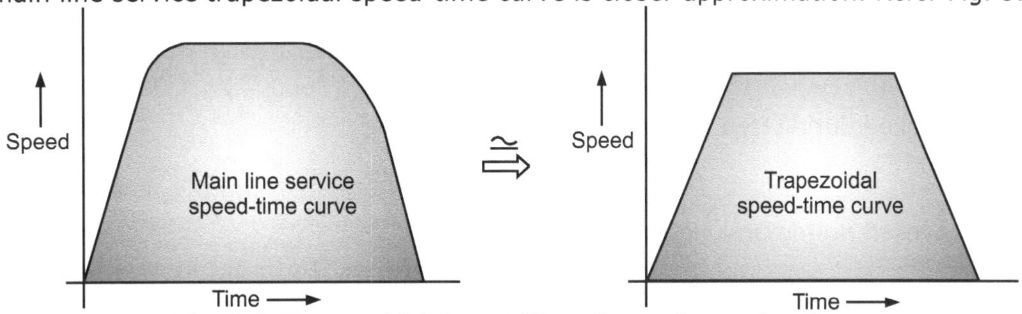

Fig. 5.8: Trapezoidal Speed-Time Curve Approximation

- For urban and suburban service, quadrilateral speed-time curve is a closer approximation.
- Refer Fig. 5.9.

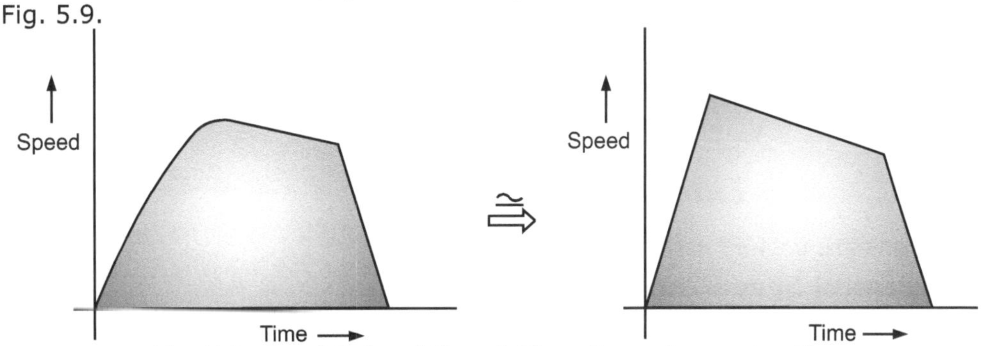

Fig. 5.9: Quadrilateral Speed-Time Curve Approximation

5.4.8.1 Trapezoidal Speed-Time Curve

Let, V_m = Crest speed or maximum speed attained in kmph

t_1 = Acceleration time

t_2 = Free running time

t_3 = Retardation time

α = Acceleration in kmphps

β = Retardation in kmphps

S = Total distance of run in km

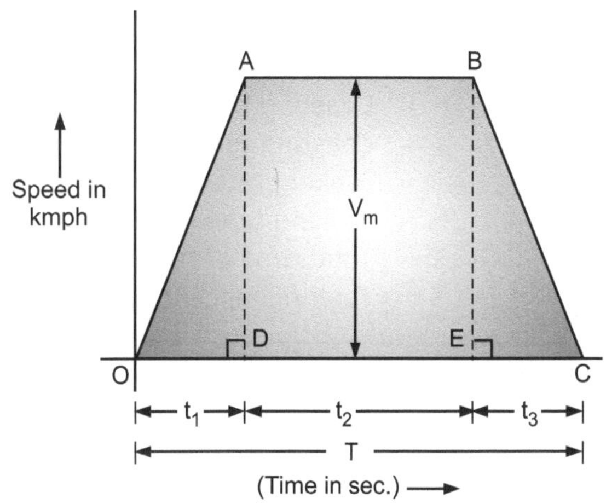

Fig. 5.10

$$\text{Acceleration} = \frac{\text{Change in velocity during acceleration}}{\text{Time for acceleration}}$$

$$\therefore \quad \alpha = \frac{V_m}{t_1}$$

$$\therefore \quad t_1 = \frac{V_m}{\alpha}$$

$$\text{Retardation} = \frac{\text{Change in velocity during retardation}}{\text{Time for retardation}}$$

$$\beta = \frac{V_m}{t_3}$$

$$\therefore \quad t_3 = \frac{V_m}{\beta}$$

Time for free running $= T - (t_1 + t_3)$

$$t_2 = T - \left(\frac{V_m}{\alpha} + \frac{V_m}{\beta}\right)$$

The area under curve gives total distance travelled.

So Area of (OABC) = S

Distance travelled during acceleration (in km)

$$= \text{Area of } \Delta \text{ AOD} = \frac{1}{2} \times t_1 \times V_m \times \frac{1}{3600} = \frac{1}{2} V_m \frac{t_1}{3600} \text{ ... km}$$

Distance travelled during free run (in km)

$$= \text{Area of rectangle DABE} = V_m \times t_2 \times \frac{1}{3600} = V_m \frac{t_2}{3600}$$

Distance travelled during braking (in km)

$$= \text{Area of } \Delta\text{CBE} = \frac{1}{2} \times t_3 \times V_m \times \frac{1}{3600} = \frac{1}{2} V_m \frac{t_3}{3600}$$

Following equation can be obtained after substitution (derivation is not expected as mentioned in the syllabus).

$$\frac{1}{\alpha} + \frac{1}{\beta} = \frac{7200 \text{ S}}{V_m^2}\left[\frac{V_m}{V_a} - 1\right]$$

$$V_m = \frac{T - \sqrt{T^2 - 14400 \text{ KS}}}{2K}$$

where, $$K = \frac{1}{2\alpha} + \frac{1}{2\beta}$$

Above equations are useful for solving the numericals.

5.4.8.2 Quadrilateral Speed-Time Curve

Let, V_1 = Maximum speed at the end of acceleration

t_1 = Acceleration time in seconds

α = Acceleration in kmphps

V_2 = Maximum speed at the end of coasting in kmphps

t_2 = Coasting time in seconds

β_c = Coasting retardation in kmphps

β = Braking retardation in kmphps

t_3 = Braking time in seconds

T = Total time of run in sec.

S = Total distance travelled in km

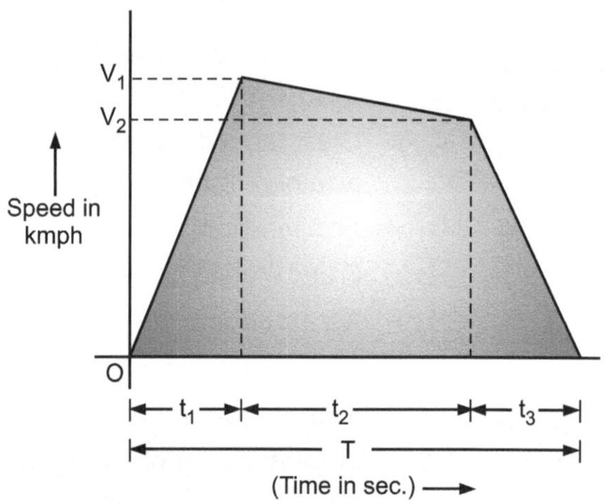

Fig. 5.11

$$\text{Acceleration} = \frac{\text{Change in velocity during acceleration}}{\text{Time for acceleration}}$$

$$\alpha = \frac{V_1}{t_1}$$

$$\therefore \quad t_1 = \frac{V_1}{\alpha}$$

$$\text{Coasting retardation} = \frac{\text{Change in velocity during coasting}}{\text{Time for coasting}}$$

$$\beta_c = \frac{V_1 - V_2}{t_2}$$

$$t_2 = \frac{V_1 - V_2}{\beta_c}$$

$$\text{Braking retardation} = \frac{\text{Change in velocity during braking}}{\text{Time for braking}}$$

$$\beta_2 = \frac{V_2}{t_3}$$

$$\therefore \quad t_3 = \frac{V_L}{\beta}$$

- Rearranging the terms we get following expressions (derivation is not expected as mentioned in the syllabus).

$$V_2 = V_1 - \beta_c t_2$$

$$V_2 = \frac{\beta_c}{\beta} V_2 = V_1 - \beta_c \left(T - \frac{V_1}{\alpha}\right)$$

$$V_2 = \frac{V_1 - \beta_c T + \frac{\beta_c}{\alpha} V_1}{1 - \frac{\beta_c}{\beta}}$$

Also

$$t_2 = \frac{V_1 - V_2}{\beta_c}$$

SOLVED EXAMPLES

Example 5.1: *The speed-time curve of a train consists of:*

(i) Uniform acceleration of 6 kmphps for 21 sec.

(ii) Free running for 10 min.

(iii) Uniform retardation of 6 kmphps for stopping the train.

(iv) A stop of 5 min.

Calculate the distance between the stations, average speed, schedule speed.

Solution: Given:

$$\alpha = 6 \text{ kmphps}$$

$$t_1 = 21 \text{ sec}$$

$$t_2 = 10 \times 60 = 600 \text{ sec.}$$

$$\beta = 6 \text{ kmphps}$$

$$\text{Maximum speed } V_m = \alpha \cdot t_1 = 6 \times 21 = 126 \text{ kmph}$$

$$\text{Time of retardation } t_3 = \frac{V_m}{\beta} = \frac{126}{6} = 21 \text{ sec.}$$

To calculate distance between stations i.e. total distance travelled(s), we should find distance travelled during – acceleration, free running and braking.

$$\text{Distance travelled during acceleration} = \frac{1}{2} V_m \frac{t_1}{3600} = \frac{1}{2} \times 126 \times \frac{21}{3600} = 0.3675 \text{ km}$$

$$\text{Distance travelled during free running} = V_m \cdot \frac{t_2}{3600} = 126 \times \frac{600}{3600} = 21 \text{ km}$$

$$\text{Distance travelled during braking} = \frac{1}{2} V_m \frac{t_3}{3600} = \frac{1}{2} \times 126 \times \frac{21}{3600} = 0.3675 \text{ km}$$

$$\text{Total distance between stations (S)} = 0.3675 + 21 + 0.3675$$

$$\boxed{S = 21.735 \text{ km}}$$

$$\text{Actual time of run} = t_1 + t_2 + t_3 = 21 + 600 + 21 = 642 \text{ sec.}$$

$$\text{Average speed in kmph} = V_a = \frac{\text{Total distance between stations in km}}{\text{Actual time of run in hours}}$$

$$= \frac{S}{\frac{642}{3600}} = \frac{21.735 \times 3600}{642}$$

$$\boxed{V_a = 121.878 \text{ kmph}}$$

Schedule speed in kmph, $V_s = \dfrac{\text{Distance between stations in km}}{[(\text{Actual time of run}) + (\text{Stop time})] \text{ in hrs.}}$

$$= \frac{21.735}{\frac{[642 + (5 \times 60)]}{3600}} = \frac{21.735 \times 3600}{942}$$

$$\boxed{\mathbf{V_s = 83.063\ kmph}}$$

Example 5.2: *A train has schedule speed of 60 kmph between stops which are 6 km apart. Determine the crest speed over the run assuming:*

(i) Duration of stops as 60 sec.

(ii) Acceleration as 2 kmphps.

(iii) Retardation as 3 kmphps.

The speed-time curve is trapezoidal.

Solution: Given:

$$S = 6 \text{ km}$$

$$\text{Schedule speed} = 60 \text{ kmph}$$

$$\alpha = 2 \text{ kmphps}$$

$$\beta = 3 \text{ kmphps}$$

Distance between the stations = S = 6 km

$$\text{Schedule speed in kmph} = \frac{\text{Distance between stations/Stops in km}}{(\text{Actual time of run + Stop time}) \text{ in hrs.}}$$

$$\therefore \quad 60 = \frac{6}{\frac{(\text{Actual time of run}) + (\text{Stop time})}{3600}}$$

$$\therefore \quad \text{Actual time of run} = 360 - 60$$

$$T = 300 \text{ seconds}$$

We know the expression, $V_m = \dfrac{T - \sqrt{T^2 - 1400\ KS}}{2\ K}$

where, $K = \dfrac{1}{2\alpha} + \dfrac{1}{2\beta}$

Note: The value of T is to be put in seconds for above expression and V_m comes in kmph.

$$K = \frac{1}{2 \times 2} + \frac{1}{2 \times 3} = \frac{1}{4} + \frac{1}{6} = 0.4166$$

$$V_m = \frac{300 - \sqrt{(300)^2 - (14400 \times 0.4166 \times 6)}}{2 \times 0.4166}$$

$$\boxed{\mathbf{V_m = 81.13\ kmph}}$$

Example 5.3: *A train has scheduled speed of 30 kmph over level track between the stations 1 km apart. Station stopping time is 20 sec. Assuming braking retardation of 3 kmphps and maximum speed 25% greater than average speed, calculate the acceleration.*

Solution: Given:

$$\text{Scheduled speed} = 30 \text{ kmph}$$

$$S = 1 \text{ km}$$

$$\beta = 3 \text{ kmphps}$$

$$V_m = 1.25\ V_a$$

$$\text{Schedule speed in kmph} = \frac{\text{Distance in km}}{\text{Schedule time in hr}}$$

$$\text{Schedule time} = (\text{Actual time of run}) + (\text{Stop time})$$

$$\therefore \quad \text{Scheduled speed} = 30 = \frac{1}{\frac{\text{Schedule time}}{3600}} = \frac{3600}{\text{Schedule time}}$$

$$\therefore \quad \text{Schedule time} = \frac{3600}{30} = 120 \text{ sec.}$$

$$\text{Actual time of run} = (\text{Schedule time}) - (\text{Stop time}) = 120 - 20$$

$$T = 100 \text{ seconds}$$

$$\text{Average speed in kmph} = \frac{\text{Distance travelled in km or Distance between stops in km}}{\text{Actual time of run in hr}} = \frac{1}{\frac{100}{3600}}$$

$$V_a = 36 \text{ kmph}$$

$$V_m = 1.25\, V_a \ldots \text{given}$$

$$\therefore \quad V_m = 45 \text{ kmph}$$

We know the expression,

$$\frac{1}{\alpha} + \frac{1}{\beta} = \frac{7200\, g}{V_m^2}\left[\frac{V_m}{V_a} - 1\right]$$

$$\therefore \quad \frac{1}{\alpha} + \frac{1}{3} = \frac{7200 \times 1}{(45)^2}\left[\frac{45}{36} - 1\right]$$

Solving, we get, $\boxed{\alpha = 1.8 \text{ kmphps}}$

Example 5.4: *A train is required to run between two stations 1.6 km apart at an average speed of 40 kmph. The run is to be made to a simplified quadrilateral speed-time curve. If the maximum speed is to be limited 64 kmph, acceleration to 2 kmph and coasting and braking retardation to 0.16 kmphps and 3.2 kmphps respectively, determine the duration of acceleration, coasting and braking period.*

Solution: Given: S = 1.6 km, V_a = 30 kmph, V = 64 kmph

α = 2 kmphps, β_c = 0.16 kmphps, β = 3.2 kmphps

$$\text{Duration of acceleration} = t_1 = \frac{V_1}{\alpha} = \frac{64}{2} = \boxed{32 \text{ seconds}}$$

$$\text{Actual time of run} = T = \frac{3600 \times S}{V_a} = \frac{3600 \times 1.6}{40} = 144 \text{ seconds}$$

$$\text{Duration of coasting} = t_2 = \frac{V_1 - V_2}{\beta_c} = \frac{64 - V_2}{0.16} \text{ seconds}$$

$$\text{Duration of braking} = t_3 = \frac{V_2}{\beta} = \frac{V_2}{3.2} \text{ seconds}$$

$$\text{Actual time of run} = T = t_1 + t_2 + t_3$$

$$\therefore \quad 144 = 32 + \frac{64 - V_2}{0.16} + \frac{V_2}{3.2}$$

$$\therefore \quad V_2 = 48.5 \text{ kmph}$$

$$t_3 = \frac{V_m - V_2}{\beta_c} = \frac{64 - 48.5}{0.16} = \boxed{96.85 \text{ seconds}}$$

and

$$t_2 = \frac{V_2}{\beta} = \frac{48.5}{3.2} = \boxed{15.15 \text{ seconds}}$$

5.4.9 Tractive Effort

1. Define: Tractive effort.
2. Define with units: Tractive effort.

- It is the effective force on the wheel rims of a locomotive which is required for its propulsion. The tractive effort is a vector quantity which is tangential to the wheel. It is measured in Newtons. The notation used for tractive effort is F_t.

- In order to propel the train/push forward the train, the tractive effort or force should be equal to sum of the following forces:
  (i) Tractive effort required for linear and angular acceleration of train (F_a).
  (ii) Tractive effort to overcome effect of gravity (F_g).
  (iii) Tractive effort to overcome frictional resistance of train (F_r).

$$F_r = F_a \pm F_g + F_r$$

Positive sign for motion up the gradient, negative sign for motion down the gradient.

Mechanics of train movement:

- Refer Fig. 5.12, the motor armature is connected to pinion whose diameter is 'd'. The pinion transfers its motion to a general wheel with diameter d.
- Let the torque developed by the motor is T Nm and tractive effort at the edge of pinion is F'. The relation between these two quantities is given by

$$T = F' \frac{d'}{2} \text{ Nw} \qquad \text{... (Torque = Force} \times \text{Radius)}$$

$$F' = \frac{2T}{d'} \text{ Nw}$$

- The tractive effort (F') at the edge of the pinion is transferred to the wheel of the locomotive. At the wheel, tractive effort available is F.

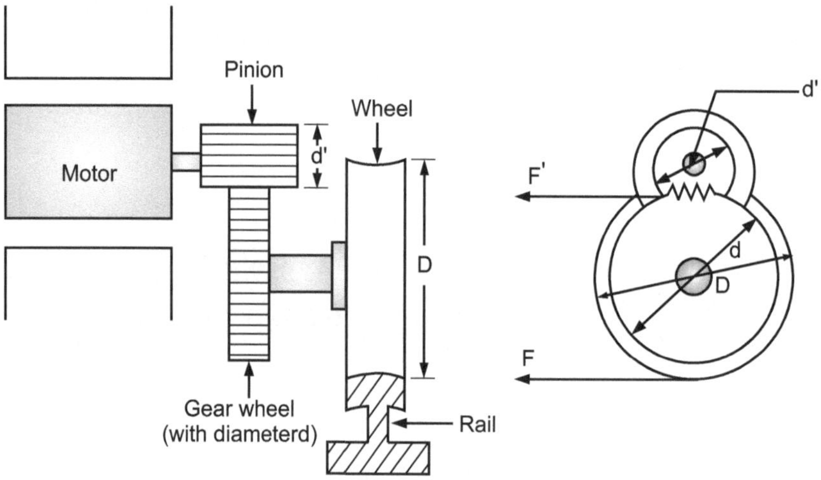

Fig. 5.12

- The relation between these two is given by

$$F = \eta F' \left(\frac{d}{D}\right)$$

where, η = Efficiency of transmission gear
d = Gear wheel diameter
D = Wheel diameter

Substituting value of F', we get

$$F = \eta \left(\frac{2T}{d'}\right) \left(\frac{d}{D}\right) = \eta T \left(\frac{2}{D}\right) \left(\frac{d}{d'}\right)$$

$$F = \eta T \frac{2\gamma}{D} \text{ where, } \frac{d}{d'} = \gamma = \text{gear ratio}$$

5.4.10 Coefficient of Adhesion

1. Define: Coefficient of adhesion.
2. Define with units: Coefficient of adhesion.

- The tractive effort can be increased by increasing the torque of motor. However, there is limit upto which the tractive effort can be increased. After that limit even if torque of motor is increased, the tractive effort does not increase but causes slipping of driving wheel on track.
- It has been found that the maximum value of tractive effort F_t at which driving wheels will not slip depend upon the dead weight W over the driving axle.
- The relationship between tractive effort and dead weight is given by,

$$\frac{F_t}{9.81\ \mu_a} = 1000\ W$$

$$\mu_a = \frac{F_t}{9.81\ (1000\ W)}$$

$$= \frac{\text{Maximum tractive effort that can be applied without slipping of wheels}}{\text{Adhesive weight}}$$

where, F_t = Total tractive effort in Nw

W = Dead weight on axles in tonne (1 tonne = 1000 kg)

μ_a = Coefficient of adhesion ($\mu_a < 1$)

- Thus, in order to increase tractive effort it is just not sufficient to increase the torque or HP of traction motor alone but at the same time weight on driving axle has to be increased.
- As seen from Fig. 5.13 the coefficient of adhesion decreases with increase in speed. The variation is hyperbolic.

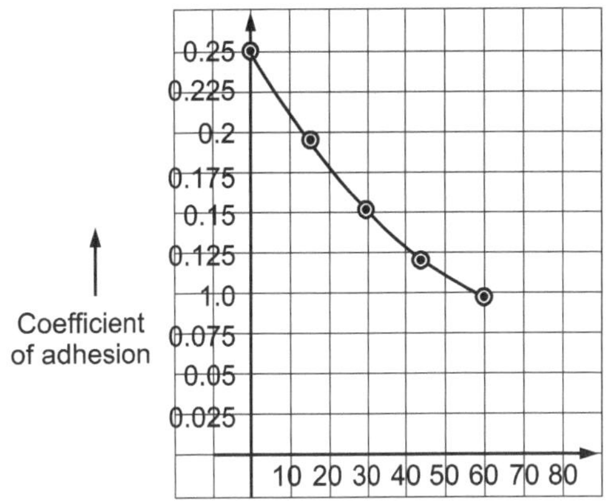

Fig. 5.13: Variation of co-efficient of adhesion with speed

5.4.10.1 Factors Affecting Coefficient of Adhesion

1. What are the factors affecting the coefficient of adhesion?

1. Effect of track condition of

(i) Presence of oil, dew etc. on track reduces the coefficient of adhesion (μ_a). Sometimes sand treatment is given to track for improving coefficient of adhesion.

(ii) Unevenness in tract surface reduces μ_a.

2. Influence of mechanical parts

(i) Uneven distribution of axle load reduces μ_a.

(ii) Bogie pitching oscillations cause unequal distribution of axle load, so μ_a decreases.

(iii) Better suspension arrangements can improve μ_a.

(iv) Sudden increase in tractive effort affect μ_a.

3. During notching up period, tractive effort varies considerably, this also affects μ_a.

5.4.10.2 Total Tractive Effort for Moving the Train

1. Elaborate the parts of total tractive effort with usual notations.

- The total tractive effort to move or propel the train is given by,

$$F_t = F_a \pm F_g + F_r$$

1. F_a (Tractive effort for linear acceleration):

Tractive effort for linear acceleration is given by

$$F_a = 1000\ W\ \alpha\ \dots\ (\text{Force} = \text{Mass} \times \text{Acceleration})$$

where, W = Dead mass in tonne (1 tonne = 1000 kg)

α = Acceleration in kmphps

$$\alpha \text{ in m/s}^2 \text{ will be } = \alpha \times \frac{1000}{3600} = 0.2778 \text{ a m/s}^2$$

$$\therefore \quad F_a = 1000 \text{ W} \times 0.2778\, \alpha = 277.8 \text{ W}\alpha \text{ ... Nw}$$

- When train starts running, the rotating parts of train such as wheel, motors have angular acceleration. So tractive effort for linear acceleration alone should not be considered. The effective accelerating weight (W_e) of train has to be considered.
- This effective weight (W_e) is about 8 to 15% greater than dead weight (W).

Thus, $F_a = 277.8\, W_e\, \alpha$ ... Nw

2. F_g (Tractive effort required to balance gravitational pull):

W = Dead weight of train in tonne

θ = Angle of slope

g = Gravity constant = 9.81 m/s²

- When train moves on an upgradient, the gravity component of dead weight parallel to track i.e. W sin θ will cause the train to come down. So a tractive effort in opposite direction will be required to overcome this gravitational pull. The force due to gradient

$$F_g = 1000 \text{ W} \cdot \sin\theta \text{ ... kg.}$$

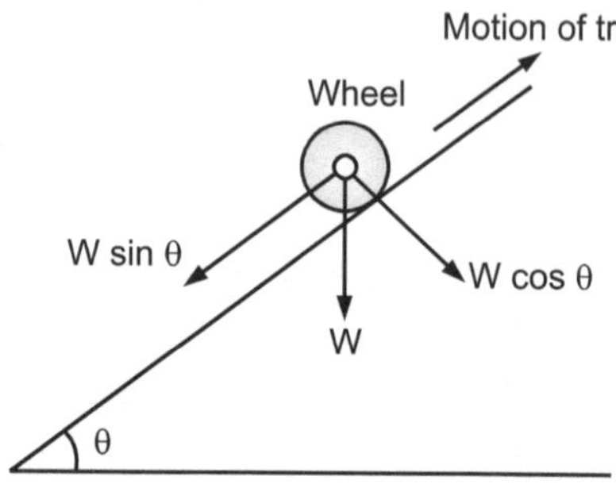

Fig. 5.14

- The gradient is expressed as rise in meters in a track in a distance of 100 meters and it is known as percent gradient (%G).

i.e. $G = \sin\theta \times 100$

$$\sin\theta = \frac{G}{100}$$

$$\therefore \quad F_g = 1000 \text{ W} \cdot \sin\theta = 1000 \text{ W} \frac{G}{100}$$

$$F_g = 10 \text{ WG kg} = 10 \times 9.81 \text{ WG Nw}$$

$$F_g = 98.1 \text{ WG Nw}$$

3. F_r (Tractive effort to overcome train resistance)

- The train resistance is due to
  (i) Aerodynamic drag.
  (ii) Mechanical friction due to moving parts.
  (iii) Friction at track.
- Tractive effort required to overcome train resistance is given by

$$F_r = W{\cdot}r \text{ Nw}$$

where, W = Dead weight of train in tonne

r = Tractive resistance in Newton per tonne

- Tractive effort to overcome train resistance is also required to be taken into account. Greater the radius of curvature, lesser is the resistance. This curvature resistance is added in the train resistance if mentioned.

Thus the total tractive effort,

$$F = F_e \pm F_g + F_r$$

$$F_t = 277.8\, W_e\, \alpha \pm 9.81 \text{ WG} + W{\cdot}r \text{ ... Putting values}$$

+ ve sign for motion up the gradient

− ve sign for motion down the gradient

Power output from driving axles:

$$\text{Power} = \frac{\text{Work done}}{\text{Time}} = \text{Force} \times \frac{\text{Distance}}{\text{Time}} = \text{Force} \times \text{Velocity} = F_t \times V_m$$

$$\therefore \quad P = F_t \times V_m \times \frac{1000}{3600} \text{ in watt}$$

or

$$P = F_t \times \frac{V_m}{3600} \text{ in kW}$$

List of important formulae for solving problem:

Sr. No.	Parameter and notations	Units
1.	W = Dead weight	Tonne
2.	W_e = Effective weight (e.g. in problem it is mentioned that rotational weight is 10% that means $W_e = 1.1$ W)	Tonne
3.	G = Gradient e.g. in problem if it is mentioned that gradient is of 1 in 500 then $\%G = \frac{1}{500} \times 100 = 0.2\%$ or if gradient is of 30 in 100 then $\%G = \frac{30}{1000} \times 100 = 3\%$	Percentage
4.	Total tractive effort $F_t = (277.8\ W_e \cdot \alpha) \pm 9.81\ WG + W \cdot r$ ... Positive sign for up gradient ... Negative sign for down gradient	F_t in Newton W_e in tonne α in kmphps W in tonne G in percentage r in Newton per tonne
5.	Maximum output from driving axle $= \frac{F_t \times V_m}{3600}$ kW	F_t in Newton V_m in kmph
6.	Total torque $T = \frac{F_t D}{2\eta\gamma}$ Nw meter ... D = diameter of train wheel η = overall efficiency γ = gear ratio	F_t in Nw D in meter

F_t is in Newton and velocity in meter/second.

Example 5.5: *A train weighing 203 tonnes accelerates uniformly from rest to the speed of 45 kmph up a gradient of 1 in 500, the time taken is 30 second. The power is then cut off and train coasts down a uniform gradient of 1 in 1000 for a period of 40 seconds when brakes are applied for a period of 15 seconds so as to bring the train uniformly to rest on this gradient. Calculate the tractive effort required and maximum power output from driving axles. Assume train resistance to be 44 Newtons per tonne and allow 10% for rotational inertia.*

Solution: W = 203 tonnes = 203 × 1000 kg

It is given that allow 10% for rotational inertia i.e. $W_e = 1.1\ W = 1.1 \times 203 = 223.3$ tonnes.

Gradient, $G = \frac{1}{500} \times 100 = 0.2\%$

$V_m = 45$ kmph

$t_1 = 30$ seconds

r = Train resistance
= 44 Newtons/tonne

$$\alpha = \text{Acceleration} = \frac{V_m}{t_t} = \frac{45}{30} = 1.5 \text{ kmphps}$$

Total tractive effort required,

$$F_1 = (277.8\ W_e\ \alpha) + (9.81\ WG) + (Wr)$$

$$F_t = (277.8 \times 233.3 \times 1.5) + (9.81 \times 203 \times 0.2) + (203 \times 44) = 105964 \text{ Newton}$$

Maximum power output from driving axle

$$= \frac{F_t \times V_m}{3600} = \frac{105964 \times 45}{3600} = 1324.55 \text{ kW}$$

Example 5.6: *A electric train weighing 200 tonnes has eight motors geared to driving wheel, each wheel is 90 cm diameter. Determine the torque developed by each motor to accelerate the train to a speed of 48 kmph in 30 seconds up a gradient of 1 in 200. The tractive resistance is of 50 N/tonne. The effect of rotational inertia is 10% of the train weight, the gear ratio is 4 to 1 and gearing efficiency is 80%.*

Solution: Given: W = 200 tonnes, D = Diameter of driving wheel = 90 cm = 0.9 m

$$\% \text{ Gradient } G = \frac{1}{200} \times 100 = 0.5\%$$

Tractive resistance, r = 50 N/tonne

Gear ratio, r = 4

Gradient, η = 0.8

Equivalent accelerating weight of the train (W_e):

$$W_e = 1.10\ W = 1.10 \times 200 = 220 \text{ tonnes}$$

Maximum speed V_m = 48 kmph

t_1 = 30 sec.

$$\text{Acceleration } \alpha = \frac{V_m}{t_1} = \frac{48}{30} = 1.6 \text{ kmphps}$$

Tractive effort required:

$$F_r = 277.8\ W_e \times \alpha + 98.1\ W{\cdot}G + W{\cdot}r$$

$$= 277.8 \times 220 \times 1.6 + 98.1 \times 200 \times 0.5 + 200 \times 50 = 117596 \text{ N}$$

$\therefore$ Total torque developed,

$$T = \frac{F_t \times D}{\eta \times 2r} = \frac{117596 \times 0.9}{0.8 \times 2 \times 4} = 16{,}537 \text{ Nm}$$

$\therefore$ Torque developed by each motor

$$= \frac{16537}{8} = \mathbf{2067\ Nm}$$

Example 5.7: *A 200 tonne motor coach having 4 motors each developing 6000 Nm torque during acceleration, starts from rest. If the gradient is 30 in 1000, gear ratio 4, transmission efficiency 90%, wheel radius 45 cm, train resistance 50 N/tonne, addition of rotational inertia 10%, calculate the time taken to attain a speed of 50 kmph. If the line voltage is 3000 volt DC and efficiency of motor is 85%, find the current taken during notching (acceleration) period.*

Solution: W = 200 tonnes

Each motor develops a torque of 6000 Nm

$\therefore$ Total torque developed by 4 motors = T = 6000 × 4 = 24000 Nm

η = 90%

γ = 4

Wheel radius = 45 cm

$\therefore$ Wheel diameter (D) = 45 × 2 = 90 cm = 0.9 m

$$\text{Percentage gradient } \% \text{ G} = \frac{30}{1000} \times 100 = 3\%$$

$$r = 50 \text{ N/tonne}$$

Addition of rotational inertia is 10%.

i.e. $W_e = 1.1 \times 200 = 220$ tonnes

$$F_t = \eta T \frac{2\gamma}{D} = \frac{0.9 \times 24000 \times 2 \times 4}{0.9} = 192000 \text{ Nw}$$

$$F_t = 277.8\, W_e\, \alpha + 98.1\, WG + Wr$$

$$192000 = (277.8 \times 220 \times \alpha) + (98.1 \times 200 \times 3) + (1000 \times 50)$$

Solving above equation we get, $\alpha = 2.015$ kmphps

The train accelerates from rest i.e. zero speed to 50 kmph. The time required for this acceleration is to be determined.

$$\alpha = \frac{V_m}{t_1}$$

$$\therefore \quad t_1 = \frac{V_m}{\alpha} = \frac{50}{2.015} = 24.82 \text{ second}$$

We know that power output from,

$$\text{Driving axles or output of motors} = \frac{F_t V}{3600} \text{ kW} = \frac{192000 \times 50}{3600} = 2666.7 \text{ kW}$$

Efficiency of motor is 85%.

$$\therefore \quad \text{Input power of motors} = \frac{2666.7}{0.85} = 3137 \text{ kW}$$

$$\text{Total input power to motors (in kW)} = \frac{V \cdot I}{1000}$$

$$\therefore \quad 3137 = \frac{3000 \times I}{1000}$$

$$\therefore \quad I = 1046 \text{ Amp}$$

Note: This current is current drawn by 4 motors.

$$\therefore \text{ Current drawn by one motor} = \frac{1046}{4} = 261.5 \text{ amp.}$$

Example 5.8: *A 200 tonnes motor coach having 5 motors each developing 5000 Nm torque during acceleration starts from rest. If up gradient is 30 in 1000, gear ratio is 4, gearing efficiency is 90%, wheel radius is 45 cm, train resistance is 50 N/T, additional inertia is 10% then calculate time taken to attain speed of 55 km/hr. If line voltage is 3k V_{dc} and efficiency of motor is 87%, find the current taken during notching period by each motor.*

Solution: W = 200 tonnes; Torque of each motor = 5000 Nm

$\therefore$ Total torque developed by 5 motors

$$\therefore \quad T = 5000 \times 5 = 25000 \text{ Nm}$$

Given: $\eta = 0.9$, Gear ratio $\gamma = 4$

$$\text{Wheel diameter (D)} = 45 \times 2 = 90 \text{ cm} = 0.9 \text{ m}$$

$$\% \text{ Gradient G} = \frac{30}{1000} = 3\%$$

$$r = \text{train resistance} = 50 \text{ N/tonne}$$

Additional rotational inertia = 10%

$$\therefore \quad W_e = 1.1 \times 200 = 220 \text{ tonnes}$$

Using the relation $F_t = \eta \cdot T \cdot \dfrac{2\gamma}{D} = 0.9 \times \dfrac{25000 \times 2 \times 4}{0.9} = 2{,}00{,}000$ Nw

Also, $F_t = 277.8\, W_e\, \alpha + 98.1\, WG + W \cdot r$

$$\therefore \quad 2{,}00{,}000 = 277.8 \times 220 \times \alpha + (98.1 \times 200 \times 3) + (1000 \times 50)$$

$$\therefore \quad \alpha = 1.4896 \text{ kmphps}$$

The train accelerates from rest (zero) to 55 kmph.

∴ The time required for this acceleration is to be found.

$$\therefore \quad t_1 = \frac{V_m}{\alpha} = \frac{55}{1.4896} = 36.92266 \text{ sec.}$$

Knowing that power output from driving axles (i.e. o.p.g. motors)

$$= \frac{F_t V}{3600} \text{ kW} = \frac{200000 \times 55}{3600} = 3055.55 \text{ kW}$$

$$\text{Efficiency of motor} = 0.87$$

$$\therefore \quad \text{Input power of motor} = \frac{\text{Output power}}{\text{Efficiency}} = \frac{3055.55}{0.87} = \mathbf{3152.132822 \text{ kW}}$$

$$\text{Line voltage} = 3 \text{ kV}$$

$$\therefore \text{Total current of all motors} = \frac{3512.132822}{3} = \mathbf{1170.71 \text{ Amp}}$$

$$\therefore \quad \text{Current by each motor} = \frac{1170.71}{5} = \mathbf{234.142 \text{ A}}$$

Example 5.9: *An electric train weighing 200 tonnes has 8 motors geared to driving wheel. Each wheel has diameter of 90 cm. Determine the torque developed by each motor to accelerate the train to a speed of 48 kmph in 30 seconds up a gradient of 1 in 200. The tractive resistance is 50 Nw/tonne. The effect of rotational inertia is 10% of train weight. The gear ratio is 4 to 1 and gearing efficiency is 80%.*

Solution: W = 200 tonnes, D = 90 cm = 0.9 m, V_m = 48 kmph, t_1 = 30 seconds

Gradient is 1 in 200.

$$G = \text{Gradient} = \frac{1}{200} \times 100 = 0.5\%$$

Tractive resistance r = 50 Nw/tonne

$$W_e = 1.1 \times W = 1.1 \times 200 = 220 \text{ tonnes}$$

From trapezoidal speed-time curve,

$$\alpha = \frac{V_m}{t_1} = \frac{48}{30} = 1.6 \text{ kmphps}$$

Total tractive effort required,

$$\begin{aligned} F_t &= (277.8\, W_e\, \alpha) + (98.1\, WG) + (Wr) \\ &= (277.8 \times 220 \times 1.6) + (98.1 \times 200 \times 0.5) + (200 \times 50) \\ &= 117595.6 \text{ Nw} \end{aligned}$$

In mechanics of train movement, we have seen the expression

$$\text{Force} = \eta T \frac{2\gamma}{D}$$

For our case, we can write it as

$$F_t = \eta T \frac{2\gamma}{D}$$

$$\therefore \quad T \cdot \frac{F_t D}{2\eta\gamma} = \frac{117595.6 \times 0.9}{2 \times 0.8 \times \left(\frac{4}{1}\right)}$$

$$T = 16536.88 \text{ Nw m}$$

$$\therefore \text{ Torque produced by each motor} = \frac{16536.88}{8} = \mathbf{2067.11 \text{ Nw m}}$$

5.4.11 Energy Output from Driving Axle

1. Obtain an expression for energy output from driving axle.

Consider that train is running as per the trapezoidal speed-time curve (Fig. 5.15).

S_1 = Distance travelled in km during acceleration

S_2 = Distance travelled in km during free running in km

S_3 = Distance travelled in km during braking.

During braking the train resistance matters. The energy is spent for overcoming train resistance.

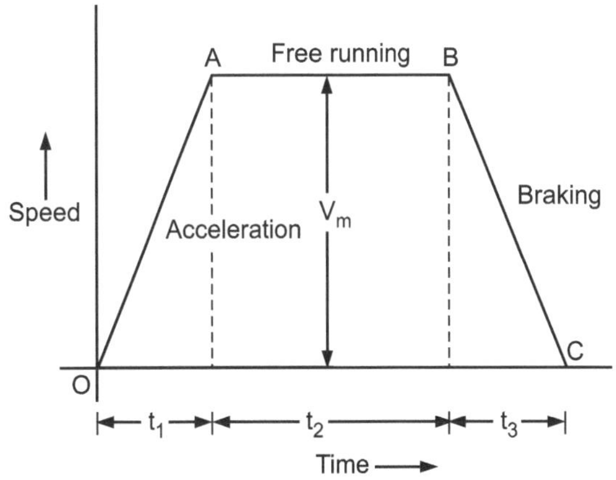

Fig. 5.15: Trapezoidal speed-time curve

The total energy required for the train to run will be

$$E = \begin{pmatrix}\text{Energy required by the} \\ \text{train for acceleration}\end{pmatrix} + \begin{pmatrix}\text{Energy required by the} \\ \text{train for overcoming} \\ \text{gravity component}\end{pmatrix} + \begin{pmatrix}\text{Energy required by the} \\ \text{train for overcoming} \\ \text{train resistance}\end{pmatrix}$$

$$\therefore \quad E = E_a + E_g + E_r$$

1. Energy required by train for acceleration (E_a):

We know that,

$F_a = 277.8\ W_e\ \alpha$ where W_e is effective weight in tonne and α is acceleration in kmphps.

$F_a = 277.8\ W_e\ \alpha \times 1000$ ... if effective weight is in kg

Energy required during acceleration.

$$E_a = (\text{Average power during acceleration}) \times (\text{Time of acceleration})$$

$$E_a = \left(\frac{1}{2}\frac{F_a V_m}{3600}\right) \times \left(\frac{t_1}{3600}\right) \ \ldots \text{kWh} = \left(\frac{V_m}{2} \times \frac{F_a}{3600} \times \frac{t_1}{3600}\right)$$

Substituting value of F_a we get,

$$E_a = \frac{V_m}{2} \times \frac{(277.8 \times W_e \times \alpha \times 1000)}{3600} \times \frac{t_1}{3600}$$

We know that

$$\alpha = \frac{V_m}{t_1}$$

$$\therefore \quad t_1 = \frac{V_m}{\alpha}$$

Substituting value of t_1 in the above equation, we get

$$E_a = \frac{V_m}{2}\left(\frac{277.8 \times W_e \times \alpha \times 1000}{3600}\right) \times \frac{V_m}{3600 \times \alpha} = \frac{V_m^2 \times 277.8 \times W_e \times 1000}{2 \times 3600 \times 3600}$$

$$E_a = 0.01072\ V_m^2\ W_e \quad \ldots \text{ where } V_m \text{ is in kmph and } W_e \text{ is in tonne}$$

2. Energy required by train to overcome gravity component (E_g):

We know the expression,

$$F_g = 98.1\ WG$$

From which we can write,

$$E_g = 98.1\ WG \times (S_1 + S_2) \ \ldots \text{Newton km}$$

where, $S_1 + S_2$ is the distance travelled in km during acceleration and free running with gradient of G%.

It is assumed that the train has to overcome gradient during acceleration and free running also, therefore, $S_1 + S_2$ is taken

$$= 98.1 \text{ WG } (S_1 + S_2) \times 1000 \text{ Nw/m or watt sec.}$$

$$= 98.1 \text{ WG } (S_1 + S_2) \times \frac{1000}{3600} \text{ watt/hr}$$

$$E_g = 98.1 \times 0.2778 \times \text{WG } (S_1 + S_2) \ldots \text{ watt hr}$$

3. Energy required by train to overcome train resistance (E_r):

We know the expression,

$$F_r = W{\cdot}r$$

From which we can write,

$$E_r = W{\cdot}r \ (S_1 + S_2) \text{ Nw km,}$$

where, W = Dead weight of train in tonne

r = Train resistance in Newton/tonne

S_1 = Distance travelled during acceleration

S_2 = Distance travelled during free running

The train resistance is present during acceleration and free running, therefore, S_1 and S_2 is taken.

$$= W{\cdot}r \ (S_1 + S_2) \times 1000 \text{ Nw km or watt sec.} = W{\cdot}r \ (S_1 + S_2) \times \frac{1000}{3600} \text{ watt hr.}$$

$$E_r = 0.2778 \ W{\cdot}r \ (S_1 + S_2) \times \frac{1000}{3600} \text{ watt hr.}$$

4. Total energy output from driving axle (E):

$$= E_a + E_g + E_r$$

$$= \left[0.01072 \ V_m^2 \ W_e\right] + [98.1 \times 0.2778 \times \text{WG } (S_1 + S_2)] + [0.2778 \ W{\cdot}r \ (S_1 + S_2)]$$

$$= 0.01072 \ V_m^2 \ W_e + 0.2778 \ (98.1 \text{ WG} + W{\cdot}r) \ (S_1 + S_2)$$

$$E = 0.01072 \ V_m^2 \ W_e + 0.2778 \ (98.1 \text{ WG} + W{\cdot}r) \ S_{12}$$

where, $S_{12} = S_1 + S_2$

i.e. distance during acceleration + free running

5.4.12 Specific Energy Output and Specific Energy Consumption

1. Derive the expression for specific energy consumption using respective notations.
2. What is meant by specific energy consumption? Derive the expression for specific energy consumption.
3. Derive the expression for specific energy, output on level track using simplified speed-time curve.
4. Define with units: Specific energy consumption.

- In case of electric tractions, the tractive power is obtained from electric motors.
- The power flow diagram is shown in Fig. 5.16.
- The output available at the wheel is evaluated by considering following factors:
  (i) Power/energy required for acceleration.
  (ii) Power/energy required for overcoming gravity component.
  (iii) Power/energy required for overcoming train resistance.

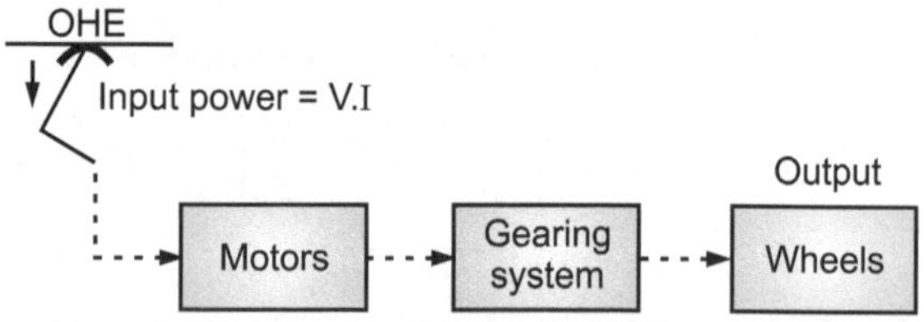

Fig. 5.16: Power flow in electric traction

- Once output is known, the input can be obtained by considering efficiency of motor and gearing system. If this efficiency (η) is known, we can find input power or energy.

  i.e. $\dfrac{\text{Output}}{\eta} = \text{Input}$

- This input is known as specific energy consumption/specific input power.

- This is discussed in detail below:

$$\text{Specific energy output} = \frac{\text{Energy output in watt hours}}{(\text{Weight of train in tonne}) + (\text{Distance of train in km})}$$

$$= \frac{0.01072\, V_m^2\, W_e + 0.2778\,(98.1\, WG + W\cdot r)S_{12}}{W + S}$$

$$= \left(\frac{0.01072\, V_m^2}{S} \times \frac{W_e}{W}\right) + 0.2778\, \frac{S_{12}}{S}\,(98.1\, G + r) \text{ watt hrs/tonne-km.}$$

- This equation is useful for solving numericals.
  Note for level track gradient, G = 0

$$\therefore \quad \text{Specific energy output} = \left(\frac{0.01072\, V_m^2}{S} \times \frac{W_e}{W}\right) + 0.2778\, \frac{S_{12}}{S}\, r \text{ watt. hrs/tonne-km}$$

- Let overall efficiency of gearing system and motor = η.
- If we consider this efficiency, we can find specific energy input (or) specific energy consumption.

$$\eta = \frac{\text{Specific energy output}}{\text{Specific energy input or specific energy consumption}}$$

$$\therefore \quad \text{Specific energy consumption} = \frac{\text{Specific energy output at driving axle}}{\eta}$$

$$\therefore \quad \text{Specific energy consumption} = \frac{0.01072\, V_m^2 \times \frac{W_e}{W}}{S\eta} + 0.2778\, \frac{S_{12}}{S} \times \frac{r}{\eta} \text{ watt hrs./tonne km}$$

$$\text{Specific energy consumption} = \frac{\text{Specific energy output at driving axles}}{\text{Overall efficiency of motors and gearing}}$$

5.4.12.1 Factors Affecting Specific Energy Consumption

1. What are the factors affecting specific energy consumption ?

(i) Energy consumption depends on the efficiency of whole system.
(ii) Energy required for accelerating the train is proportional to the square of maximum velocity.
(iii) Energy required during free running depends on dead weight of the train, train resistance and distance of free run.
(iv) Energy required to overcome gravity effect depends on dead weight of train and percentage gradient.

For electric train following factors affect specific energy consumption:

(i) Efficiency of the whole system.
(ii) Moment of inertia of rotating parts.
(iii) Crest speed/Maximum speed.
(iv) Duration of acceleration.
(v) Train resistance.
(vi) Percentage gradient.

List of important formulae for solving problems:

Sr. No.	Parameter and notations	Units
1.	$S_{12} = S_1 + S_2$ S_{12} = Distance travelled during acceleration and free running	S_{12} in km S_1, S_2 in km
2.	$\text{Schedule speed in kmph} = \frac{\text{Distance between stops}}{\frac{(\text{Schedule time})}{3600}}$ $\text{Schedule speed in kmph} = \frac{S}{\frac{(\text{Schedule time})}{3600}}$	S in km, schedule speed in kmph, schedule time in seconds

contd. ...

3.	Specific energy output from driving axle $= \left(\frac{0.01072\,V_m^2}{S} \times \frac{W_e}{W}\right) + 0.2778\,\frac{S_{12}}{S}$ r... watt hr/tonne-km In problems many times W and W_e are not mentioned separately but the effect of rotational inertia is given e.g. If it is mentioned that effect of rotational inertia or mass is 10% then we should take $\frac{W_e}{W} = 1.1$.	V_m in kmph S in km W_e in tonne W in tonne
4.	Distance travelled during acceleration or notching up period $= \frac{1}{2}\,t_1\,\frac{V_m}{3600}$	t_1 in second V_m in kmph
5.	Distance travelled during braking $= \frac{1}{2}\,t_3\,\frac{V_m}{3600}$	t_3 in second V_m in kmph
6.	Specific energy consumption = $\left[\frac{0.01072\,V_m^2 \times \frac{W_e}{W}}{S\eta}\right] + \left[0.2778\,\frac{S_{12}}{S} \times \frac{r}{\eta}\right]$ watt hrs/tonne km ... η = Overall efficiency of motors and gearing system Generally $\frac{W_e}{W}$ is directly known.	V_m in kmph W_e in tonne W in tonne S in km S_{12} in km r in Nw/tonne

Example 5.10: *An electric train has crest speed of 52 kmph and average speed of 42 kmph on a level track between stops 1400 m apart. It is accelerated at 1.7 kmphps and is braked at 3 kmphps with a braking time of 15.8 seconds. Estimate the energy consumption at axles of the train per tonne-km. Taken constant track resistance of 50 Nw/tonne and allow 10% for rotational inertia.*

Solution: Given: V_m = 52 kmph, V_a = 42 kmph.

Level track, therefore G = 0

Distance between station = S = 1400 m = 1.4 km

α = 1.7 kmphps, β = 3 kmphps, t_3 = 15.8 seconds 10% is allowed for rotational inertia.

i.e. $W_e = 1.1\,W$

$\therefore \quad \frac{W_e}{W} = 1.1$

$$\text{Specific energy output (at axle)} = \left(\frac{0.01072\,V_m^2}{S} \times \frac{W_e}{W}\right) + 0.2778\,\frac{S_{12}}{S}\,r \text{ watt hrs./tonne-km.}$$

We need to find S_{12} ($S_1 + S_2$) i.e. distance travelled during acceleration and free running.

Distance travelled during braking

$$= \frac{1}{2} \times V_m \times \frac{t_3}{3600} = \frac{1}{2} \times 52 \times \frac{15.8}{3600}$$

$$S_3 = 0.1141 \text{ km}$$

and $S = 1.43$ km given

$$S_1 + S_2 + S_3 = S$$

$$S_{12} = S_1 + S_2 = S - S_3$$

$$= 1.4 - 0.1141$$

$$= 1.285 \text{ km}$$

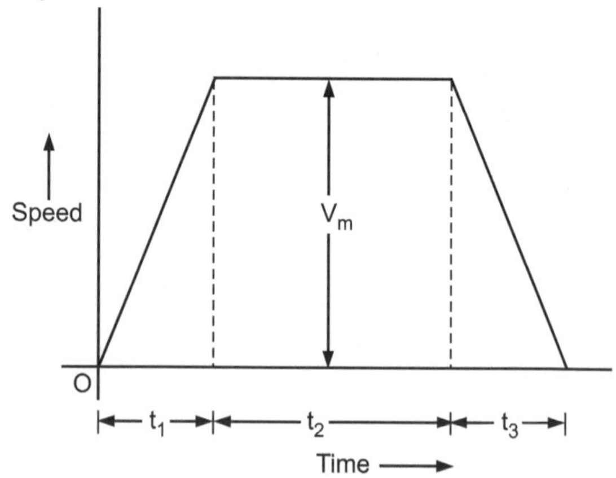

Fig. 5.17

$\therefore$ Energy consumption at the axle of train per tonne

$$= \left(\frac{0.01072 \times (52)^2}{1.4} \times 1.1\right) + \left(0.2778 \times 50 \times \frac{1.286}{4}\right) = \mathbf{35.534 \text{ watt hr.}}$$

Example 5.11: *A 250 tonnes train with 10% rotational inertia effect is started with uniform acceleration and reaches a speed of 50 kmph in 25 seconds on a level track. Find specific energy consumption if journey is to be made according to simplified trapezoidal speed-time curve. The acceleration in 2 kmphps, braking retardation is 3 kmphps and distance between the stations is 2.4 km. Efficiency of motors is 0.9, track resistance is 5 kg/tonne.*

Solution: W = 250 tonnes, 10% rotational inertia effect,

i.e. $W_e = 1.1\ W$

$\therefore \quad \frac{W_e}{W} = 1.1$

$V_m = 50$ kmph

$t_1 = 25$ sec.

Level track, therefore, G = 0, $\alpha = 2$ kmphps, $\beta = 3$ kmphps

$r = 5$ kg/tonne $= 5 \times 9.81$ Nw/tonne

$S = 2.4$ km

$\eta = 0.9$

$$\text{Specific energy output at axle} = \left(\frac{0.01072\ V_m^2}{S} \times \frac{W_e}{W}\right) + 0.2778\ \frac{S_{12}}{S}\ r \text{ watt hr./tonne-km}$$

We need to find S_{12} i.e. $S_1 + S_2$.

Duration of braking $= t_3$

$$t_3 = \frac{V_m}{\beta} = \frac{50}{3} = 16.66 \text{ sec.}$$

Distance travelled during braking (S_3)

$$= \frac{1}{2} V_m \frac{t_3}{3600} = \frac{1}{2} \times 50 \times \frac{16.66}{3600}$$

$$S_3 = 0.1157 \text{ km}$$

and $S = 2.4$ given

$\therefore \quad S_1 + S_2 + S_3 = S$

$\therefore \quad S_{12} = S_1 + S_2 = S - S_3 = 2.4 - 0.1157 = 2.284$ km

$$\therefore \text{Specific energy output at axle} = \left(\frac{0.01072 \times 50^2}{2.4} \times 1.1\right) + \left(0.2778 \times \frac{2.284}{2.4} \times 9.81\right)$$

$$= 12.283 + 12.967 = 25.25 \text{ watt hr./tonne km}$$

$$\text{Specific energy consumption} = \frac{\text{Specific energy output at axle}}{\eta} = \frac{25.25}{0.9}$$

$$= \mathbf{28.05 \text{ watt hr./tonne km}}$$

Example 5.12: *An electric train has quadrilateral speed-time curve as follows:*

(i) Uniform acceleration from rest at 2 kmphps for 30 seconds

(ii) Coasting for 50 seconds

(iii) Braking period of 20 seconds

The train is moving a uniform upgradient of 1% tractive resistance in 40 newtons per tonne, rotational inertia effect 10% of dead weight, duration of station stop 15 seconds and overall efficiency of transmission gear and motor as 75%. Calculate the value of its sphedule speed and specific energy consumption of run.

Solution: Maximum speed $V_1 = \alpha t_1 = 2 \times 30 = 60$ kmph

Coasting retardation $\beta_C = 0.452$ kmphs

$V_2 = 37.4$ kmph

Braking retardation $\beta = 1.87$ kmphps

Total distance travelled $S = 1.03$ km

Schedule time, $T_s = t_1 + t_2 + t_3 + \text{stop duration}$

$= 30 + 50 + 20 + 15 = 115$ seconds

Schedule speed, $$V_s = \frac{3600 \times S}{T_s} = \frac{3600 \times 1.03}{115} = 32.34 \text{ kmph}$$

Distance travelled when power is on,

$$S_1 = \text{Distance travelled during acceleration period}$$
$$= \frac{V_1 t_1}{7200} = \frac{60 \times 30}{7200} = 0.25 \text{ km}$$

$$\text{Specific energy output} = \frac{0.01072\, V_m^2}{S} \cdot \frac{W_e}{W} + 0.2778\,(98.1\,G + r)\frac{S_1}{S}$$
$$= \frac{0.01072 \times 60^2}{1.03} \times 1.1 + 0.2778\,(98.1 \times 1 + 40) \times \frac{0.25}{1.03}$$
$$= \mathbf{50.5\ Wh/tonne - km}$$

$$\text{Specific energy consumption} = \frac{50.5}{0.75} = \mathbf{67.34\ Wh/tonne\text{-}km}$$

Example 5.13: *It is proposed to put an electric trolley service in a city. The schedule speed is to be 45 kmph. The distance between stops is 2.8 km. The track is assumed to be level. Each stop is of 30 seconds duration. Using simplified speed-time curve, calculate the maximum speed assuming the acceleration to be 2 kmphps and retardation to be 3:2 kmphps. The dead weight of the car is 16 tonnes, rotational inertia is 10 per cent of the dead weight and track resistance is 40 newtons per tonne. If the overall efficiency is 80%, calculate (a) maximum power output from driving axle, (b) the specific energy consumption in watt hours per tonne-km.*

Solution: Schedule speed, $V_s = 45$ kmph

Schedule time of run, $T_s = \frac{S \times 3600}{V_s} = \frac{2.8 \times 3600}{45} = 224$ seconds

Actual time of run, $T = 224 - 30 = 194$ seconds

$$K = \frac{1}{2\alpha} + \frac{1}{2\beta} = \frac{1}{4} + \frac{1}{6.4} = \frac{13}{32} = 0.40625$$

Maximum speed, $V_m = \frac{T}{2K} - \sqrt{\left(\frac{T}{2K}\right)^2 - \frac{3600\,S}{K}}$

$$= \frac{194}{2 \times 0.40625} - \sqrt{\left(\frac{194}{2 \times 0.40625}\right)^2 - \frac{3600 \times 2.8}{0.40625}} = \mathbf{59\ kmph}$$

Acceleration time, $t_1 = \frac{V_m}{\alpha} = \frac{58}{2} = 29$ seconds

Duration of braking, $t_3 = \frac{V_m}{\beta} = \frac{58}{3.2} = 18.125$ seconds

Time of free run, $t_2 = T - (t_1 + t_3) = 194 - (29 + 18.125) = 146.875$ seconds

Tractive effort during acceleration,

$$F_1 = 277.8\,W_e\,\alpha + W_r = 277.8 \times 16 \times 1.1 \times 2 + 16 \times 40$$
$$= 10418.56 \text{ Newtons}$$

(a) Maximum power output $= \frac{F_t V_m}{3600}$ kW $= \frac{10418.56 \times 58}{3600} = \mathbf{167.855\ kW}$

(b) Distance travelled during braking period

$$= \frac{1}{2}\frac{V_m \times t_3}{3600} = \frac{1}{2} \times \frac{58 \times 18.125}{3600} = 0.146 \text{ km}$$

Distance travelled with power on, $S_1 = 2.8 - 0.146 = 2.654$ km

$$\text{Specific energy output} = \frac{0.01072\, V_m^2}{S} \times \frac{W_e}{W} + 0.2778\, r\,\frac{S_1}{S} \text{ Wh/tonne-km}$$
$$= \frac{0.01072 \times 58^2}{2.8} \times 1.1 + 0.2778 \times 40 \times \frac{2.654}{2.8}$$
$$= \mathbf{24.7\ Wh/tonne\text{-}km}$$

$$\text{Specific energy consumption} = \frac{24.7}{0.8} = \mathbf{30.9\ Wh/tonne\text{-}km}$$

5.5 TRACTION MOTORS

5.5.1 Desirable Characteristics of Traction Motor

1. Write six desirable characteristics of traction motor.

1. **Suitable speed-torque characteristics:** The torque required at start or tractive effort required at start is high or maximum. During the constant speed coasting, the tractive effort requirement is not high because kinetic energy is developed and the tractive effort required is for overcoming the track resistance and gravity component.

   Thus the requirement is that motor should develop very high starting torque which should fall off at high speeds.
2. **Parallel running:** In traction work, usually more than one motor operate at a time. Hence, these motors should be capable of operating in parallel. There occurs a small difference in rotational speed of various motors because of uneven wear and tear of wheels. This should not produce wide variations in torques developed and current drawn by various motors.
3. **Voltage fluctuations:** In traction work, on account of heavy current in rush at starting, considerable voltage fluctuation of supply line is a normal feature. Traction motor should be capable of withstanding voltage fluctuations without adverse effect on their performance.
4. **Temporary interruption of supply:** This occurs when section insulators and cross overs are crossed with the controller ON. Motor should withstand these fluctuations without heavy in rush of current.
5. **Overload capacity:** The traction motors are subjected to heavy loads which cause large rush of current. This may produce large armature reaction and bad commutation. The arcing produced on commutator surface may exceed over the whole periphery and flashover may cocur.

   This is to be avoided at all costs. Thus, motors should be capable of taking heavy loads without flashover.
6. **Self-relieving property:**

   $T \propto \frac{1}{N}$ and T.N. $\rightarrow$ constant

   H.P. or O/P $\propto$ T.N. $\rightarrow$ constant

   This gives motor a self-protective property against excessive overload.
7. **Amenability to speed control:** Motor should be amenable to simple speed control methods.
8. **Amenability to electric braking:** Motor should be amenable to easy and simple methods of rheostatic and/or regenerative braking.
9. **Limitation of weight and size:** The traction motors should have high power to weight ratio. The physical size of motor depends on type of insulation employed. Traction motors are wound with class H insulation.
10. **Robustness:** Motors are subjected to severe conditions. These motors should be robust enough to withstand such severe conditions. Motors should be further provided with mechanical protection so as to prevent dirt, water, mud etc.
11. **High efficiency:** Motor should have high efficiency.

5.5.2 Featuers and Advantages of Traction Motors

Mechanical Features:

- Must be robust.
- Vibration sustainable.

- Less weight.
- High speed motor.
- Small diameter, small size to be accommodated in small space.
- Totally enclosed type.

Electrical Features:

- Capable to develop high starting torque.
- Speed control be simple.
- Capable of voltage fluctuations.
- Capable of withstanding temporary interruption of supply.
- Should have overload withstanding capability.
- Equal load sharing capability when operated parallel with other motors.

Therefore suitable motors for traction purpose are:

1. D.C. series motor.
2. D.C. compound motor.
3. A.C. single-phase series motor.
4. 3-phase induction (slip ring type) motor.

5.5.3 Advantages, Suitability of Motors for Traction Purpose

5.5.3.1 Suitability of D.C. Series Motor for Traction

1. D.C. series motor is used for traction purpose. Justify the answer giving its characteristics.

- The D.C. series motor has features that make it suitable for traction work. These are as follows:
  1. D.C. series motors develop high starting torque which is required for traction purpose.
  2. The high torque is produced at low speeds and low torque is produced at high speeds.
  3. D.C. series motor is ideally suitable for parallel running.

Suppose two motors are connected to different driving axles and these are operating simultaneously. Ideally when more than one motors are operating, they should rotate at same speed with equal load on them i.e. torque should be same.

But in actual practice there is uneven wear and tear of individual wheels, the wheels are connected to axles and axles are driven by motors. Hence, motors also run at different speeds which is not desired.

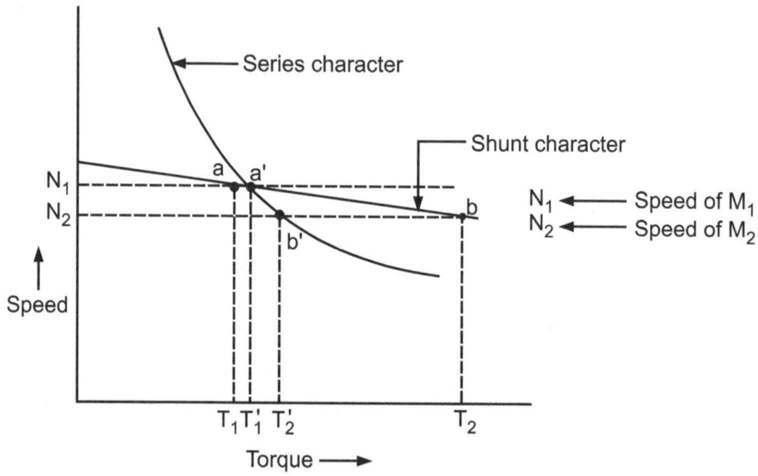

Fig. 5.18 (a)

Now consider that the motors M_1 and M_2 are operating simultaneously and their speeds are N_1 and N_2 i.e. there is difference in their speeds.

If motors are having shunt characteristics, with motor M_1 (with worn wheels) running at a higher speed N_1 and motor M_2 running at a lower speed of N_2. From the characteristics it is seen that motor M_1 exerts a torque T_1 which is approximately half of T_2 developed by M_2.

But it is desired that all motors of train should share load equally i.e. torque developed by them should be approximately same even if small speed variation is there. Thus, with shunt characteristics desired performance requirements are not met.

But with series characteristics, however, it can be seen that small difference in the speeds has a very little effect on relative torques (T_1' and T_2'). Hence, these motors can meet the performance requirement effectively.

Induction motors have characteristics similar to shunt characteristics. These motors are also used for traction purpose. But to meet above requirement (i.e. variation in speed should not cause drastic difference in torque), the driving axles are coupled by a connecting rod so that there is no possibility of difference in speed.

4. In case of D.C. series motors:

$$T \propto \phi \cdot I_a$$

$$T \propto I_a \cdot I_a \text{ ...since, } \phi \text{ is produced by } I_a$$

$$\therefore \quad \phi \propto I_a$$

$$T \propto I_a^2 \text{ ... upto magnetic saturation.}$$

For shunt motor,

$$T \propto \phi \cdot I_a$$

$$T \propto I_a \text{ only since } \phi \text{ is constant}$$

Thus, for series motor, $T \propto I_a^2$

and for shunt motor, $T \propto I_a$

If torque requirement increases, the series motor will draw less current as compared to shunt motor i.e. drain in electric current is less in case of series motor.

5. Commutating properties of series motor are good. At heavy loads, the armature speed decreases due to which commutation is better and sparking is less.
6. Speed of series motor can be controlled easily and effectively.
7. The torque is unaffected by variations in supply voltage.
8. The series motor speed torque characteristics gives stability in control.
9. The output of motor is given in HP.

$$O/P = \frac{2\pi NT}{60}$$

$$\therefore \quad O/P \propto T \cdot N$$

or $\quad HP \propto T \cdot N$

For series motor, $\quad N \propto \frac{V}{I_a}$

$$\therefore \quad HP \propto T \frac{V}{I_a}$$

Also, $\quad T \propto I_a^2$

$$\therefore \quad I_a \propto \sqrt{T}$$

$$\therefore \quad HP \propto T \frac{V}{\sqrt{T}}$$

$$\therefore \quad \mathbf{HP \propto \sqrt{T}\, V}$$

- From above equation we can say that for given increase in torque there is not much increase in HP rating. This is called self-relieving property.

  Following are negative points about series motor w.r.t. traction work:

  (iii) Regenerative braking is difficult to achieve. Special arrangements are essential for it.

  (iv) Due to temporary interruption of supply, the heavy in rush current is more.

5.5.3.2 Suitability of A.C. (Single-Phase) Series Motor for Traction

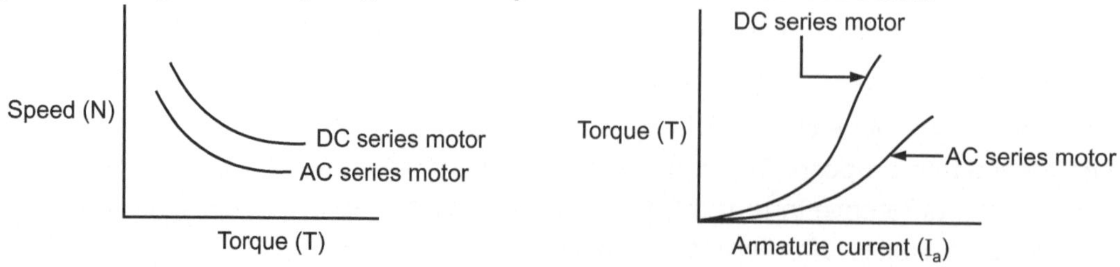

Fig. 5.18 (b): Comparison of characteristics of D.C. and A.C. Series Motor

- The traction application requires steep speed torque (N/T) characteristics.
- D.C. series motor fulfill this requirement, hence can be used in traction.
- A.C. series characteristic is similar to D.C. series motor's characteristic.
- Construction of A.C. series motor is almost similar to D.C. motor but needs following modifications:
  (i) Compensating winding is needed.
  (ii) To reduce iron losses motor operated at low flux density.
  (iii) To reduce magnetizing current, the air gap be as minimum as possible.
  (iv) Core (field system) is laminated to reduce eddy current losses.
  (v) To improve commutation and efficiency motor is operated at lower frequency.
  (vi) Shunted compoles are used.

  Note: Size of A.C. series motor is almost twice the size of D.C. series motor of same capacity.
- Power factor of A.C. series motor is poor during starting, so starting torque is somewhat less.
- The maximum operating voltage is 400 volts.
- Moderate acceleration is required for main line services. So A.C. series motor is well suited in this.
- For urban and suburban services, this A.C. series motor is not suitable as higher acceleration is required in such lines.

5.5.3.3 Suitability of Three-Phase Induction Motor

1. How three-phase induction motor is suitable for traction service?

- Three-phase induction motor have the following features:
  1. Robustness.
  2. Less maintenance.
  3. Trouble free operation.
  4. Three-phase induction motor can be used for high voltage. The current requirement is less. Because P = V.I. hence for same power, voltage is high, current required is less.
  5. Automatic regenerative braking is possible.
  6. It can sustain overload.

Following are the drawbacks w.r.t. traction work:

1. Speed-torque characteristic is flat. (Fig. 5.19).
2. The overhead conductor system is complicated because three conductors are needed.
3. If single-phase supply is tapped from OHE then single-phase to three-phase converter is needed in the locomotive.
4. The speed control method is complicated than DC motors.

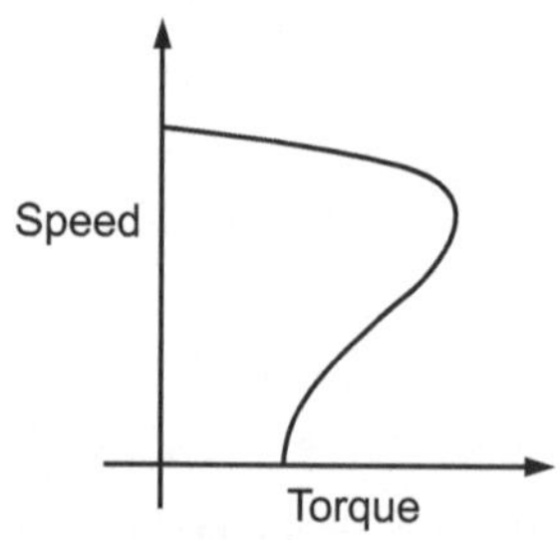

Fig. 5.19: Speed-torque characteristics of three-phase induction motor

5.5.3.4 Suitability of Linear Induction Motor for Traction

1. Explain with construction and working linear induction motor.

5.5.3.4 (A) Linear Induction Motor

- The ordinary induction motor is designed to produce a rotary motion. The same motor with a special design and certain changes in the construction can be made to produce linear motion. Such a motor is called as Linear Induction Motor (LIM). So LIM is a developed version of a cylindrical induction motor. It is a device in many different configurations.
- Small capacity LIM is designed to operate sliding doors automatically. It is also used for metallic belt conveyors. Medium capacity motors are designed to operate cranes. It is also designed to operate medium and high speed vehicles.
- Also the LIM is tried in traction work i.e. trains etc. Even then the use of LIM is very limited due to design constraints and economic considerations.

5.5.3.4 (B) Construction of Linear Induction Motor

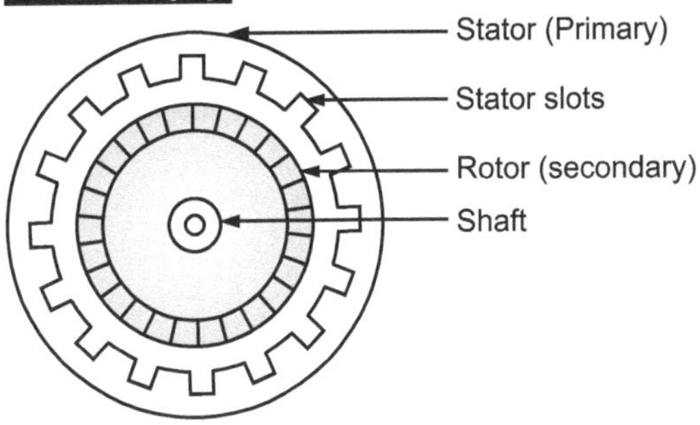

(a) Ordinary induction motor but having rotor conductor in the form of a strip or plate

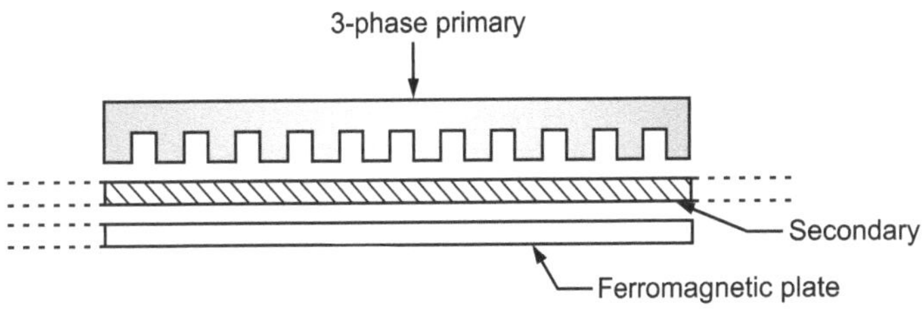

(b) If a ordinary induction motor is cut axially and spread out flat it corresponds to LIM

Fig. 5.20

- The ordinary induction motor is imagined to be cut axially and if made flat, then, it is a simple LIM as shown above in its ordinary form.
- The field system is accommodated in the primary slots (three-phase star connected winding similar to ordinary induction motor) and supplied with a three-phase balanced supply to produce the necessary moving field. In this case it moves linearly from one to other end of the stator at a constant synchronous speed (v_s).
- The secondary is a conductor of aluminium or copper in the form of a sheet or strip.
- In the single primary type construction, a ferromagnetic plate is also provided to reduce the reluctance for the flux.
- The drawback in providing the ferromagnetic plate is that, when conductor is attracted, it creates uneven air gap between primary and secondary.
- The above difficulty is avoided by providing primaries on both sides of secondary as shown in Fig. 5.21.

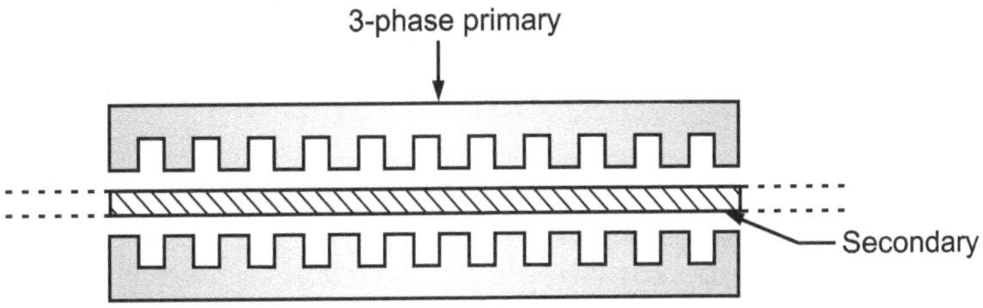

Fig. 5.21: Double primary

- When the movement required is smaller i.e. distance between the two limits is shorter then primary is made longer than secondary. Refer Fig. 5.22.

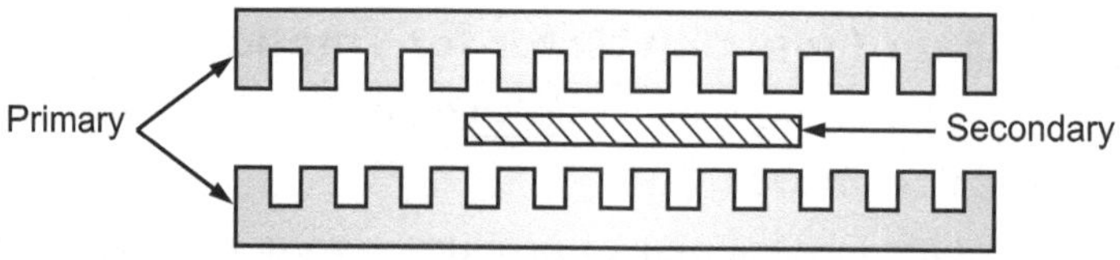

Fig. 5.22: Primary longer than secondary

- When the distance is large then primary is made shorter (i.e. in electric traction) than secondary. Refer Fig. 5.23.

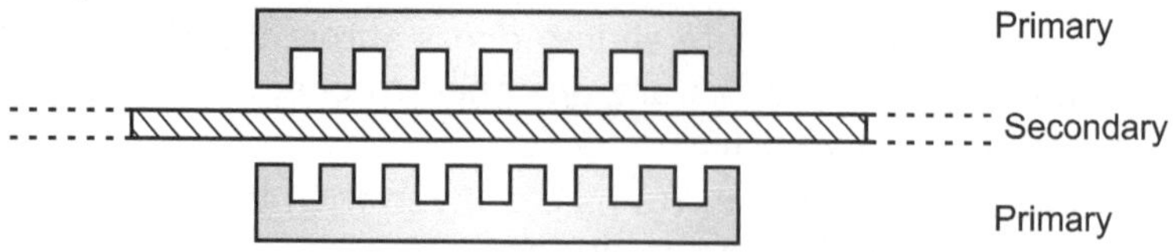

Fig. 5.23: Secondary is longer than primary

5.5.3.4 (C) Working of Linear Induction Motor

- When primary is energized i.e. supply is put on, the magnetic field is produced. This field moves linearly with a definite speed called as linear synchronous speed (v_s). The currents are induced in the secondary due to electromagnetic induction action.
- These secondary currents interact with the field produced by primary and a "linear force" is produced.
- Due to this force the secondary conductor moves in the same direction as that of the linearly moving field and motor action takes place. Secondary conductor moves with somewhat less speed than synchronous speed.
- This actual speed is noted as (v). It is possible either of the two i.e. primary or secondary can be made to move, keeping other stationary.
- In traction type motor as shown in Fig. 5.24, the stationary member is secondary and moving member is primary which is accommodated in vehicle.

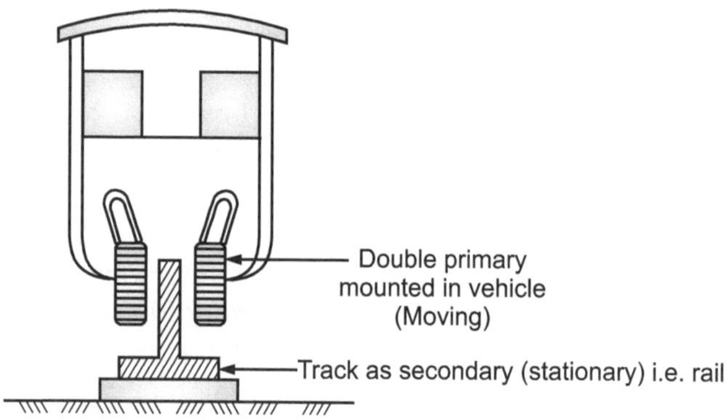

Fig. 5.24: LIM floating on track

5.5.3.4 (D) Speed Relations : Thrust, Active Power, Efficiency

- When three-phase balanced supply is connected to primary, a linearly moving field is produced. In one cycle of supply frequency (f), the field moves by a distance of two pole pitches (2τ).

  ∴ Synchronous linear speed (velocity) $v_s = 2\tau f$ m/sec.

  where τ = pole pitch in meters

  f = frequency of supply in Hz
- Similar to ordinary induction motor, the primary moves in opposite direction to the field with the secondary field fixed (on the track of rail) at a speed v.

  $$\therefore \quad \text{Slip } s = \frac{v_s - v}{v_s}$$
- It is seen from $v_s = 2\tau f$ that it is not governed by the number of poles (P) as in case of ordinary induction motor but it depends on pole pitch τ.

  ∴ Change in speed is achieved by adjusting pole pitch τ.

  $$\text{Thrust (or force)} \quad F = \frac{P_r}{v_s}$$

  where, P_r = Power supplied to the rotor in watts

  v_s = Linear synchronous speed in m/sec.

- Active power (Mechanical power) 'P_m' developed by the motor is

$$P_m = P_r (1 - s)$$

- Similar relation as in ordinary induction motor.

$$\text{Rotor copper loss} = s.P_r$$

$$\text{Efficiency} = \frac{\text{Output}}{\text{Input}} \times 100$$

- Magnetising current is large due to more air-gap and also p.f. is poor, hence efficiency is very poor of this motor.

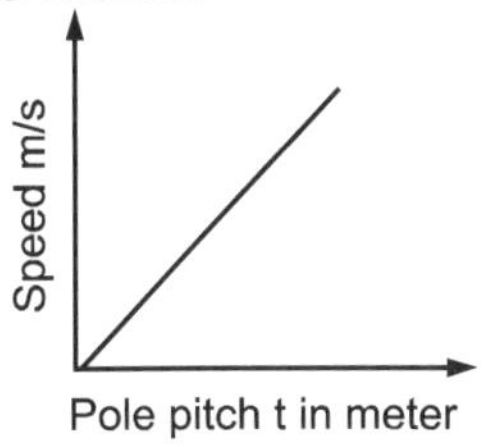

(a) Pole pitch τ versus speed characteristics

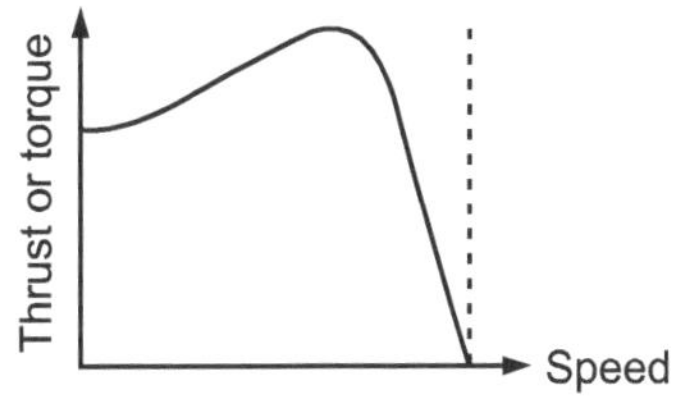

(b) Torque versus speed characteristics

Fig. 5.25: Linear motor characteristics

5.5.3.4 (E) Edge and End Effect

1. Explain edge and end effect.

(i) Edge effect:

- This is a typical phenomenon observed in LIM. Due to this effect the primary sways in a transverse manner. Due to this effect, energy loss takes place. In Fig. 5.26 the edge effect has been shown.

The secondary current distribution path over one pole has been shown in Fig. 5.26. In the case of LIM, the distribution current paths are somewhat like eddy currents unguided. It is seen that direction of current is almost circular. It is also seen that these currents tend to put a horizontal force in opposite directions. If the secondary is not allowed to move, primary will move along these current distribution paths. As a result primary will sway while in forward motion.

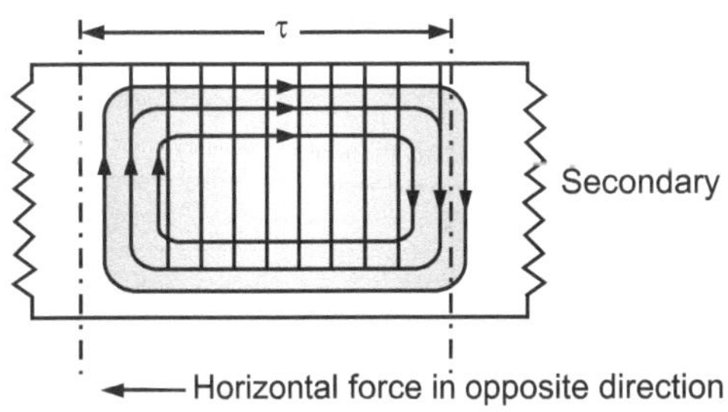

Fig. 5.26: Secondary current distribution (Edge effect)

(ii) End effect:

Unlike a rotary induction motion the path of the travelling flux is not continuous. The core starts from one end and finishes at the other end. The revolving flux sheet starts from one end and terminates on the other end. Again a fresh sheet starts from one end and finishes at the another end. This process continues, suggesting that the travelling flux is not continuous as in the case of a rotary induction motor whose rotating magnetic field is continuous. This effect is called 'End Effect'.

Advantages:

1. State the advantages of linear induction motor.

1. Simple construction, low cost, no overheating, less maintenance.
2. No limits for maximum speed due to centrifugal forces.
3. No rotating parts.
4. No limitations of tractive effort due to adhesion between the wheel and the rail.
5. Tractive effort is independent of speed as it is dependent on rotary motion.
6. We can dispense with current collection gear, if stator on the track and rotor plate is carried on the vehicle.
7. In comparison with rotary motor the linear induction motor has better power to weight ratio.

Disadvantages:

1. State the disadvantages of linear induction motor.

1. Due to edge effect and end effect, utilization of motor is inferior.
2. Capital cost of reaction rail fixed along the centre line of the track is very high. It also creates complications.
3. Due to axle movements on curves, adequate clearance at points and crossing may be difficult to maintain.
4. A lot of attractive force is exerted between the stator and rotor if single field system and ferromagnetic reactions rail is employed.
5. Power factor and efficiency of motor is less due to requirement of more magnetizing current as there is a large air gap and non-magnetic reaction rail.

Applications:

1. Give the applications of linear induction motor.

1. As a booster accelerator for moving heavy trains from rest or up the inclines or on curves.
2. Can be used as traction motors.
3. Linear induction motor forms excellent sources of motive power for magnetically suspended trains where the other motor which depends for torque conversion to linear tractive force upon the adhesive weight on driving wheels fails.

- Their applications rather than traction are
  - (i) Can be used as propulsion unit in marshaling yards in place of shunting locomotives.
  - (ii) Can be used for metallic belt conveyors.
  - (iii) Can be used for crane-movements, halters.
  - (iv) Medium-high speed vehicles on road for transport.
  - (v) Linear induction motor can be used for puller machines, sliding door operations and also for electromagnetic pumps.

5.6 TRACTION MOTOR CONTROL

5.6.1 Requirements

1. State the requirements for traction motor control.

- The traction motor control requirements are as follows :
  - (i) Connecting the motor to supply without taking excessive starting current.
  - (ii) Providing smooth acceleration without sudden jerk to avoid damage of couplings.
  - (iii) Adjusting the speed according to the type of services and route conditions.
  - (iv) Providing rheostatic or regenerative braking.
- So accordingly control equipments are used to get desired performance of traction motors.

5.6.2 Traction Control of D.C. Locomotives and EMUs

- D.C. series motors are used for traction works. The current taken by motor is given by following expression: $I = \frac{V - E_b}{R}$

  where, I = Current taken by motor

  V = Supply voltage

  E_b = Back emf

  R = Total resistance of motor (armature + field)

- At start E_b is zero, so motor takes very high current (because R is low). This high starting current is to be limited by some means.

- This can be done by
  - (i) Applying reduced voltage at start.
  - (ii) Adding some resistance in series with armature.
- For D.C. motors the second method is adopted because it is difficult to get variation in supply voltage.
- After safe starting of motor, the speed of motor is to be controlled as per the requirement. The various methods of starting and speed control are as follows:

1. List and explain methods of starting and speed control of D.C. locomotives.

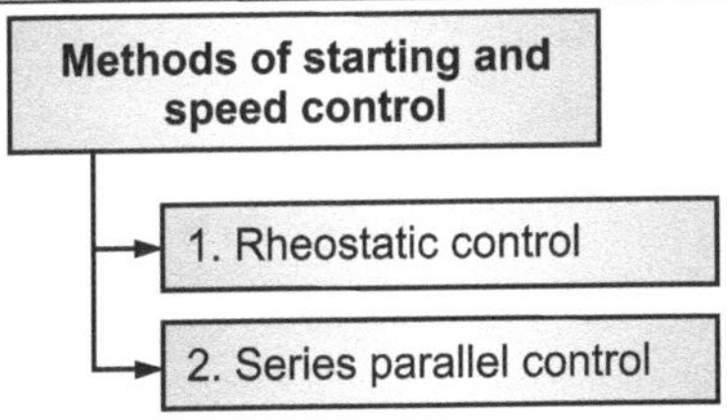

Fig. 5.27: Methods of Starting and Speed Control

1. Rheostatic Control:

- This is used to limit the starting current. Function is similar to that of starter. A rheostat resistance is connected in series with the motor as shown in Fig. 5.28.
- At start **1** is closed and remaining contacts are open, so motor starts with full resistance in series.

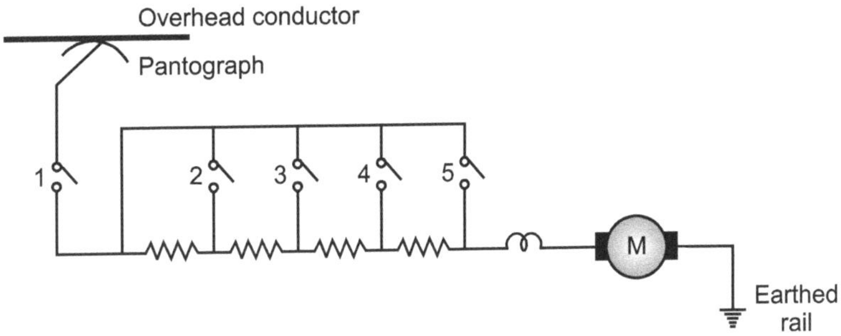

Fig. 5.28

- Thus, starting current is limited. As the motor gathers speed, **2** is closed so as part of resistance is cut down. As the motor accelerates further **3, 4, 5** are closed, thus, full resistance is cut down.
- The resistance is required only during starting of motor and it is not supposed to carry current when motor is working normally. So this resistance is designed for short time rating and for continuous rating. Considerable amount of power (I^2R) is wasted during starting.

2. Series Parallel Control:

- The motors are run in series and then in parallel. In each case, the voltage applied to motor will be different and the motors can thus run at different speeds.
- There are three methods of series parallel control:
  - (a) Open circuit transition.
  - (b) Shunt transition or short circuit transition.
  - (c) Bridge transition.

5.6.3 Series Parallel Control, Steps Involved

- In traction more than one motor operates at a time and these are connected in series or parallel (depending upon the requirement).

5.6.3.1 Open Circuit Transition

- In this method, two motors are initially in series. This series connection is disconnected and power is switched OFF. Refer Fig. 5.29.

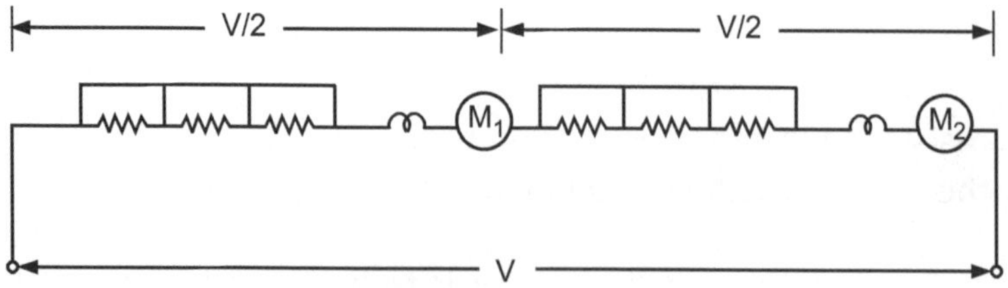

(a) Motors in series

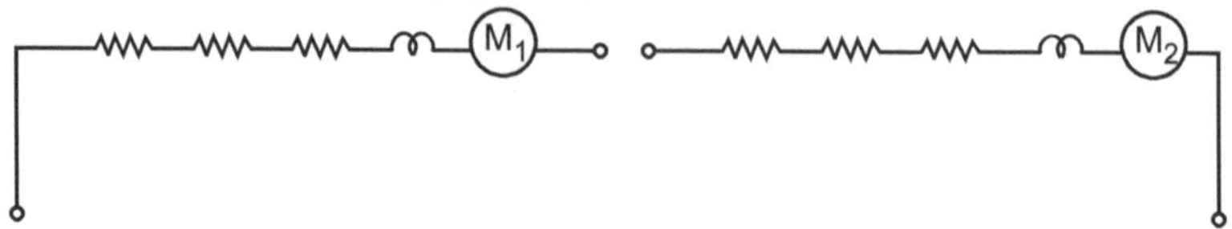

(b) Power OFF

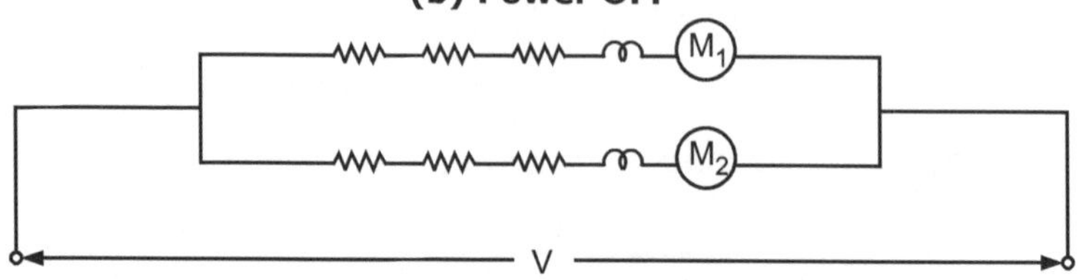

(c) Motors in parallel

Fig. 5.29: Open Circuit Transition

- Then starting resistance is reinserted in the circuit, the two motors are connected in parallel and supply is restarted.

Advantages:

(i) Simple method.

(ii) Easy operation.

Disadvantages:

(i) During series-parallel process, there is interruption of supply which is not desired.

(ii) There is loss of tractive effort i.e. torque becomes zero during transition.

(iii) Jerk on two motors is more.

5.6.3.2 Shunt Transition / Short Circuit Transition

- The supply is given through starting resistance and this resistance is gradually cut down and motors are connected in series. The voltage across each motor is $\frac{V}{2}$ and they run at half rated speed. Now to run the motors at rated speed, the motors are disconnected and then reconnected in parallel, again supplied through starting resistance. The resistance is gradually cut down.

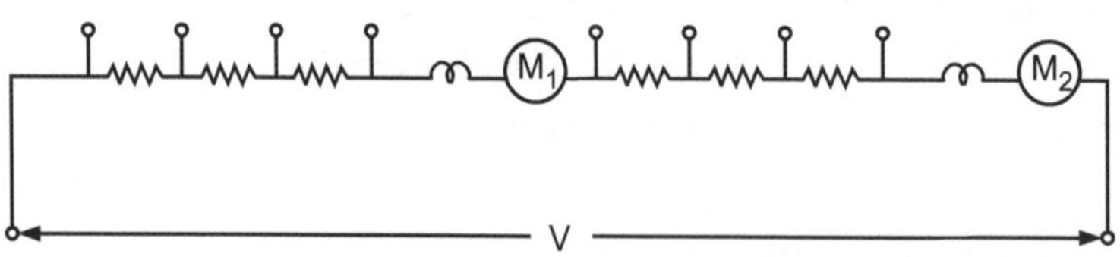

(a) Both motors are connected in series through full starting resistance

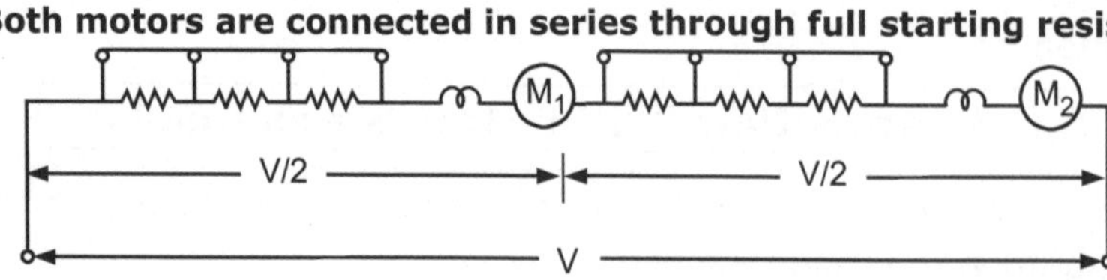

(b) Starting resistance is gradually cut down and both motors get connected across supply and receiving half the rated voltage $\left(\text{i.e. } \frac{V}{2}\right)$ and run at $\frac{1}{2}$ rated speed

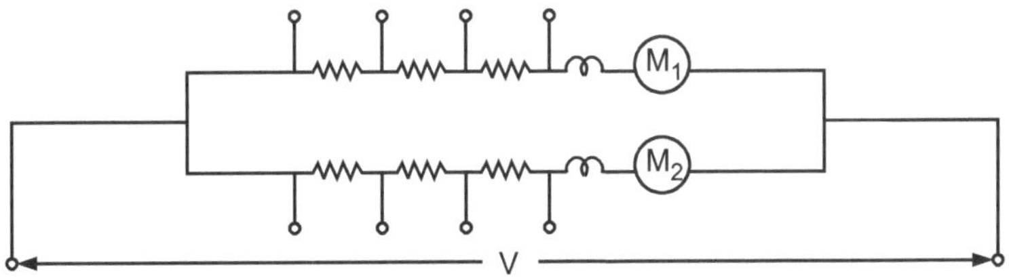

(c) Both motors are put in parallel alongwith their respective external resistance across supply

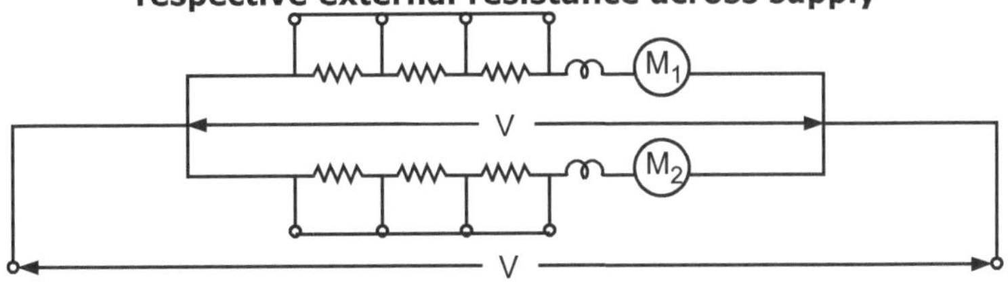

(d) External resistance is gradually cut down and both motors get connected across supply and receiving voltage = V and run at rated speed

Fig. 5.30: Series-Parallel Transition

- However various steps are involved between (b) and (d) which are illustrated in following Fig. 5.31.

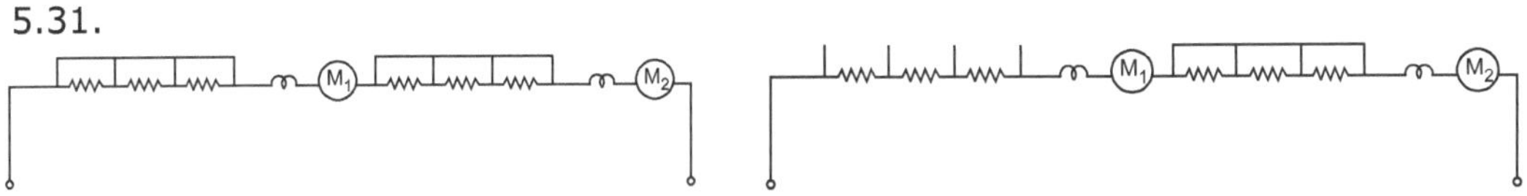

(a) Motors running in series with no starting resistance

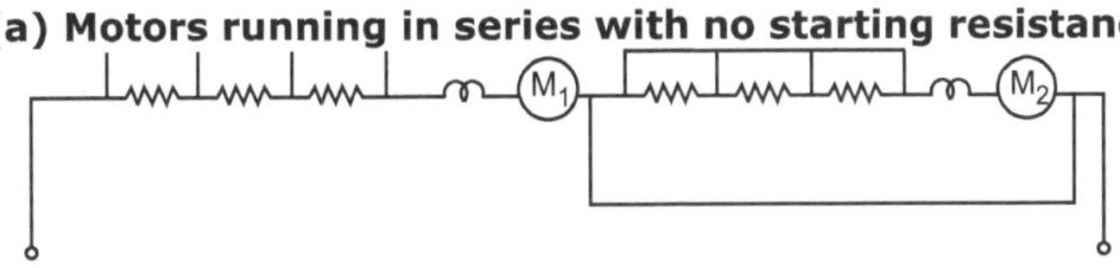

(b) M_2 is short-circuited

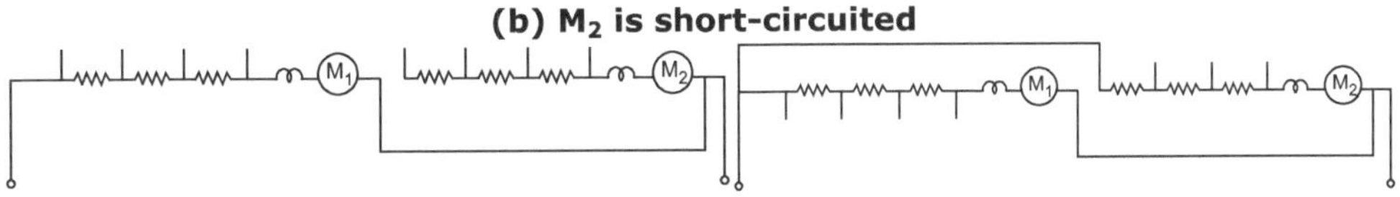

(c) Open end is connected to corresponding terminal of other motor

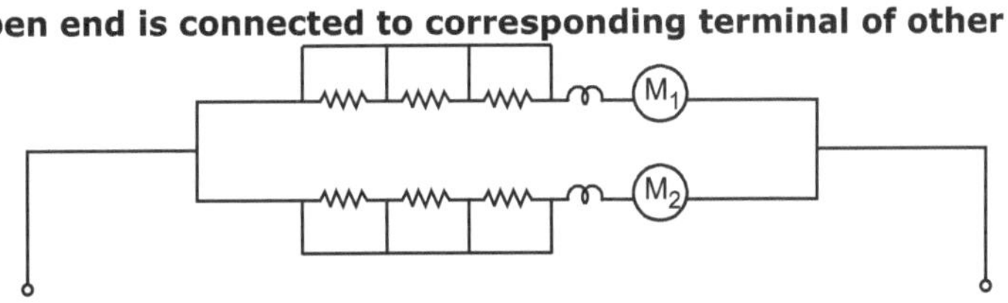

(d) Motors connected in parallel

Fig. 5.31: Steps in Series-Parallel Transition

- The above process is also known as shunt transition / short circuit transition.
- Note that the resistance inserted in individual motor circuits is for limiting starting current and it is also used during transition from series to parallel. This resistance is not for the purpose of speed control.
- It is observed that during shunt transition motor, M_2 is short circuited [Fig. 5.31 (b)] and then it is disconnected from supply and afterwards reconnected in parallel. Thus only one motor is in operation during transition.
- This results in loss of tractive effort during transition, i.e. total torque reduces by about 50% during transition period causing noticeable jerk in the motion of the vehicle. Inspite of this, this method of transition is preferred because the time during which motor remains shorted (transition time) is quiet small and this method operates satisfactory over wide range of loads, thereby making it suitable for locomotive application. Generally, this method is used for voltages above 600 volt.

Advantages:

(i) The supply remains connected to the motor during series-parallel transition.

(ii) Simple operation.

(iii) Comparatively smooth transition than open circuit transition.

Disadvantages:

(i) During transition, only one motor is in the system. So load on this motor is more.

(ii) There is loss of tractive effort.

(iii) There is jerk on one motor.

5.6.3.3 Bridge Transition

- The main advantage of bridge transition method is that none of the motor is disconnected from the supply during transition.
  1. All starting resistances are shorted and motors are in series.
  2. & 3. Bridge link is moved, till a portion of starting resistance is connected in parallel with each motor.
  4. Bridge link is removed and motors are in parallel with individual starting resistances which will be gradually cut down.
- It is seen that during transition, the two motors remain connected to the supply. And the resistances are varied such that motor current remains almost same. Hence, torque remains the same and jerk is avoided. So uniform acceleration is obtained.

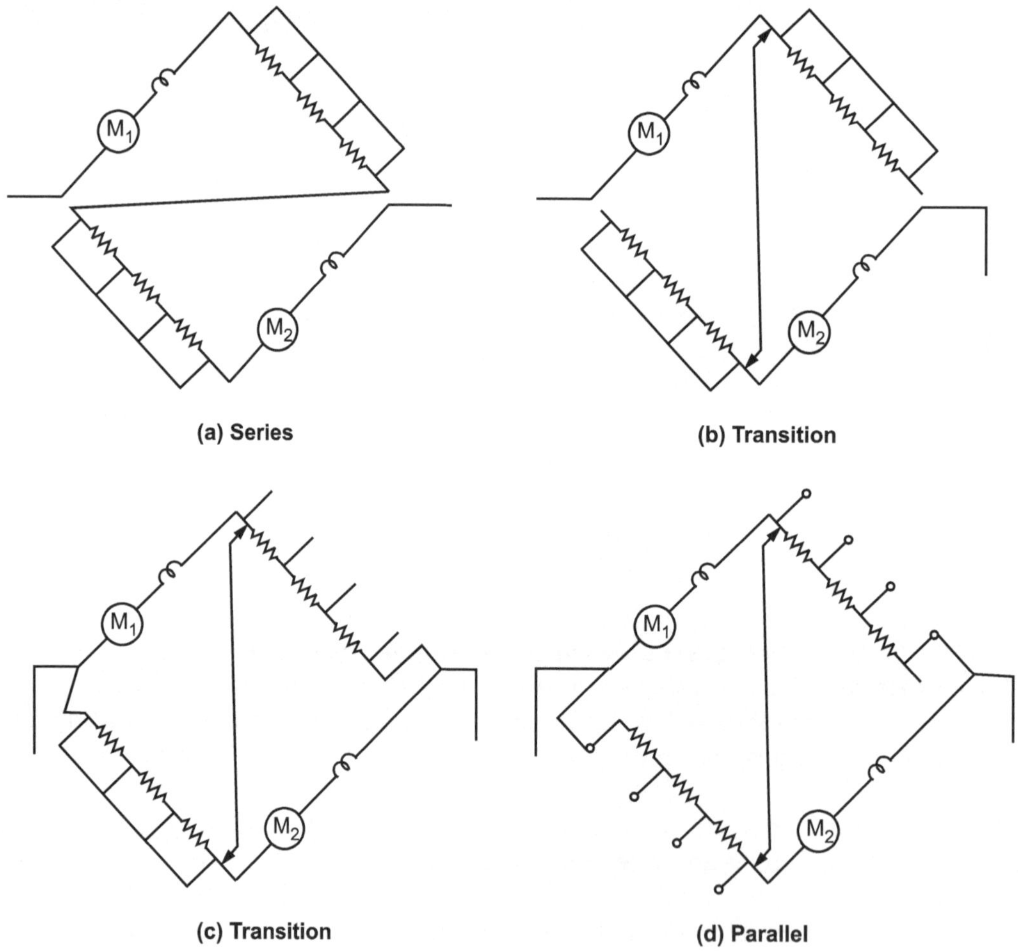

Fig. 5.32: Bridge Transition

Advantages:

(i) Simple operation, the steps involved are easy.

(ii) Both motors remain connected to the supply during transition.

(iii) No motor is overloaded.

(iv) There is no jerk, so operation is smooth.

Disadvantages:

(i) Circuit is somewhat complex than other methods.

5.6.3.4 Block Diagram of A.C. Locomotive and Function of each Part

1. Draw a neat labelled block diagram of A.C. locomotive and state function of each part.

- The block diagram of electric locomotive is as shown in Fig. 5.33. The pantograph collects the current through sliding contact with the overhead conductor or catenary. Air circuit breaker is provided at the top of the locomotive which disconnects the supply in case of over voltage. Step-down transformer with tap charger on HV side is provided in the locomotive. The taps can be charged as per the speed requirement of the traction motor. The traction motor is D.C. so the D.C. supply is provided through rectifier unit. A filter or smoothing choke is also provided so that pure D.C. supply can be fed to the motors. The filter reduces the undesirable effects such as bad commutation and heating (which are produced because of alternating component of rectified A.C.).

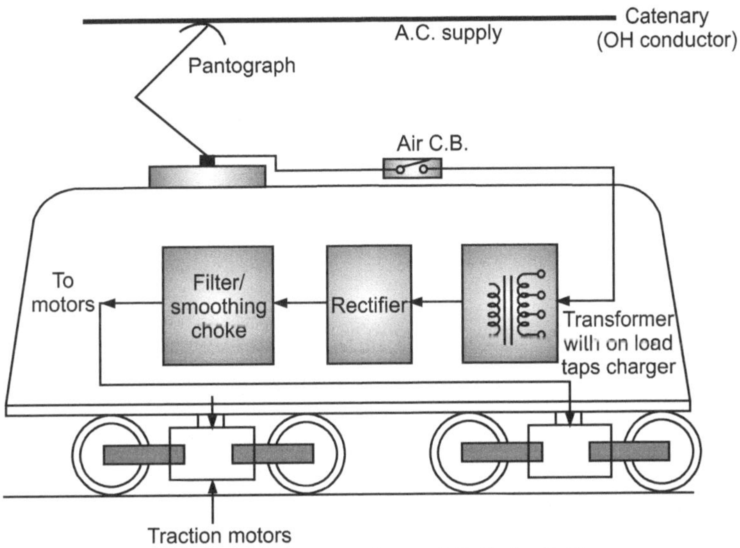

Fig. 5.33: Conversion of A.C. to D.C.

5.6.3.5 Comparison of Open Circuit (Series) and Short Circuit (Shunt) Transition

1. Compare open circuit transition and shunt circuit transition.

Sr. No.	Point of Comparison	Open Circuit Transition	Shunt (Short) Circuit Transition
1.	Inrush current during transition	More	Less
2.	Loss of tractive effort	More	Less (50%)
3.	Diagram during transition	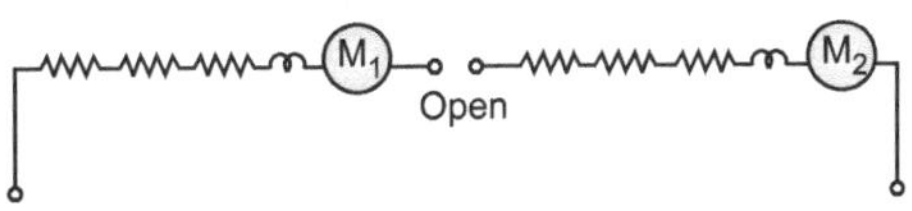	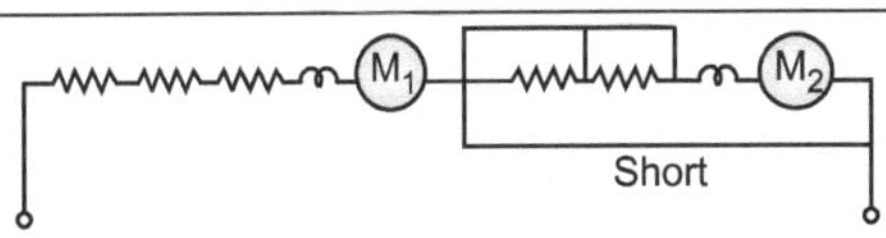
4.	Jerk during transition	More	Less
5.	Complexity of operation	Less	More

Comparison between Shunt Transition and Bridge Transition:

1. Compare shunt transition and bridge transition.

Sr. No.	Point of Comparison	Shunt Transition	Bridge Transition
1.	Inrush current during transition	Moderate	No inrush current
2.	Loss of tractive effort	50%	No
3.	Jerk during transition	Less	No jerk
4.	Complexity of operation	More	Less
5.	Diagram	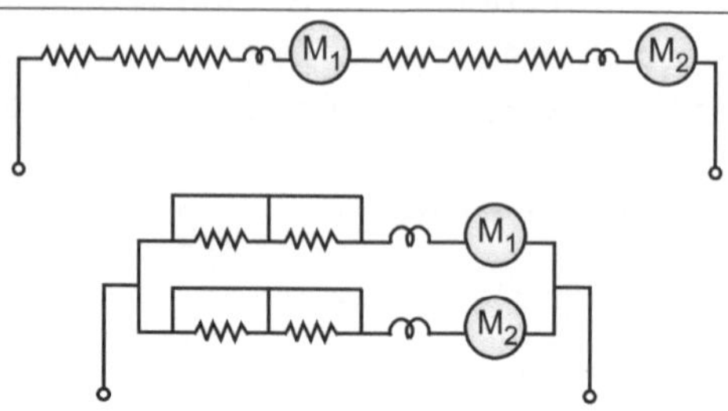 **Fig. 5.34**	Series : Refer Fig. 5.32 (a) Parallel : Refer Fig. 5.32 (d)
6.	Connections of resistances in the circuit	Series-Parallel	In the form of Wheatstone's bridge
7.	Acceleration achieved	Not as smooth as bridge transition	Smooth
8.	Continuity of supply to motor during transition	One more is shorted and other motor receives supply.	All motors receive continuous supply.

5.6.4 Energy Loss and Efficiency Calculations of Series-Parallel Transition

1. Derive the expression for energy loss and efficiency for series-parallel control of two D.C. series motors.

- Consider a DC motor started with a controller resistance. The variation in back emf (E_b), variation in voltage drop of controller resistance and variation in motor resistance drop (I_aR_a) is to be observed and plotted.
- At start, $E_b = 0$. Therefore, the supply voltage is dropped in the controller and armature resistance (series field resistance is neglected) i.e. $V = I_aR_a$ + Voltage drop in controller. As the motor speed increases, E_b increases and controller resistance is cut down in such a way that motor does not take more than full-load current during starting. The motor accelerates till all the resistance is cut out and at this instant, $V = E_b + I_aR_a$.
- The above process is represented in Fig. 5.35 (a).
- On X-axis time is plotted and on Y-axis all voltages and voltage drops i.e., I_aR_a, E_b, voltage drop in controller are plotted.
- The process starts at point O. Here motor is just switched ON. So $E_b = 0$. The motor draws current (I_a) and I_aR_a drop takes place. I_aR_a = OA.
- Same current flows through entire controller resistance and creates voltage drop in controller = BA.

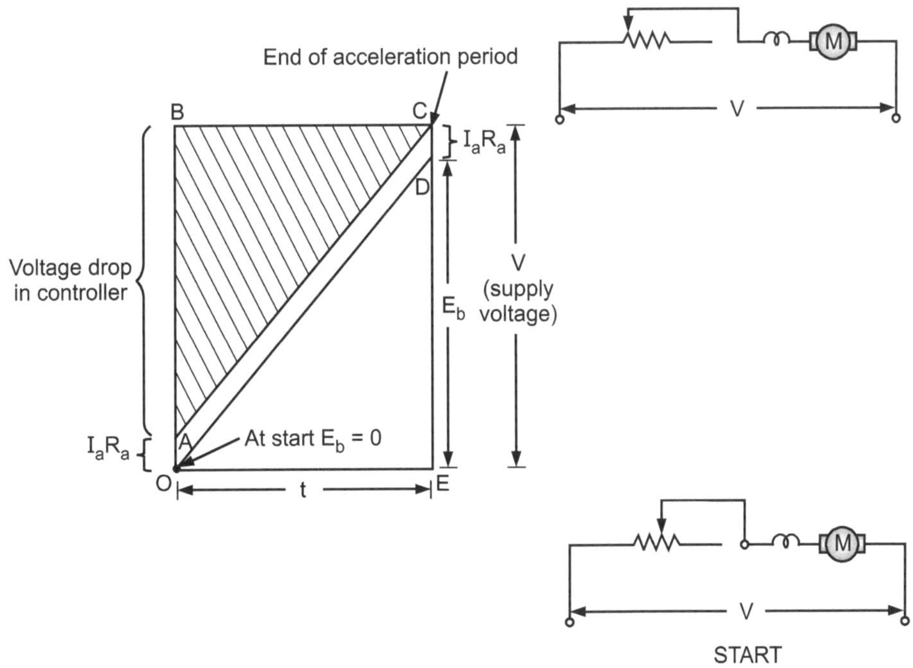

Fig. 5.35 (a)

- Now controller resistance is gradually cut down. The motor speed increases, hence E_b increases linearly as shown by line OD. So the magnitude of E_b at start is zero (point O) and it is equal to ED at the end of acceleration period when entire controller resistance is cut down.
- As the controller resistance is cut down, I_a remains reasonably constant. Therefore I_aR_a drop (=OA) remains constant. Therefore, OA = CD. This drop from start to end of acceleration period remains constant.
- Since controller resistance is gradually cut down the voltage drop in controller resistance will linearly decrease. It will decrease from BA to zero i.e. point C.
- So at start the voltage drop in controller was maximum (BA), since entire controller resistance is in series with the motor and it decreases to zero (point C) at the end of acceleration period.
- This is how the energy diagram for simple rheostatic starting is plotted. If we neglect I_aR_a drop, the intercepts OA, CD will no longer exist and hence the energy diagram will be as per Fig. 5.35 (b).

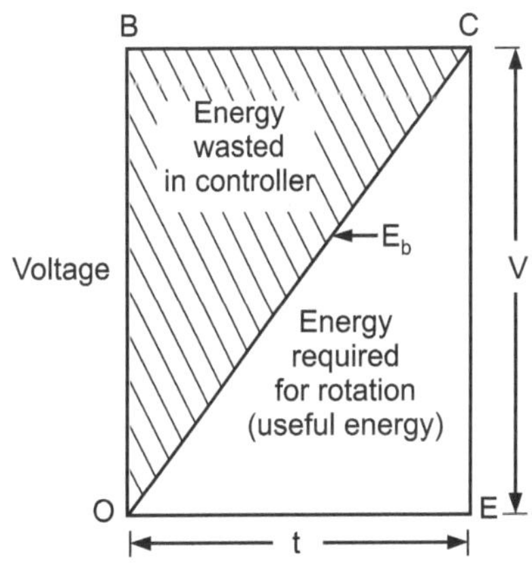

Fig. 5.35 (b)

OC = Represents variation of E_b

CE = V

OE = Time (t) required to cut down resistance or acceleration time.

The area of OBCE = $t \times V$

- If we multiply it by current of motor, we get

  $I_a \times$ (Area of OBCE) = $I_a (t \times V) = V I_a t$ watt second
- Thus the area of OBCE when multiplied by current gives total energy input to the system. Hence Fig. 5.35 (a) and (b) are known as energy diagram.
- Now we need to find energy efficiency of this process. This will be worked out on average basis. We consider two instants:

  (i) Point O (start of acceleration)

  (ii) Point E (end of acceleration)

The average useful energy at these instants is worked out.

$$\text{The average useful energy will be} = \frac{(\text{Energy input at point E}) - (\text{Energy input at point O})}{2}$$

$$\text{Average useful energy} = \frac{VIt - 0}{2} = \frac{VIt}{2} = \frac{VI_a t - 0}{2} = \frac{VI_a t}{2}$$

$$\text{The input energy} = VIt$$

$$\text{Energy efficiency, } \eta = \frac{\text{Average useful energy}}{\text{Input energy}} = \frac{\left(\frac{VI_a t}{2}\right)}{VI_a t}$$

$$\eta = \frac{1}{2} = 50\%$$

So 50% energy is wasted in controller resistance, it can be seen in Fig. 5.35 (b) that 50% area is shown with shaded portion that represents loss in controller and 50% blank area represents useful energy.

$$\text{Energy loss in controller resistance} = \text{Area } (\Delta \text{ OBC}) \times I_a = \left(\frac{1}{2} tV\right) \times I_a$$

$$= \frac{VI_a t}{2} \text{ watt second which is 50\% of total input energy}$$

$$\text{Useful energy} = \text{Area } (\Delta \text{ OEC}) \times I_a = \left(\frac{1}{2} tV\right) \times I_a$$

$$= \frac{VI_a t}{2} \text{ watt second which is 50\% of total input energy}$$

Thus, for single motor started with controller resistance, the efficiency of starting is 50%.

5.6.4 (A) Series Parallel Transition with Two Motors

1. Describe series parallel transition with two motors.

- Now consider series parallel transition with two motors.

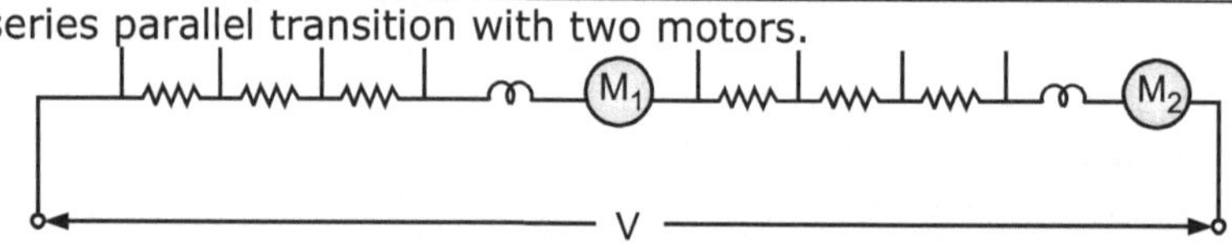

Fig. 5.36

- E_b of both motors at start is zero, the voltage drop in the individual regulator or controller will be V - (Armature drop in both motors). The voltage drop in controller goes on reducing till whole resistance is cut out and voltage across each motor is half the supply voltage. Motors are then put in parallel. The voltage drop in controller resistance is half of the supply voltage.

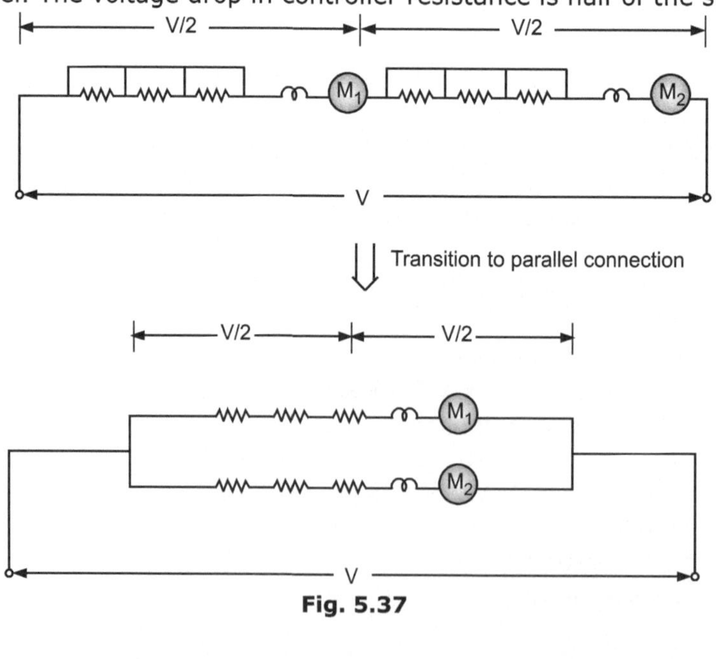

Fig. 5.37

- The voltage drop in controller will gradually reduce till the end of acceleration period (whole controller resistance is cut). This process is illustrated in Fig. 5.38.
- For energy diagram of two motors, started with series parallel operation, note following points:
- Energy diagram is divided into two sections – series operation, parallel operation.
- The voltage drops, E_b, voltage across motors are included in energy diagram.
- For series operations, all above drops will be plotted for both motors.
- For parallel operation, all above drops will be plotted for one motor. Since both motors are identical, the variation in all these parameters will be same.
- Three instants are important:
  (i) Starting instant when both motors are in series.
  (ii) Transition instant when motors are switched from series to parallel.
  (iii) And last instant when both motors operate in parallel with complete controller resistance cut down.

  Refer Fig. 5.38.

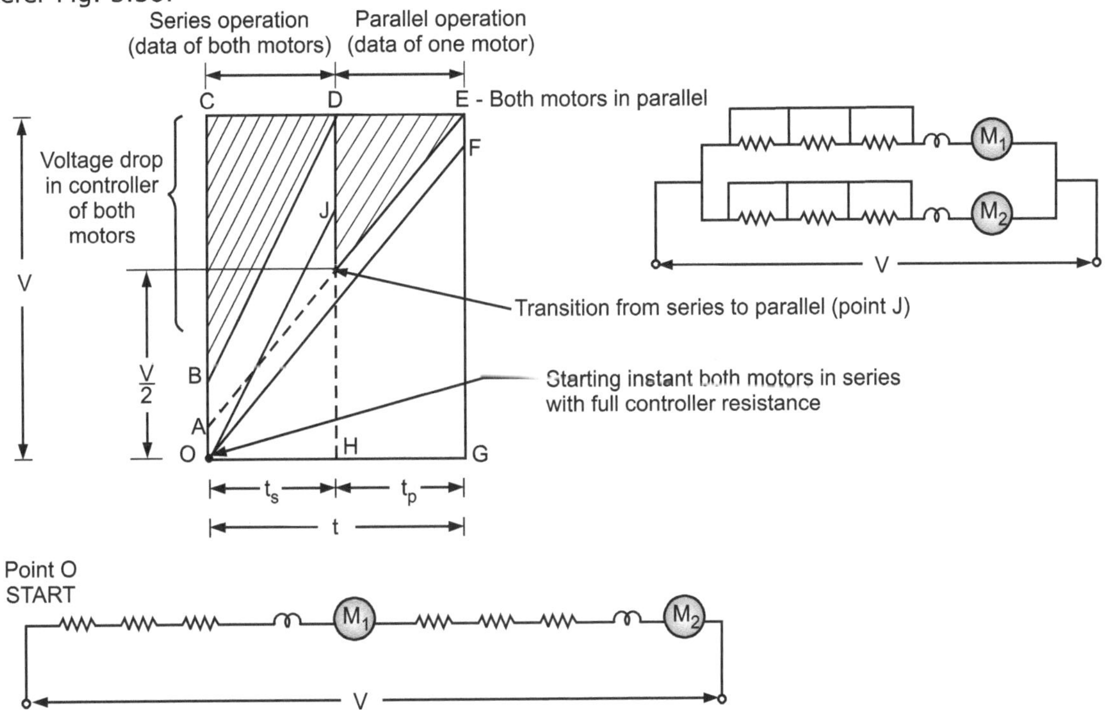

Fig. 5.38

5.6.4 (B) Limitations of Series Parallel cum Rheostatic Control

1. State the limitations of series parallel cum rheostatic control.

1. In plain rheostatic control, considerable amount of energy is wasted in starting resistance.
2. In series-parallel transition - for shunt transition method, during transition period one motor is short circuited, hence there is loss of tractive effort and jerk is experienced. So acceleration is not smooth.
3. In both methods of series-parallel transition, considerable amount of energy is wasted in starting resistance and uniform acceleration is somewhat difficult to obtain.
4. The current taken from supply is utilized to accelerate the train to 20-25 mph if full voltage is applied and remaining power or current is wasted in resistance.
5. The speed ranges obtained with different connections (series-parallel) are limited. So use of solid state devices became popular. The chopper and PWM techniques can control the speed of D.C. motor efficiency.

5.6.5 Chopper Control of Motors in D.C. Traction System

1. Explain chopper control of motors in D.C. traction system.

D.C. Chopper:

- It is a static device that converts fixed D.C. input voltage to a variable D.C. output voltage directly (Refer Fig. 5.39). A chopper may be thought as a D.C. equivalent of an A.C. transformer since they behave in an ideal manner.
- A chopper can be considered as a high speed ON-OFF semiconductor switch. It connects and disconnects the supply to the load at a very high speed.

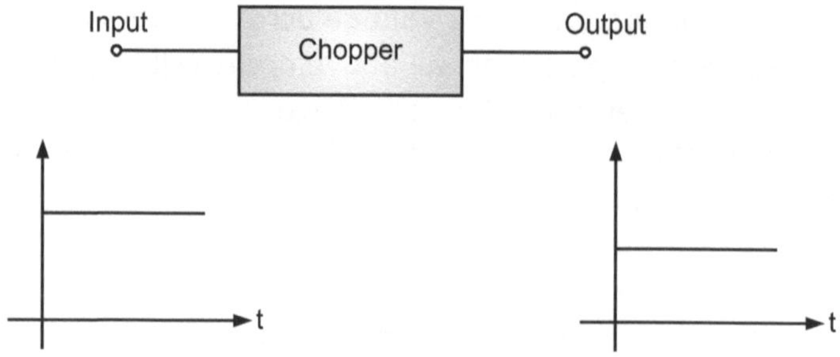

Fig. 5.39: Function of D.C. chopper

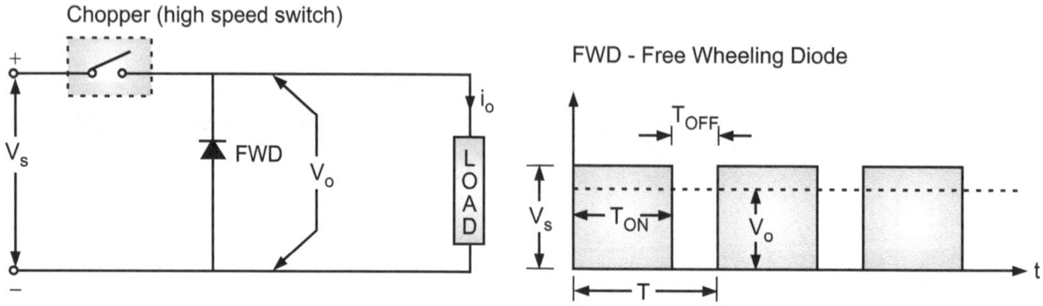

Fig. 5.40: Chopper operation

- The chopped voltage waveform is shown in Fig. 5.40. The chopper is represented by a high speed switch. Its switching time can be decided as per the requirement i.e. T_{ON} and T_{OFF} can be controlled as per the requirement.
- During T_{ON}, the chopper is ON and load voltage is equal to source voltage V_S.
- During T_{OFF}, the chopper is OFF (switch is open), the load current flows through Free Wheeling Diode (FWD). Thus load terminals are shorted through FWD, hence load voltage is zero during T_{OFF}. But load current continues to flow during T_{OFF}.
- The average load voltage comes out to be V_o shown by dotted line.

$$V_o = \frac{T_{ON}}{T_{ON} + T_{OFF}} \cdot V_s = \frac{T_{ON}}{T} V_s = \alpha V_s$$

where, T = Chopping period

α = Duty cycle

$\therefore \quad V_o = \alpha V_s$

Thus, load voltage can be controlled by controlling duty cycle (α).

- A separately excited motor is convenient to use i.e. field winding connected separately across the supply.
- Both motor's armature and field receive supply through thyristor (chopper). By varying duty cycle (α), the voltage applied to armature and current through field winding can be controlled independently. Thus speed control is possible.
- For switching OFF the thyristor during T_{OFF}, a thyristor commutating circuit is used. The freewheeling diode maintains the motor current during T_{OFF}.
- The thyristor acts as a fast switching device (chopper), this may cause a lot of electrical interference with neighboring circuits. So a line filter is used which prevents this interference.

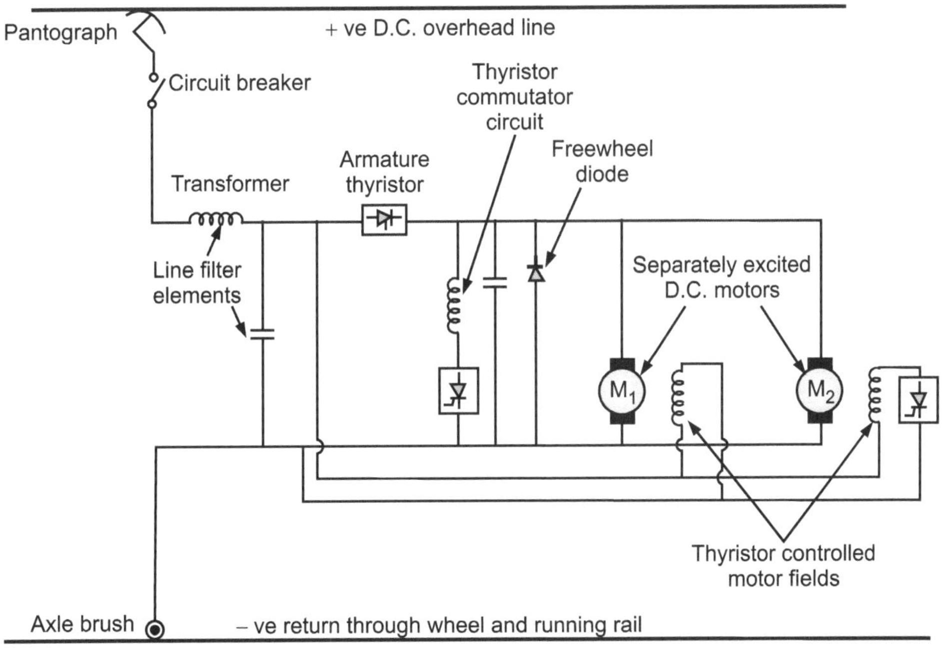

Fig. 5.41: A.C. locomotive with tap changer control of transformer output

5.6.6 Speed Control by Tap Changing Transformer

1. With the help of neat diagram, explain speed control by tap changing transformer.

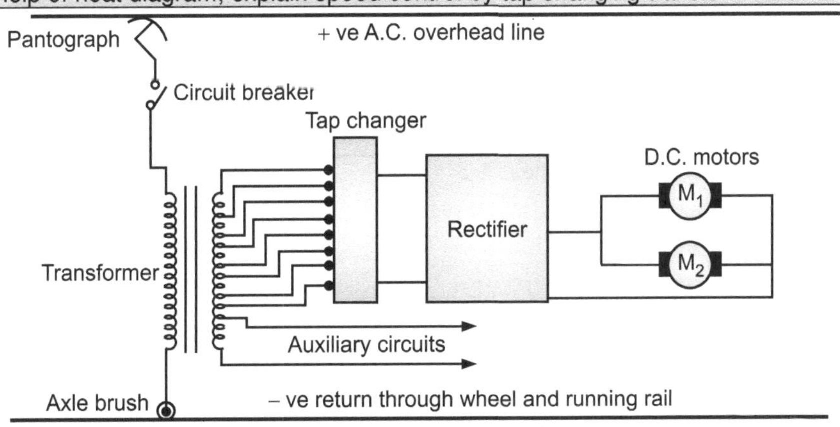

Fig. 5.42: Speed control by tap changing transformer

- This system is popular for 25 kV A.C. supply. A.C. supply is collected by pantograph from overhead conductor and it is given to the transformer. The transformer has got number of tappings. The tap changer device does the tap changing as per the requirement. This A.C. voltage is given to the rectifier which converts it to D.C. This D.C. voltage is fed to traction motors. The control part is an A.C. side i.e. controlled A.C. voltage is fed to rectifier and thus output D.C. voltage magnitude can be controlled. The tapping sequence can be decided and accordingly contactors can be used for tap changing operation.

5.6.7 PWM Control of Induction Motors

1. Describe PWM control of induction motors.

- A device that converts D.C. power into A.C. power at desired output voltage and frequency is called as an inverter. The output of inverter can be controlled by a technique known as Pulse Width Modulation (PWM).

Pulse Width Modulation:

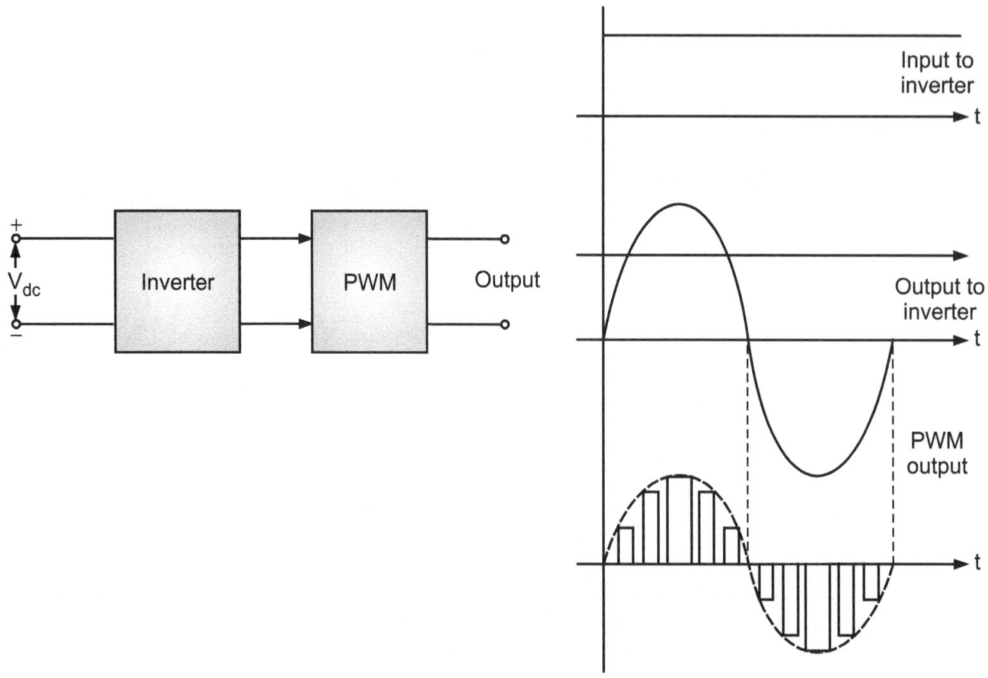

(a) (b)

Fig. 5.43: PWM Technique

- In this method, a fixed D.C. input voltage is given to the inverter and a controlled A.C. output voltage is obtained by adjusting T_{ON} and T_{OFF} periods of inverter components.
- The SCRs are used for this purpose. The stator voltage and frequency of induction motor can be controlled by using PWM technique and thus speed control of induction motor is possible.

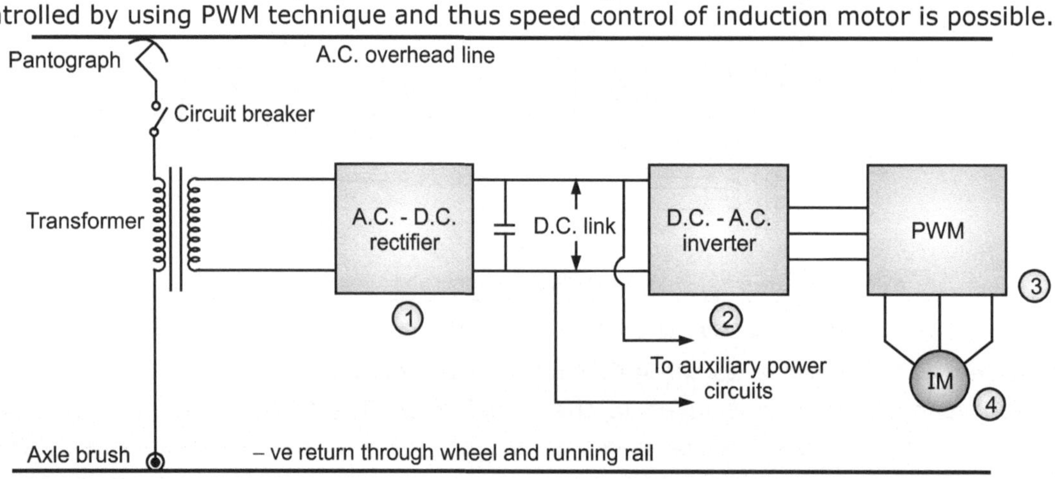

Fig. 5.44: PWM for induction motor

- The above circuit shows PMW scheme for induction motor used for traction work.
- Single-phase supply is collected from overhead conductor by means of pantograph and is passed on to step down transformer.
- Rectifier unit (1) converts A.C. into D.C.
- This D.C. is converted into 3-phase A.C. by inverter unit (2).
- 3-phase A.C. goes to PWM unit (3). This unit's technique is used to supply controlled voltage and frequency to induction motor (4).
- In this way, the speed transaction of induction motor can be controlled.

5.7 BRAKING

- Fast trains are the necessity of today. This is due to increased traffic density. With modern traction equipments, desired acceleration can be achieved easily and it does not matter if high acceleration is not achieved i.e. there is no risk involved. However, if stopping system fails, it will cause a disaster. So braking system of trains is important.

5.7.1 Requirements of Braking System

1. What are the requirements of braking systems?

- Requirements of braking system are:
  (i) Braking system should be robust.
  (ii) Simple to operate and control.
  (iii) It should be reliable and should require less maintenance.
  (iv) The brake actuation time should be as small as possible.
  (v) System should apply brakes simultaneously over all the vehicles.
  (vi) The brakes should be very gradual and smooth. However in case of emergency braking, safety is prime most consideration and hence, train should be stopped within minimum possible distance.
  (vii) To get uniform deacceleration, generally the braking force applied is proportional to the axial load.
  (viii) The braking system should respond effectively for repeated operation.
  (ix) The kinetic energy of train should be stored during braking and later on it should be used to accelerate the train.

5.7.2 Types of Braking

1. State the types of braking.

Types of **Braking**
- (A) Mechanical braking
  - (a) Air brake system/compressed air brake system
  - (b) Vacuum brake system
- (B) Hydraulic braking
- (C) Electric braking
  - (a) Eddy current braking
  - (b) Magnetic braking
  - (c) Regenerative braking
    - (i) French method
    - (ii) Regenerative braking using exciter

Fig. 5.45: Types of Braking

5.7.2 (A) Mechanical Braking

1. Explain in brief mechanical braking.

- The mechanical brakes are applied by brake block shoes which are pressed with force against wheels. The braking force is generally produced by means of piston-cylinder arrangement. The movement of piston is transmitted to the brake block shoes through system of levers.

(a) Air Brake System/Compressed Air Brake System

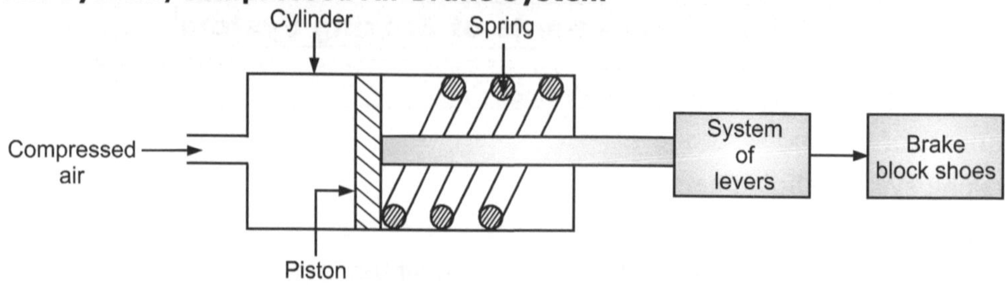

Fig. 5.46: Air brake system

- Air at high pressure enters the cylinder, causing the movement of piston against the spring tension and motion is transmitted to brake block shoe via system of levers.

(b) Vacuum Brake System:

- In this system, the vacuum is maintained inside the cylinder, then air is admitted to one side of piston which causes movement of piston.
- The vacuum is created inside the cylinder by means of vacuum pump. The air is sucked from both sides of piston. At the same time, vacuum is created in train pipe. When chain is pulled by any passenger, the air injector valve operates and vacuum in the system is destroyed and air rushes to the downside of piston through the coupler. This causes upward movement of piston which is transmitted to the system of lever and then to brake block shoes and thus braking is achieved. The resetting of system is done by operating the vacuum pump again.

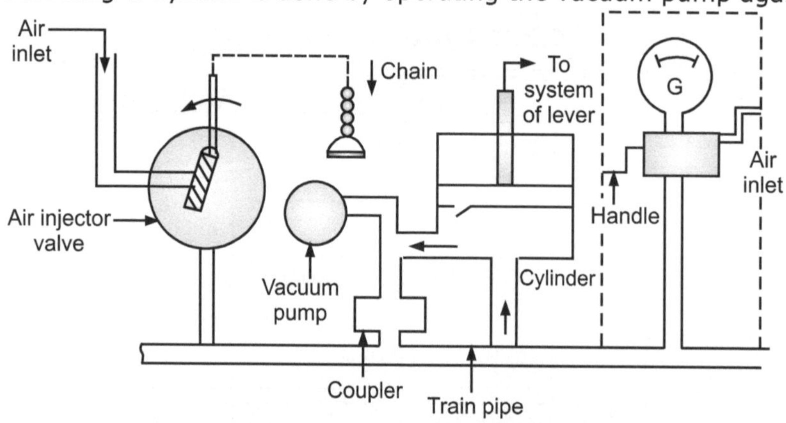

Fig. 5.47: Vacuum brake system

(c) Handbrake for Parking:

- This is provided in the driver's cabin and guard's cabin. It is a mechanical brake. It is operated by means of system of levers. A handle is provided to operate the system. When this handle is pulled the system of levers operate the braking mechanisms and brakes are applied simultaneously to all wheels of the engine.

5.7.2 (B) Hydraulic Braking

- An assembly of circulating fluid is fixed to the rotating member, the circulation of oil or fluid in the assembly generates the necessary braking torque.

5.7.2 (C) Electric Braking

- It has to be supplemented by mechanical braking.

Comparison between Mechanical Braking and Electrical Braking

1. Compare mechanical and electrical braking.

Sr. No.	Mechanical Braking	Electrical Braking
(i)	Mechanical braking produces excessive wear and tear of brake block shoes of brake lining. Therefore, frequent replacement is necessary.	No such need.
(ii)	Due to constant wear and tear of brake lining, frequent adjustments are necessary to compensate the offset error, therefore maintenance is costly.	Less maintenance.
(iii)	Produces metal dust causing wearing of bearings.	No such dust is formed.
(iv)	If they are not correctly adjusted then sudden braking action produces shock. This results in discomfort to passengers.	Adjustment is easy and braking action is gradual and smooth.
(v)	Heat is produced in brake shoes of brake lining, which may damage the lining and may result in failure of braking system.	Heat is produced at a convenient place which is is no way harmful to braking system.
(vi)	Regenerative braking is not possible.	60 to 80% of kinetic energy of apparatus can be fedback to line in regenerative brake.
(vii)	Mechanical braking alone can be used.	Electrical braking is alone not sufficient, it is to be supplemented by mechanical braking.
(viii)	Low cost	High initial cost.

Types of Electric Braking:

(a) Eddy Current Braking: Traction motors are made to work as generator and they supply power to shoe of eddy current brakes.

In rotary eddy current braking, rotor is attached to motor shaft and stator is supplied with D.C. excitation. The currents produced in rotor tend to reduce relative speed between rotor and stationary field (dynamic braking action) and braking is obtained.

(b) Magnetic Braking: Passage of current through coil of brake shoe produces magnetic flux in perpendicular direction to rail. The force of attraction is produced which provides braking.

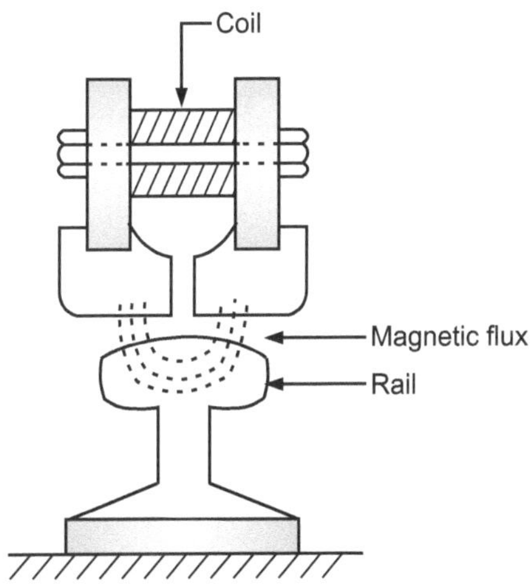

Fig. 5.48: Magnetic braking

Sequence of Braking:

- The sequence will be mechanical brakes supplemented by electric braking preferably regenerative electric braking.

5.7.3 Regenerative Braking

1. What are the difficulties in case of regenerative braking in case of D.C. series motor?
2. Explain regenerative braking for D.C. series motor.

- In case of regenerative braking, the motor is made to run as a generator and it remains connected to the line. Thus, the kinetic energy is converted into electrical energy which is fedback to the line.

Conditions for regenerative braking:

1. The necessary condition for regenerative braking to take place (in case of D.C. motors) is that the back e.m.f. or generated voltage must exceed the supply voltage.
2. The braking system should have mechanical stability i.e. braking system should be in a position to apply more and more braking torque in case of powerful overhauling (steep down gradient).
3. Braking system should have electrical stability Le braking torque applied should be independent of supply voltage fluctuations.
4. If the energy fedback to line is surplus i.e. supply is greater than demand then the generating station should be able to consume the energy, otherwise, regenerative braking is ineffective.

- The regenerative method is obviously preferable to other two methods because the braking energy is not wasted but it is fedback to the line. This fedback energy can be further utilized by other electrical load connected to the line.

D.C. Shunt Motor:

- Suppose decrease in load on motor causes increase in speed above normal with field current remaining the same. Thus, the back e.m.f. will increase and it may exceed supply voltage. When this happens, the power is fedback to the line and kinetic energy (KE) decreases gradually, thereby, decreasing the speed and thus braking is achieved.

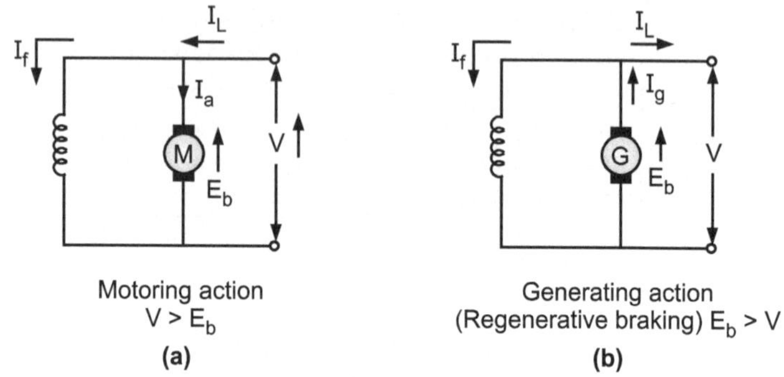

Fig. 5.49: Regenerative braking - Shunt motor

- Alternatively the field current can be increased and same effect can be achieved. Regenerative breaking can thus easily be applied with shunt motor.

  Note: Direction of field current should remain same.

D.C. Series Motor:

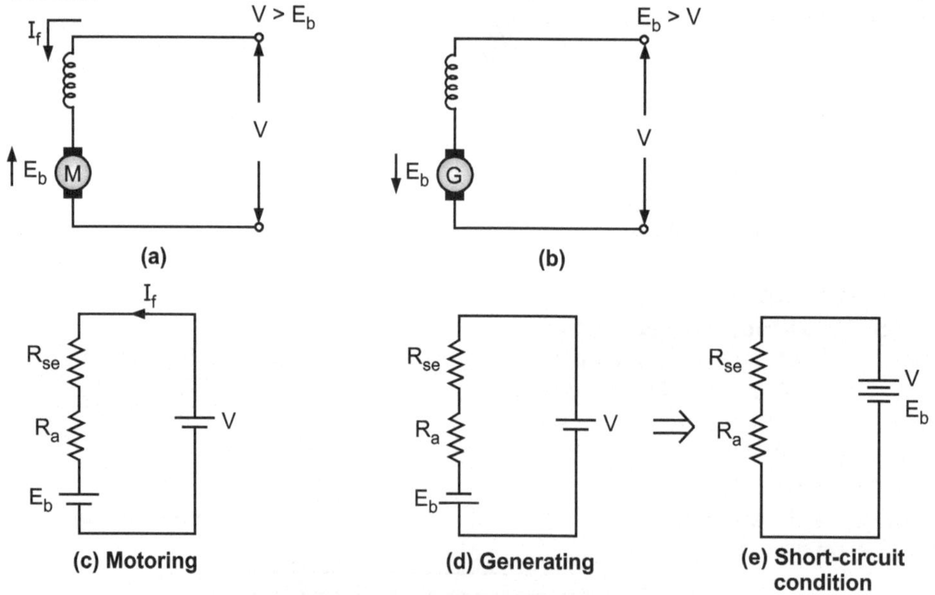

Fig. 5.50: Regenerative braking - Series motor

Difficulties in Regenerative Braking:

- The conditions with a series motor for regenerative braking are not so simple, since the armature current necessary to produce regeneration also flows through field winding. Therefore field current reverses causing direction of E_b as shown in Fig. 5.50. Thus E_b and V are added and the resistance of circuit is (R_a + R_{se}) which is very low. This causes short circuit condition (Fig. 5.50 (e)).

- Other possibility is, the field flux will be neutralized during the transition.
- Thus, the requirement is to reverse the field winding connection at the exact instant but the method is still useless because at the instant of reversal, the e.m.f. generated will be less i.e. supply voltage will be greater than generated e.m.f. Hence regenerative braking is not possible.
- Thus, the main trouble associated with regenerative braking by series motors is seen to be due to the lack of the control of the field. There are various methods of overcoming this difficulty, either by modification of the windings or by supplying the machine with separate excitation.
- These methods described below are mainly used in electric traction.
- In following methods, the field circuit is modified to avoid short circuit condition. The modification also gives compensation for voltage fluctuations occurring in line during regenerative braking.

(i) French method:

1. Explain French method of regenerative braking.

- In this method, auxiliary field windings are used.
- During motoring action, the main series field winding and auxiliary field windings are connected in parallel and this whole combination is in series with the armature (Refer Fig. 5.51).
- In generating mode, the auxiliary field winding group is connected in series and this combination is connected to motor circuit as shown. Thus the machine now behaves as a compound generator with a slightly differential action. This eliminates the short circuit condition.
- The shunt combination is sensitive to voltage fluctuations. Note that slight changes in supply voltage frequency occur in the traction system. The shunt combination provides necessary balancing action and thus maintains potential difference between V and voltage generated (with $E_b > V$).

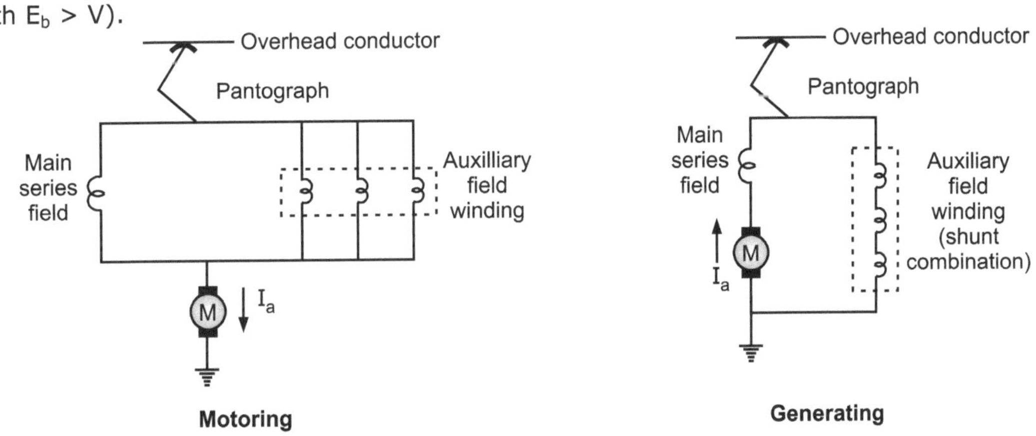

Fig. 5.51: French method

- The idea of French method can be implemented for more than one motor also. In locomotives 4 to 6 motors are used at a time. In that case the main field winding of motor is retained and other is made to function like auxiliary winding.

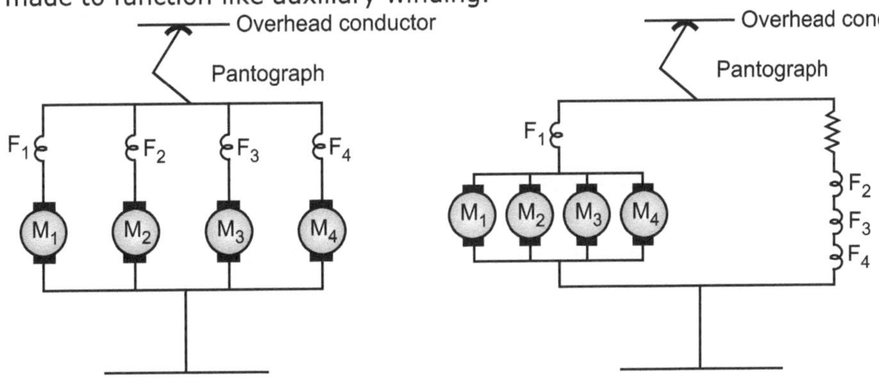

Fig. 5.52

- Suppose during regenerative braking the line voltage increases beyond the generated voltage. Same increased voltage is present across the shunt combination, thus it will cause increase in exciting current through it. Hence, as voltage is developed the generator also increases and again becomes greater than supply voltage. Reverse will happen when line voltage falls. The method is applicable for trolley bus and trams.

(ii) Regenerative Braying using Exciter :

1. Explain regenerative braking system using exciter.

- The series field windings of motor are connected in series and these are excited by the exciter. The exciter unit has got its own field windings F_1 and F_2. F_1 is again connected to a separate D.C. source. The field current of F_1 is to be controlled by the operator. F_1 and F_2 are connected in such a manner that their fluxes act in opposite direction.
- Suppose during regenerative braking, the line voltage decreases, hence (E_b−V) increases, hence regenerative current increases and same current is flowing through F_2, therefore excitation or flux of F_2 increases. Since, F_1 and F_2 oppose each other, the e.m.f. E falls on account of reduced excitation. (Action is similar to differential compound generator). As soon as the e.m.f. of the exciter falls, the current in the field windings of M_1 and M_2 decreases and therefore regenerative e.m.f. also decreases. Thus compensation for decrease line is provided.
- In another arrangement (Fig. 5.53), armature of exciter is connected in the circuit of field windings and stabilizing resistance. The current stabilizing resistance is the summation of regenerated current and exciter current.
- Suppose during regenerative braking, the line voltage falls, then (E_b−V) increases, therefore, regenerative current increases. Therefore, voltage drop across stabilizing resistance increases. Therefore, the balance of voltage or net voltage available in the exciter armature circuit reduces. This causes reduction in exciting currents of field windings of motor, hence regenerated e.m.f. decreases. Thus compensation is provided.
- The function of stabilizing resistance is to prevent the current surges when the vehicle crosses one section of supply to another and to compensate for variation in line voltage.

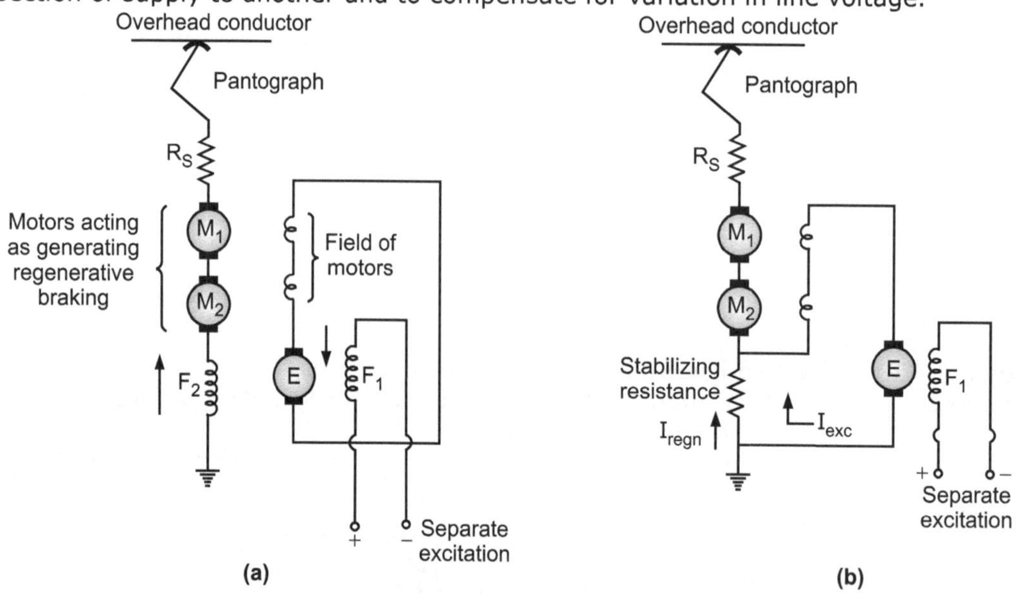

Fig. 5.53: Regenerative braking using exciter

Regenerative braking with three-phase induction motor:

- It is achieved automatically when the motor runs a speed slightly above the synchronous speed. The motor starts working as an induction generator. The induction generator however is not self-exciting and must be connected to a system supplied from synchronous generator i.e. its field winding/rotor coinciding will be supplied with D.C. by an exciter of some other synchronous generator system.

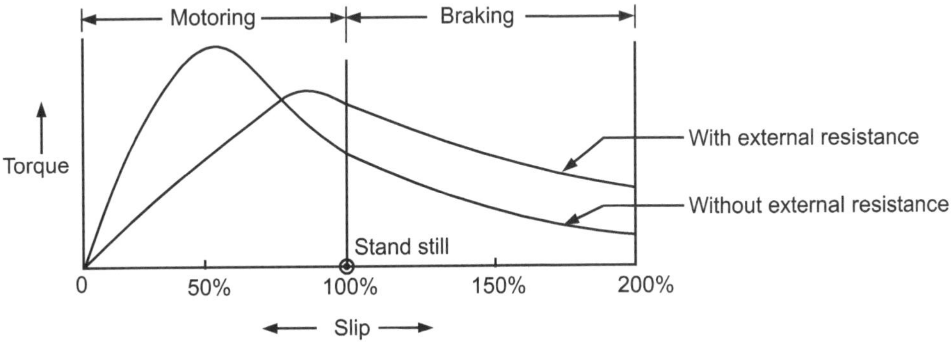

Fig. 5.54: Regenerative braking with induction motor

Example 5.14: *A 2340 tonnes train including loco proceeds down a gradient of 1 in 80 for 5 minutes during which period its speed gets reduced from 60 kmph to 36 kmph by application of regenerative braking. Find the energy returned to the lines if the tractive resistance is 5 kg/tonne, rotational inertia 10% and overall efficiency of the motor during regeneration is 70%.*

Solution: Accelerating weight of train,

$$W_e = 1.1 \times W$$
$$= 1.1 \times 2{,}340 = 2{,}574 \text{ tonnes}$$

Track resistance, $r = 5 \times 9.81$ Newton/tonne

Gradient, $G = \frac{100}{80} = 1.25\%$

Regenerative period, $t = 5 \text{ minutes} = 5 \times 60 = 300 \text{ seconds}$

Overall efficiency, $\eta = 0.7$

Energy available due to reduction in speed

$$= 0.01072\, W_e \left(V_1^2 - V_2^2\right)$$
$$= 0.01072 \times 2{,}574 \times (60^2 - 36^2)$$
$$= 63{,}575 \text{ Wh or } 63{,}575 \text{ kWh}$$

Tractive effort required while going down the gradient

$$= Wr - 98.1\, WG$$
$$= 2{,}340 \times 5 \times 9.81 - 98.1 \times 2{,}340 \times 1.25$$
$$= -172{,}165.5$$

i.e., tractive effort of 172,165.5 Newtons is available.

Distance moved during regeneration

$$= \frac{V_1 + V_2}{2} \times \frac{1{,}000}{3{,}600} \times t = \frac{60 + 36}{2} \times \frac{1{,}000}{3{,}600} \times 300 = 4{,}000 \text{ m}$$

Energy available on account of moving down the gradient over distance of 4,000 m

$$= \frac{172{,}165.5 \times 4{,}000}{1{,}000 \times 3{,}600} = 191.3 \text{ kWh}$$

Total energy available $= 63{,}575 + 191.3 = 254.875$ kWh

Energy returned to the line $= 0.7 \times 254.875 =$ **178.4 kWh**

Average speed $= \frac{40 + 20}{2} = 30$ kmph

Average time taken to cover 2 km $= \frac{2}{30} = \frac{1}{15}$ hours

Average power, $P = \frac{\text{Energy returned to the line}}{\text{Time}} = \frac{30.28}{1/15} =$ **454 kW**

Example 5.15: *A train weighing 400 tonnes has speed reduced by regenerative braking from 40 kmph to 20 kmph over a distance of 2 km on a down gradient of 2%. Calculate the electrical energy returned to the line. Tractive resistance is 40 N/tonne and allowed rotational inertia of 10% and efficiency of conversion is 75%.*

Solution: Accelerating weight of the train $W_e = 1.1 \times W = 1.1 \times 400 = 440$ tonnes

Distance travelled during period, S = 2 km

Gradient, G = 2%

Efficiency of conversion, $\eta = 75\% = 0.75$

Energy available due to reduction in speed

$$= 0.01072\ W_e \left(V_1^2 - V_2^2\right) = 0.01072 \times 440 \times (40^2 - 20^2)$$

$$= 5{,}660 \text{ watt-hours} = 5.66 \text{ kWh}$$

Tractive effort required while going down the gradient

$$F_t = Wr - 98.1\ WG = 400 \times 40 - 98.1 \times 400 \times 2 = -62{,}480 \text{ Newtons}$$

i.e. Tractive effort of 62,480 Newtons is available.

Hence energy available on account of moving down the gradient a distance of 2 km

$$= \frac{F_t \times S \times 1{,}000}{1{,}000 \times 3{,}600} \text{ kWh} = \frac{62{,}480 \times 2 \times 1{,}000}{1{,}000 \times 3{,}600} = \mathbf{34.7\ kWh}$$

Total energy available = 5.66 + 34.7 = **40.37 kWh**

Energy returned to the line = 0.75 × 40.37 = **30.28 kWh**

5.7.4 Dead Man's Handle

- The function of Dead Man's Handle is to stop the train in case of driver faints or becomes incapicitated. The dead man's handle is attached to the controller.
- If driver releases this handle, all the relay contacts are opened and motors are disconnected from the supply.
- Thus train stops.
- This dead man's handle causes opening of contacts of master controller (C_1, C_2, C_3, ...).
- Thus supply of all the motors is cut-off and this causes stopping of the motor.

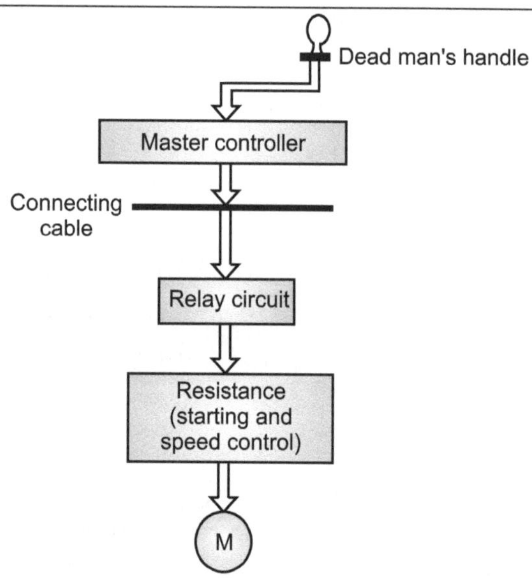

Fig. 5.55: Function of Dead Man's Handle

Important Points

- **Traction:** The act of drawing or pulling by power is known as traction.
- **Non-electric traction system:** This traction system does not use electricity. e.g. steam engine drive, direct internal combustion engine drive (direct IC engine drive).
- **Electric traction system:** This traction system uses electricity partly or completely. e.g. straight electric drive, battery electric drive, diesel electric drive.
- **Speed-time or Speed-distance curve:** The movement of trains and their energy consumption is most conveniently studied by means of speed-time and speed-distance curves.
- **Maximum or Crest speed:** It is the maximum speed attended by a train during a run.
- **Average speed:** It is the mean speed maintained by a train during a run.
- **Schedule speed:** It is the ratio of the distance between the stops and the total time taken including time for stops to cover the distance.

Practice Questions

1. What are the requirements of traction system?
2. Describe in brief the following and state their advantages and disadvantages:
   (a) Steam engine drive
   (b) Electric drive
   (c) Diesel electric drive
   (d) Direct internal combustion engine drive.
3. What are the different types of track electrification systems?
4. Give advantages of 25 kV A.C. system over D.C. system of traction.
5. Draw a diagram of A.C. electric locomotive and explain function of each part in it.
6. Write a short note on speed-time curve.
7. What is specific energy consumption? Write total expression for it.
8. What are the different factors affecting specific energy consumption?
9. What is coefficient of adhesion? What are the factors affecting the coefficient of adhesion?
10. What are the desired characteristics and traction motor?
11. Discuss the suitability of D.C. series motor for traction work.
12. With a suitable diagram, describe series parallel control of D.C. series motor.
13. With a suitable diagram explain bridge transition method.
14. List the type of traction system and write in brief about each.
15. Write short notes on:
    (i) Kando system of track electrification.
    (ii) D.C. system of track electrification.
16. Write short notes on:
    (i) Third rail system of current collection.
    (ii) Overhead current collection system.
17. What are different types of catenary? Describe each type in brief with suitable diagram.
18. Describe features for following services for traction in brief:
    (i) Urban service.
    (ii) Suburban service
    (iii) Main line service.
19. Draw simplified trapezoidal and quadrilateral speed-time curve and describe meaning of each notation related to the figure.
20. Explain the following terms:
    (i) Crest speed or maximum speed.
    (ii) Average speed.
    (iii) Schedule speed.
21. What are the various factors that affect the schedule speed? Explain each in brief.
22. Compare shunt and series motor performance for parallel running.
23. What are the requirements of traction control?
24. What is the drawback of series parallel control?
25. Why series parallel transition is known as short circuit transition?
26. Define unit of "tractive effort".
27. Define coefficient of adhesion, state its unit.
28. Explain the factors which affect co-efficient of adhesion.
29. Derive the relation of specific energy consumption.
30. State the factors which affect specific energy consumption.

31. List out desirable characteristics of traction motor.
32. State the mechanical and electrical features of traction motor.
33. How three-phase induction motor is suitable for traction service?
34. Explain with diagrams the working of linear induction motor.
35. Explain edge and end effect of linear induction motor.
36. State advantages and disadvantages of linear induction motor.
37. State the requirements of traction motor control.
38. Explain open circuit traction.
39. Explain short circuit traction.
40. Write on bridge transition.
41. Draw the block diagram of A.C. locomotive.
42. Compare open circuit with short circuit transition.
43. Explain chopper control of motors in D.C. systems.
44. With the help of neat diagram, explain speed control by tap changing transformer.
45. Explain PWM control of induction motor.
46. What are the requirements of braking system?
47. Compare mechanical and electrical braking.
48. Explain regenerative braking of D.C. series motor.
49. Write a note on Dead Man's Handle.

Problems with Answers

Numericals on Speed-time Curve:

1. The trapezoidal time curve of a train consists of:
   (i) Uniform acceleration of 6 kmphps for 25 sec.
   (ii) Free running for 10 min.
   (iii) Uniform deacceleration of 6 kmphps to stop the train.
   (iv) A stop time of 5 min.

   Find the distance between the stations, the average and schedule speed.

   [**Ans.:** S = 26.042 km, V_a = 144.23 kmph, V_s = 98.68 kmph]
2. A schedule speed of 45 kmph is required between the stations 1.5 km apart. Find the maximum speed over the run if the stop time is 20 sec. duration. The values of acceleration and retardation are 2.4 kmphps and 3.2 kmphps respectively. Assume simplified trapezoidal speed time curve.

   [**Ans.:** V_m = 74 kmph]
3. The distance between two stations is 1 km and schedule speed is 30 kmph with station stopping time of 20 sec. Assuming braking retardation of 3 kmphps and maximum speed 1.25 times average speed ($V_m = 1.25\ V_a$). Determine the acceleration required with trapezoidal speed time curve.

   [**Ans.:** α = 1.8 kmphps]
4. A train runs with an average speed of 73 kmph. Distance between the stops is 6 km. The train accelerates at 2 kmphps and retards at 3 kmphps. Find maximum speed and schedule speed. Duration of stop is 60 sec. Assume trapezoidal speed time curve.

   [**Ans.:** V_m = 81.2 kmph, Schedule speed = 60 kmph]
5. A train is required to run between two stations 3 km apart at an average speed of 40 kmph. The maximum speed is not to exceed 60 kmph. Acceleration is 2 kmphps, coasting retardation is 0.15 kmphps and braking retardation is 3 kmphps. Assuming quadrilateral speed-time curve, determine the duration of accelerating, coasting, braking periods and total running period.

   [**Ans.:** t_1 = 30 sec, t_2 = 231.6 sec, t_3 = 8.4 sec, T = 270 sec]

■■■

Chapter **6...**

Economic Aspects of Utilizing Electrical Energy

Syllabus

6.1 Economic Aspects of Utilizing Electrical Energy.

6.2 Costing of Electrical Energy: Fixed Charges, Semi-Fixed Charges and Running Charges.

6.3 Formulation of Electrical Tariffs.

6.4 Various Types of Tariffs: Tariffs in force for Domestic, Commercial and Industrial Consumers. Simple, Block rate, Two-part, Three-part, kVA, MD, P.F. Tariffs.

6.5 Power Factor Improvement: Causes of Low Power Factor, Disadvantages of Low Power Factor, Power Factor Improvement by using Static Capacitors, Location of Capacitors for Power Factor Improvement, Most Economical Power Factor. Automatic Power Factor Controller (Derivation and Simple Numericals)

6.6 Energy Conservation: Importance and Need of Energy Conservation, Measures for Energy Conservation in (i) Electric Drives, (ii) Electric Traction, (iii) Electric Heating, (iv) Refrigeration and Air Conditioning, (v) Illumination.

Objectives

- Students will understand selection of relevant tariff for the given applications with justifications.
- Develop the ability to draw sketches, method of power factor improvements, their advantages in power savings.
- Students will be in a position to calculate the most economical power factor for a given application.
- Describe the procedure to maintain the specified former factor of the system.
- To apply/select different methods of energy conservation, their needs and advantages.

6.1 INTRODUCTION (ECONOMIC ASPECTS)

- In most of the engineering works "Cost" is of the first importance.
- Cost is the criteria for accepting the project.
- Electricity company supplies electric power to the wide varieties of consumers to meet their requirements.
- When designing any project (say generating station), efforts are made to achieve overall economy so that per unit cost of generation is as low as possible.
- This helps the supplier to supply electric energy to the consumers at reasonable rates.
- So trying at every step, the company shall keep in mind that the product to be given to supplier shall be as minimum as possible so that consumers are satisfied.
- At the same time the supplier shall get enough profit to pay the employees, maintain the project, safety future expansion and reasonable profit.
- These are the economical aspects of generation, supplies and utilization of energy.

6.2 COSTING OF ELECTRICAL ENERGY

6.2.1 Classification of Costs

Cost

1. Fixed cost 2. Semi-fixed cost 3. Running/Operating cost

1. Fixed cost:

- This cost is independent of maximum demand (MD) and energy output.
- This mainly include:
  - (i) Cost of central organization
  - (ii) Interest on capital cost of land
  - (iii) Salaries of employees.

2. Semi-fixed cost:

- This cost depends on M.D. and independent of energy output.
- This cost mainly depends on:
  - (i) Annual interest and depreciation on the capital cost of generating plant, transmission and distribution network.
  - (ii) Buildings and civil work.
  - (iii) All types of taxes, insurance charges, salaries of lower class employees, clerical staff.
  - (iv) It also includes yearly compensation given to the workers.
  - (v) Semi-fixed cost and M.D.

3. Running cost/Operating cost:

- This cost includes the following in running of the plant:
  - (i) Annual cost of fuel.
  - (ii) Annual cost of lubricating oil.
  - (iii) Annual cost of water.
  - (iv) Maintenance.
  - (v) Repairs of equipments.
  - (vi) Wages/Salaries of operational and maintenance staff.
  - (vii) Salaries of supervisory staff.
- Running cost is approximately proportional to the units of energy generated (kWH/MWH).
- Prior to decide the implementation of a plant all these:
  1. Annual fixed cost
  2. Semi-fixed cost + Operating cost this sum must be assured that it is minimum.

6.3 FORMATION OF ELECTRIC TARIFF

6.3.1 Meaning of the Term Tariff

- Tariff means schedule of rates or charges. Tariff in case of electric supply means schedule or rate frames for supply of electric energy to the different classes of consumers.

6.3.2 Objectives of Tariff (Desirable Characteristics)

- Tariff must cover the following items:
  1. Recovery of cost of capital investment in Generation, Transmission and Distribution equipments.
  2. Recovery of cost of operation, supplies and maintenance of equipment.
  3. Recovery of cost of equipment for metering, billing, collection and related miscellaneous services.
  4. The satisfactory returns on capital investments.
- The aim is to distribute equally the cost of supplying energy among the various classes of consumers.

6.3.3 Formation of Tariff

- The principle factors involved in fixing of a tariff are enumerated below:
  1. Proper return is secured from each consumer. While fixing the tariff for different classes of consumers, it has to be taken into account whether the tariff will result in a revenue, meeting all the expenditure of the supply authority. In addition, collect money to enable future expansion for anticipated future load.
  2. The tariff should be simple and capable of easy explanation to the public. A complicated tariff may cause an opposition from the public.
  3. The consumers are encouraged to use more electricity.
  4. The tariff should be such as to satisfy the consumers of all categories. A big consumer should be charged at a lower rate than a small consumer.
  5. The consumers are encouraged to use power during off peak hours. They are penalised for high loads demanded at system's peak load.
  6. The consumers are penalised for poor power factor.

6.3.4 Types of Tariffs

1. State various types of tariffs. Explain suitable tariff for industrial consumers.
2. State various types of tariffs and explain 'flat demand rate' type.
3. State various types of tariffs. Explain suitable tariff for industrial consumers.
4. What are different tariffs used by electricity supply authority? Explain any one of them in brief.

1. **Block Rate Tariff:** The first block of energy is charged at higher rate and succeeding blocks of energy are charged at progressively reduced rates, for example,
   - The first 30 units may be charged at the rate of ₹ 4.0 paise/unit.
   - The next 50 units may be charged at the rate of ₹ 3.50 per unit.
   - The consumption exceeding 100 units may be charged at the rate of ₹ 3.0 per unit.
   - The advantage of such a tariff is that the consumer gets an incentive for consuming more electrical energy. This increases the load factor of the system and hence the generation cost is reduced. The drawback is that it lacks a measure of the consumer's demand. This tariff is most popular now-a-days among domestic, commercial and small industrial consumers.
2. **Power Factor Tariffs:** Since, the efficiency of plant and equipment depends upon the power factor, therefore, in order to increase the utility of plant and equipment to the maximum, the plant must be operated at the most economical power factor. That is why, sometimes consumers are penalised for poor power factor by applying the following types of power factor tariffs:
   (a) kVA Maximum Demand Tariff: This type of tariff is derived by modifying form of two part tariff. In this case, maximum demand is measured in kVA instead of in kW. This type of tariff encourages the consumers to operate their machines and other equipment at improved power factor, because low power factor will cause more demand charges.
   (b) kWh and kVARh Tariff: kWh and kVARh are measured and charged separately. Since, kVARh decreases with the increase in power factor, therefore, the consumer is encouraged to improve the power factor of his installation in order to decrease the charges on account of kVARh recorded.
3. **Two-Part Tariff**: The total charge to be made to the consumer is split into two components namely fixed charge and running charge. Since, fixed charge is independent of energy consumed and proportional to the maximum demand so the fixed charge is made at a certain amount per kW of maximum demand, which can be assessed on the basis of connected load or on the total combined kW capacity of all the consuming devices owned by a particular consumer. Running charge is made at a certain amount per kWh for the total energy consumed. This type of tariff is expressed by the expression:
   Total energy charges, e = ₹ a × kW + ₹ b × kWh where ₹ a is the charge per kW of maximum demand assessed and ₹ b is the charge per kWh of energy consumed.

This tariff is mostly applicable to medium industrial consumers.

In this tariff charge made on maximum demand recovers the fixed charges such as interest and depreciation on the capital cost of building and equipment, taxes and insurance charges and operating cost which is independent of energy supplied by it and varies with variation of maximum demand. Charge made on total energy consumption recovers the operating cost, which varies with variation in energy supplied.

The drawback is that consumer has to pay his fixed charge irrespective of the consumption. For example, if during any month any industry remains closed, the owner will be required to pay the fixed charges unnecessarily.

4. **Flat Rate Tariff:** In this type of tariff, different types of consumers are charged at different rates i.e. the flat rate for light and fan loads is slightly higher than that for power load. The rate for each category of consumers is framed by taking into account its load factor and diversity factor.

   The method is usually most popular with the public, since it can easily be understood by the consumers and the calculations at the supplier's end are simple. If the energy consumed during the billing period is x units and the flat rate is ₹ a per unit then charges are ₹ a x i.e. for 200 units @ ₹ 2.5 the bill will be 200 × 2.5 = ₹ 500.

Drawbacks of this type of tariff:

(i) Separate meters are required for different types of supply (domestic and power).

(ii) Difficulty is experienced to derive the load factor and diversity factor for various types of loads.

6.3.4.1 Compare Block Rate Tariff and Flat Rate Tariff

Sr. No.	Comparison point	Block Rate Tariff	Flat Rate Tariff
1.	Charges to consumer (rates)	First block at higher rate, successive blocks are charged at low rates.	Different types of consumers are charged at different rates.
2.	Incentive	Consumer gets incentive for consuming more electric energy.	No such incentive.
3.	Drawback and benefit	Lacks a measure of consumer's demand.	Advantageous as it is easy to be understood by consumers.
4.	Popularity	Most popular among domestic, commercial and small Industrial consumers.	Most popular in public due to its ease of understanding the tariff.

6.3.5 Time of Day Tariff (TOD)

- This tariff is also called as Time of Day metering or Time of Usage (TOU) or Seasonal Time of Day (STOD).
- The scheme of this tariff is such that the consumer has to pay less in non-peak load period of the day and to pay at higher rate during the peak-load period.
- For such tariff scheme the metering involves dividing the day, month and year into tariff slots.
- This tariff is beneficial to the consumer as well as to the electricity supply company. This is because it controls the usage on the part of consumer resulting in automatic load control on the part of the company.
- It is the consumer's responsibility to control his own usage or pay more on peak load. Naturally consumer avoids the usage during peak load period and the load is thus controlled.
- This helps the electricity company to plan transmission-distribution infrastructure accordingly.
- The peak load is generally in the day time of 1 pm to 6 pm or upto 9 pm.
- In summer days, peak-load period is generally 6 to 12 noon in winter, evening 5 to 8 pm.

- This tariff splits the rate into two segments:
  (i) In peak
  (ii) Off-peak load and charges are decided by the company and revising the rates as and when the company decides.
- The rates of tariff vary from place to place, region to region. Generally large consumers adopt this tariff.

Comparison between Two-Part Tariff and Three-Part Tariff:

Sr. No.	Comparison point	Two-Part Tariff	Three-Part Tariff
1.	Slit up of charges	Two-part tariff is divided into two parts: 1. Fixed charges 2. Running charges	Three-part tariff is divided into three parts: 1. Fixed charges 2. Semi-fixed charges 3. Running charges
2.	Dependency of charges on	1. Fixed charges depend on M.D. 2. Running charges depend on number of units consumed by consumer	1. Fixed charges depend on interest, depreciation of capital cost. 2. Semi-fixed charges depend on charge per kW of M.D. 3. Running charges depend on energy consumed.
3.	Relation of total charges	Total charges = ₹ (b × kW × c × kWh) where b = charge/kW c = charge/kWh of energy consumed	Total charges = ₹ (a + b × c kW) where a = Fixed charges b = Semi-fixed charges c = Running cost

6.3.6 Simple Tariff (Uniform Rate Tariff)

- Most simple form in which the cost of energy is charged on the basis of units consumed by the consumer.
- The cost per unit (kWh) is derived by the following relation:

$$\text{Cost/kWh} = \frac{\text{Annual fixed cost + Annual operating cost}}{\text{Total units supplied to the consumer per year}}$$

- Consumer is suffered by this tariff:
  (i) Cost per unit (kWh) delivered is higher.
  (ii) Domestic, bulk, industrial, commercial consumers are not discriminated, they have different load factors, different power factor or diversity factors. All are treated same.

  So this tariff is not generally exercised.

6.3.7 Step Rate Tariff

- It is a group of flat rate tariffs. In this decreasing unit charges for higher range of consumption.
- Explanation:

  Say

  ₹ 5/- unit upto less than 50 unit (i.e. kWh) consumption.

  ₹ 4.5/- unit for 50 to 200 kWh consumption.

  ₹ 4/- unit exceeding 200 kWh consumption.

  In this tariff the tendency of the consumer to cross and enter into the next slab.

  This drawback is overcome in block rate tariff.

6.3.8 Three-Part Tariff

- In this type of tariff the total charges are split into three parts:
  - **(i) Fixed charges:** Which recover expenses in giving supply.
  - **(ii) Semi-fixed charges:** These are lavid on the basis of maximum demand in kW or kVA which recovers plant's initial cost plus operating cost.
  - **(iii) Variable charges:** These charges are lavid on the basis of units consumed. This recovers changing rates of fuel. Following relation shows the total charge relation:

    Total charges = ₹ a + b · kW + c kWh

    (a) Constant charge.

    (b) Unit charge per kW of metered demand during billing period.

    (c) Unit charge for energy in ₹/kWh of consumption.

    This tariff is generally applicable to bulk supplies.

SOLVED EXAMPLES

Example 6.1: *A factory has a maximum demand of 250 kW with a load factor of 0.6. The following tariffs are offered.*

(a) Two part tariff ₹70 / kW of M.D. / year + 4 paise / kWh

(b) A flat rate of 10 paise / kWh.

Which tariff will be cheaper?

Solution: For part (a):

$$\text{Annual load consumption in kWh} = \text{MD in kW} \times \text{L.F.} \times 365 \times 24$$
$$= 250 \times 0.6 \times 365 \times 24 = 131400 \text{ kWh}$$

Now, total bill be, (A × kW max. demand) + B × kWh

$$= (70 \times 250) + \left(1{,}31{,}4000 \times \frac{4}{100}\right) = (17{,}500) + (52{,}560)$$
$$= \boxed{₹\ 70{,}060}$$

For part (b): Annual load consumption = 1,31,4000 kWh

$$\therefore \quad \text{Total bill will be} = 1{,}31{,}4000 \times \frac{10}{100} = \boxed{₹\ 1{,}31{,}400}$$

First tariff is cheaper.

Example 6.2: *A consumer has annual consumption of 3,40,500 units. The tariff is ₹230/kW of M.D. per year plus 17 paise per unit, L.F. = 40%. Find average cost per unit.*

Solution: From the above example, it is clear that it is a two part tariff.

$$(\text{Total units kWh}) = (\text{M.D. in kW}) \times (\text{Total hours of year}) \times (\text{L.F.})$$
$$3{,}40{,}500 = (\text{M.D. in kW}) \times (365 \times 24) \times (0.4)$$
$$\therefore \quad \text{M.D. in kW} = \frac{3{,}40{,}500}{365 \times 24 \times 0.4} = 97.17$$
$$\text{Total bill} = (\text{A} \times \text{kW max. demand}) + (\text{B} \times \text{kWh})$$
$$= (230 \times 97.17) + \left(\frac{17}{100} \times 3{,}40{,}500\right) = 80{,}234.1$$
$$\text{Average cost per unit} = \frac{\text{Total bill}}{\text{Total units}} = \frac{80{,}234.1}{3{,}40{,}500} = ₹\ 0.2356$$
$$₹\ 0.2356 = \textbf{23.56 paise/unit}$$

Example 6.3: *For the previous Example 6.1, how many hours per week the factory should work, so that both tariffs are equal?*

Solution: Total units consumed = (M.D. in kW) × (Working hours) × (L.F.)

$$= 250 \times (52\,x) \times 0.6 \ldots \text{ as 1 year} = 52 \text{ weeks} = \boxed{7800\,x}$$

(a)
$$\text{Bill will be} = (₹\ A \times \text{M.D.}) + (₹\ B \times \text{kWh})$$
$$= (70 \times 250) + (0.04 \times 7800\,x)$$
$$= (17{,}500) + (312\,x) \quad \ldots\text{(i)}$$

(b) Bill will be $= \frac{10}{100} \times 7800\ x = \boxed{₹\ 780\ x}$ ...(ii)

Equating (i) and (ii),

$$17{,}500 + 312\ x = 780\ x$$

$\therefore$ $$17{,}500 = 468\ x$$

$\therefore$ $$\boxed{x = 37.39}$$

So if the factory working hours are 37.39 per week then tariffs (a) and (b) are equal.

Example 6.4: *An industrial consumer has maximum demand of 320 kW and L.F. 60%. There are two tariffs in force:*

(i) (₹97/kW of M.D.) + (6 paise/kWh)

(ii) A flat rate of 14 paise / kWh.

Find which tariff is economical.

Assume 52 weeks for one year.

Solution: For (i):

Total units consumed in kWh = Max. demand kW × 52 × 7 × 24 × L.F.

= 320 × 52 × 7 × 24 × 0.6 = 16,77,312 kWh

Total bill = (₹ A × Max. demand in kW) + (₹ B × kWh)

$= (97 \times 320) + \left(\frac{6}{100} \times 16{,}77{,}312\right) = \boxed{₹\ 13{,}16{,}78.72}$

For (ii): Flat rate bill $= ₹\ A \times kWh = \frac{14}{100} \times 16{,}77{,}312 = \boxed{₹\ 2{,}34{,}823.68}$

$\therefore$ Two part tariff (i) is economical.

Example 6.5: *A factory has a maximum demand of 1000 kVA at power factor 0.8 lagging a load factor of 60 percent. Determine the average monthly bill by following tariff.*

1. *₹15 per kW of maximum demand per month plus 10 paise per kWh.*
2. *₹18 per kVA of demand per month plus 8 paise kWh.*

Solution: Given: M.D.= 1000 kVA

P.F. = 0.8 lag

L.F. = 60% = 0.6

Find electricity bill (monthly).

Total energy consumer per month = M.D. in kW × Load factor × Hours of month

= (1000 × 0.8) × 0.6 × (30 × 24) = 3,45,000 kWh

1. Monthly demand charges = ₹ 15 × M.D.= 15 × 1000 = ₹ 15,000

Monthly energy charges = ₹ – (0.1) × 345600 = 34560

Monthly electricity bill = 5,000 + 34,560 = **₹ 49,560**

2. Monthly demand charges = ₹ 18 × M.D.= 18 × 1000 = ₹ 18000

Monthly energy charges = ₹ (0.08) × 3,45,600 = ₹ 27,648

Monthly energy charges = ₹ 18,000 + 27,648 = **₹ 45,648**

Example 6.6: *A consumer has a maximum demand of 250 kW at 50% load factor. If the tariff is ₹100 per kW of maximum demand plus 20 paisa per kWh, find the overall cost per kWh.*

Solution: Given: Maximum demand = 250 kW at 50% load factor

Tariff = ₹ 100/kW + 20 paise/kWh

To find: Overall cost per kWh.

Units consumer/year = M.D. × L.F. × Hrs. in a year

= 250 × 0.5 × 8760 = 10,95,000 kWh

$$\text{Annual charges} = \text{Annual M.D. charges + Annual energy charges}$$
$$= ₹\ (100 \times 50) + (0.2 \times 10{,}95{,}000) = ₹\ 2{,}24{,}000$$
$$\therefore \quad \text{Overall cost/kWh} = \frac{2{,}24{,}000}{10{,}95{,}000} = ₹\ 0.2046 = \mathbf{20.46\ paise}$$

Example 6.7: *Annual consumption of a big consumer is 52,560 kWh. The charges are @ ₹ 100 kW maximum demand + ₹ 0.1/kWh. Find (i) Annual bill, (ii) Overall cost/kWh. Take load factor = 0.5.*

Solution:

$$\text{Hours of one year} = 8760$$
$$\text{Average load} = \frac{\text{Annual energy consumption}}{8760} = \frac{52560}{8760} = 6\text{ kW}$$
$$\text{Maximum demand (MD)} = \frac{\text{Average load}}{\text{Load factor}} = \frac{6}{0.4} = 15\text{ kW}$$
$$\text{Annual energy consumption bill} = \frac{\text{Demand charges}}{\text{Annum}} + \frac{\text{Energy charges}}{\text{Annum}}$$
$$= 100 \times 15 + 0.5 \times 52560 = ₹\ 27{,}780$$
$$\text{Overall cost} = \frac{27780}{52560} = \mathbf{0.528}$$

6.4 POWER FACTOR IMPROVEMENT

1. Explain methods of improving power factor.
2. State methods of p.f. improvement. List four advantages of it.
3. Write and explain two methods to improve power factor.
4. State any two causes of low power factor. State any two methods for improving power factor.

- As seen, the low power factor is almost invariably due to inductive nature of load and, therefore, it is necessary to connect such devices across the load, which take leading reactive power. The devices such as static capacitors, synchronous machines or synchronous condensers take leading reactive power.
- The leading reactive component of current drawn by power factor correcting device neutralises the lagging reactive component of current drawn by the load partly or completely. Power factor of the system will become unity when lagging reactive component of load current is completely neutralised by the leading reactive component of current drawn by power factor correcting device.

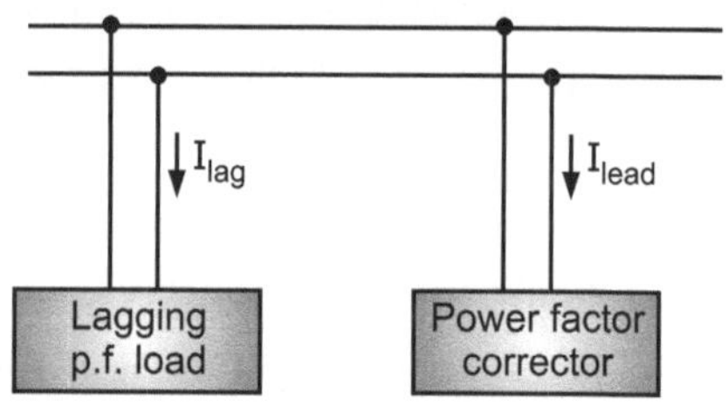

Fig. 6.1

- Let the current drawn by an inductive circuit be I lagging behind the applied voltage by an angle ϕ. The leading current required to neutralize the lagging reactive component of current drawn by the inductive circuit (equipment) to give unity power factor

$$= I \sin\phi = I\sqrt{(1 - \cos^2\phi)} = I\sqrt{1 - (\text{p.f.})^2}$$

1. By Use of Static Capacitors:

- Power factor can be improved by connecting the capacitors in parallel with the equipment operating at lagging power factor such as induction motors, fluorescent tubes. Static capacitors have small losses, low initial cost, little maintenance (static), easy installation, and has capability to operate under ordinary atmospheric conditions. However, they have short service life (8 to 10 years). They get damaged on over-voltages and uneconomical repair. The current drawn by induction motors or fluorescent tubes can be resolved into two components. The active component, which is in phase with the supply voltage and the quadrature or wattless component of constant magnitude. The capacitors draw current leading the supply voltage by 90° approximately and neutralise the quadrature or wattless component of current drawn by the equipment across which these are connected. In case of 3-phase loads, capacitors can be connected either in star or delta, as shown in Fig. 6.2 (a) and (b). These capacitors remain connected permanently across the equipment and are across the supply mains.

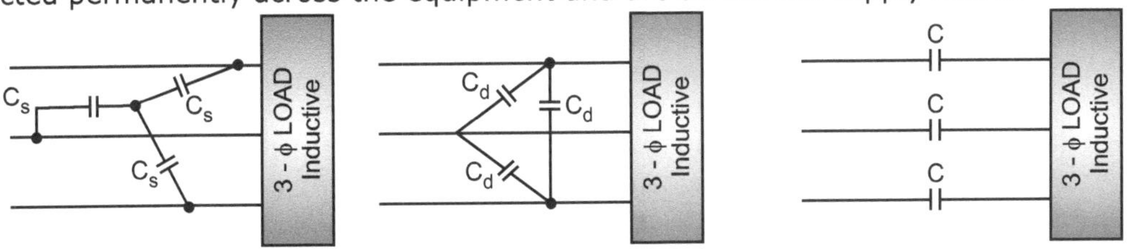

(a) Star-connected capacitors **(b) Delta-connected capacitors** **(c)**

Fig. 6.2

- The value of the static capacitors for the improvement of the power factor can be determined as follows:
- The leading current required to neutralise the lagging reactive component of the current drawn by the equipment to give unity power factor is expressed as

$$I_C = I_L = I \sin \phi = I \sqrt{(1 - \cos^2 \phi)} = I \sqrt{1 - (pf)^2}$$

- The value of capacitance in star bank is given by

$$C_S = \frac{I_C}{2\pi f V} = \frac{1}{2\pi f V} \sqrt{1 - (p.f.)^2}$$

where, V is the phase voltage, I is the phase current and f is the supply frequency.

- For given kVAR and line voltage, the delta value will be one-third of star value.
- Power factor can also be improved by connecting static capacitors in series with the line, as shown in Fig. 6.3. Capacitors connected in series with the line neutralize the line reactance. The capacitors, when connected in series with the line, are called the **series capacitors**, and when connected in parallel with the equipment, are called the **shunt capacitors**.
- Shunt capacitors are used in factories, plants and also on transmission lines.
- Series capacitors are used on long transmission lines as they provide automatic compensation with the variations in load. The capacity of the capacitors to neutralize the line reactance is given by $C = \frac{1}{(2\pi f)^2 L}$

where, f is the supply frequency and L is the inductance of the line per phase.

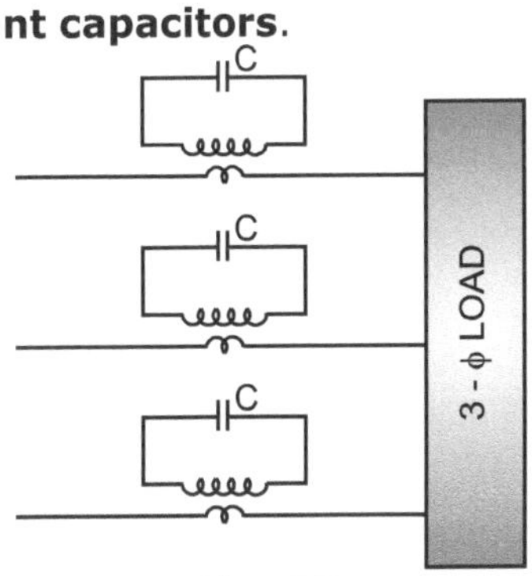

Fig. 6.3

- The reactive output of the capacitors in kVAR is given by:
1. $\frac{\text{Voltage} \times \text{Line contact}}{1,000}$ kVAR or $2\pi f\, CV^2 \times 10^{-9}$ kVAR in case of single-phase circuit where,

   V is the line voltage, f is the supply frequency and C is the capacitance in microfarads.
2. $\frac{\sqrt{3}\ \text{voltage} \times \text{line current}}{1,000}$ kVAR or $2\pi f\, CV^2 \times 10^{-9}$ kVAR in case of 3-phase star-connected circuits.
3. $6\pi f C V_L^2 \times 10^{-9}$ kVAR in case of 3-phase delta-connected circuits

   where, V_L is the line voltage, f is the supply frequency in Hz and C is the capacitance in μF between the line terminals.

2. By Use of Synchronous Machines or High Power Factor Machines:

- Synchronous machines are excited by D.C. and the power factor may be controlled by controlling the field excitation. The various synchronous machines available for power factor correction comprise synchronous motors, synchronous condensers, synchronous converters, synchronous phase modifiers, phase advancers, and synchronous induction motors.

(i) By Use of Synchronous Motors: Synchronous motors are designed for 0.8 to 1 leading power factors at full load. The unity power factor motor costs less and has a higher efficiency, but if fully loaded, it cannot furnish leading reactive kVA to compensate for lagging reactive kVA in the system.

(ii) By Use of Synchronous Condensers: An over-excited synchronous motor running on no load is called the synchronous condenser or synchronous phase advancer. It behaves like a capacitor. The capacitive reactance depends upon motor excitation. Power factor can be improved by using synchronous condensers just like shunt capacitors we connect across the supply.

In phasor diagram (Fig. 6.4), phasor I_L represents the current drawn by the industrial load, lagging behind the applied voltage V by a large angle ϕ_L and phasor I_M represents the current drawn by the synchronous condenser leading the applied voltage V by the angle ϕ_M. The resultant current I is the phasor sum of I_L and I_M and now angle of lag ϕ is much smaller than ϕ_L. Thus overall power factor is improved from cos ϕ_L to cos ϕ by the use of the synchronous condenser. In this way the power factor can be improved upto unity.

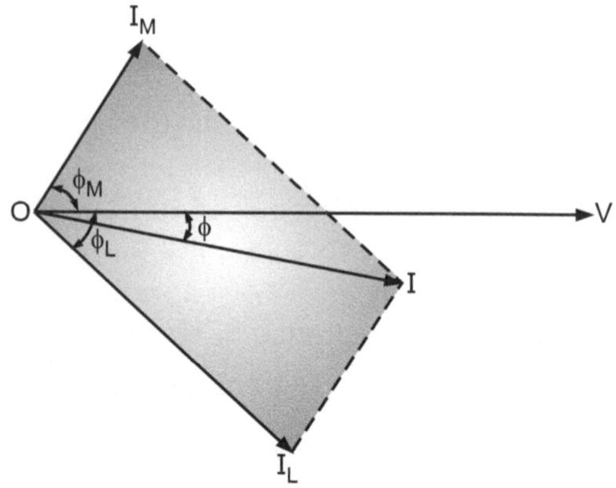

Fig. 6.4

The disadvantages of synchronous condensers over static capacitors as power factor correcting devices are: (i) except in size above about 5,000 kVAR, the cost is higher than that of static capacitors of the same rating; (ii) comparatively higher maintenance and operating costs; (iii) comparatively lower efficiency (say 97%) due to losses in rotating parts and heat losses; (iv) noise is produced in operation; (v) an auxiliary equipment is required for starting synchronous condensers; (vi) possibility of synchronous condensers falling out of synchronism causing in interruption of supply; and (vii) increase of short-circuit currents when the fault occurs near the synchronous condenser.

Synchronous condensers are used invariably at large substations for improving the power factor and voltage regulation. Machines upto 100 MVAR rating or even higher have been used. The excitation current is regulated automatically to give a desired voltage level.

(iii) By Use of Phase Advancers: The power factor of an induction motor falls mainly due to its magnetizing current drawn from the A.C. supply mains, because magnetizing current lags behind the voltage by 90°. It may be improved by connecting the set with an A.C. exciter or phase advancer which supplies this exciting current to the rotor circuit at slip frequency. Such an exciter may be mounted on the same shaft as the main motor or may be suitably driven from it. Use of phase advancer is not generally economical in connection with motors below 150 kW output but above this size, phase advancers are frequently employed. Shunt and series type of phase advancers are available according to whether the exciting winding of the advancer is connected in parallel or series with the rotor winding of the induction motor.

There are two main advantages of phase advancers (i) lagging kVAR drawn by the motor are considerably reduced due to supply of exciting ampere-turns at slip frequency and (ii) the phase advancers can be used where the use of synchronous motor is not permitted.

(iv) By Use of Synchronous Induction Motors: These are special types of motors which operate at certain loads as synchronous motors and at other loads as induction motors.

(v) By Use of High Power Factor Motors: Besides synchronous motors or synchronous induction motors there are other several types of motors which operate at a power factor of approximately unity such as compensated induction motors, and charge motors. These motors are more expensive and have higher maintenance cost than ordinary induction motors.

6.4.1 Location of Power Factor Correction Equipment

- The best location for the power factor correction equipment to be installed is where the apparatus or equipment responsible for low power factor is operating. Synchronous condensers are used at load centres where considerable corrective kVAR is required whereas static capacitors are used in smaller units and may be placed closer to the point where the load of inductive nature is installed. This relieves the distributors and feeders from excessive currents owing to low power factor.
- In case of transmission system, these should be installed at the receiving end so that not only the generators but also the transmission lines are relieved of carrying excessive current due to poor power factor. However, if synchronous condensers are installed near the generators then only generators will be relieved from the excessive current component and the transmission lines will have to carry it.

6.4.2 Advantages of Power Factor Improvement

- Modern alternators are designed to have high reactance in order that alternators may not get damaged at the time of short circuit and normally reactance is 20 times of resistance. Any change in current and power factor causes the change in terminal voltage, so voltage can be kept fairly constant by **power factor control**.
- If the power factor of the supply or power station is raised to unity, the current for the same amount of power to be supplied is reduced to minimum. This results in reduction of transmission line copper losses, and reduction of voltage drop in transmission line and in alternator windings, as copper losses are directly proportional to the square of supply current and voltage drop is directly proportional to the current.
- The terminal voltage of an alternator is given by the phasor difference of induced emf and drop, synchronous impedance drop. Neglecting the resistance (as compared to synchronous reactance), the synchronous impedance can be considered purely inductive, so synchronous impedance voltage drop leads the load current by $\frac{\pi}{2}$ i.e. 90°.

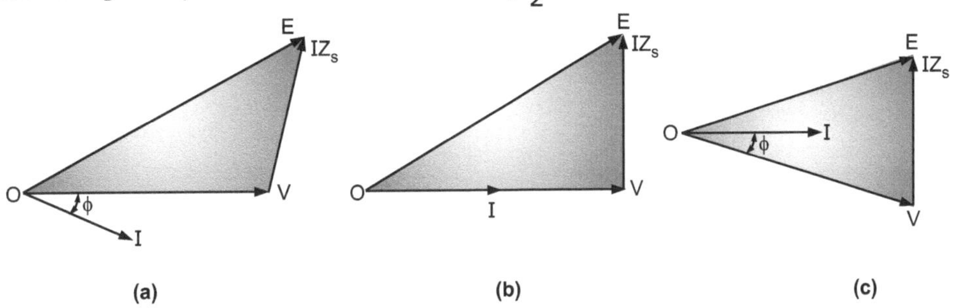

Fig. 6.5

- From the phasor diagrams shown in Fig. 6.5 (a), (b) and (c) for lagging, unity and leading power factor respectively it is obvious that as the power factor is raised from lagging to unity, the difference of terminal voltage and induced emf is reduced and this difference can be reduced to zero by making the power factor, leading and less than unity. Hence, to have zero regulation the power factor should be made leading one, so that no other regulating equipment is required.
- Why it is not economical to raise the power to unity or leading one? Reasons being:
  1. In case the power factor is improved to unity for full-load conditions, it would become leading for loads less than full load (unless some capacitors are switched off which is usually difficult). At power factors lower than unity (lagging or leading), the current for the same amount of power to be supplied is increased and thereby energy losses in generators, transformers, transmission lines and distribution lines will be increased.
  2. As the power factor approaches unity, the capacity of power factor correction device increases more rapidly i.e. the power factor of an installation can be improved from 0.7 or 0.8 to 0.8 or 0.9 by a much smaller capacitive kVAR than that required for raising the power factor from 0.9 to unity.

- The advantages of good (or improved) power factor are: (i) reduction in load current; (ii) increase in voltage level across the load; (iii) reduction in energy losses in the system (generators, transformers, transmission lines and distributors) due to reduction in load current; (iv) reduction in kVA loading of the generators and transformers which may relieve an overloaded system or release capacity for additional growth of load and (v) reduction in kVA demand charge for large consumers.

6.4.3 Most Economical Power Factor (Derivation)

1. Derive the most economical power factor equation.

A consumer takes kW = P at p.f. ϕ_1.

$\therefore$ His maximum kVA demand = $\boxed{\dfrac{P}{\cos\phi_1}}$.

He installs a unit for improvement of p.f. So now his maximum kVA demand becomes = $\boxed{\dfrac{P}{\cos\phi_2}}$.

$$\text{His saving in kVA} = \left(\frac{P}{\cos\phi_1}\right) - \left(\frac{P}{\cos\phi_2}\right)$$

$$\text{Saving in energy charges (₹)} = \boxed{₹\, A \times \left[\left(\frac{P}{\cos\phi_1}\right) - \left(\frac{P}{\cos\phi_2}\right)\right]}$$

The kVAR supplied for the improvement of p.f. by the unit is ($P \tan\phi_1 - P\tan\phi_2$).

$$\text{Cost of installation per year} = \boxed{₹\, B\,(P\tan\phi_1 - P\tan\phi_2)}$$

$$\text{Net saving in ₹} = ₹\, A\left[\frac{P}{\cos\phi_1} - \frac{P}{\cos\phi_2}\right] - ₹\, B\,(P\tan\phi_1 - P\tan\phi_2) \qquad \text{... (I)}$$

This saving is maximum when differential of equation (I) w.r.t. $\cos\phi_2 = 0$.

$\therefore$ Differentiating the equation (I) w.r.t. ϕ_2 and equating it to zero.

$$-AP\sec\phi_2\tan\phi_2 + BP\sec^2\phi_2 = 0$$

$$\therefore \quad -A\sec\phi_2\tan\phi_2 + B\sec^2\phi_2 = 0$$

$$\therefore \quad -A\tan\phi_2 + B\sec\phi_2 = 0$$

$$\therefore \quad -A\frac{\sin\phi_2}{\cos\phi_2} + \frac{B}{\cos\phi_2} = 0$$

$$\therefore \quad -A\sin\phi_2 + B = 0$$

$$\sin\phi_2 = \frac{B}{A}$$

$$\sin^2\phi_2 = \left(\frac{B}{A}\right)^2$$

$$1 - \cos^2\phi_2 = \left(\frac{B}{A}\right)^2$$

$$-\cos^2\phi_2 = \left(\frac{B}{A}\right)^2 - 1$$

$$\cos^2\phi_2 = 1 - \left(\frac{B}{A}\right)^2$$

$$\boxed{\cos\phi_2 = \sqrt{1 - \left(\frac{B}{A}\right)^2}}$$

where, B = charging rate / kVA of p.f. improvement unit

A = charging rate / kVA of maximum demand

SOLVED EXAMPLES

Example 6.8: *A factory takes 200 kW at 110 volts from a 3-phase supply and p.f. of 0.71 lagging. A synchronous motor is installed which takes an additional 100 kW. What must be the kVA rating of this motor to raise the p.f. of the system to 0.87 lagging?*

Solution: Factory load = 200 kW

Motor load = 100 kW

$\therefore$ Total load = 200 + 100 = 300 kW

$\cos \phi_1$ i.e. factory load p.f. = 0.71 lagging

$\therefore$ $\phi_1 = 44^\circ\, 46'$

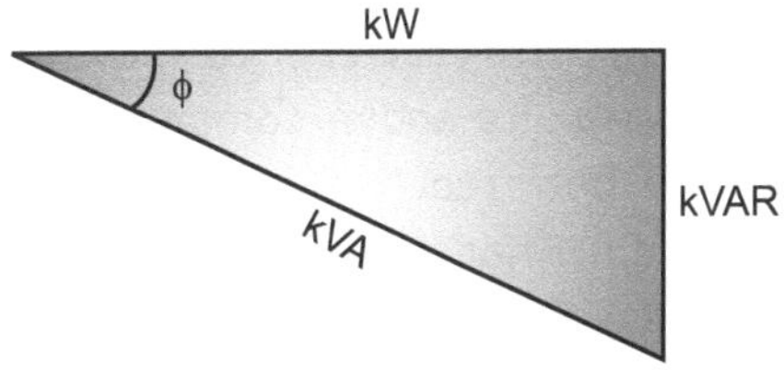

Fig. 6.6

$\cos \phi_T$ i.e. p.f. of the total system = 0.87 lagging (with synchronous motor)

$\therefore$ $\phi_T = 29^\circ\, 33'$

Now total system p.f. to be 0.87 and total load = 300 kW

Then kVAR (total) = 300 tan 20° 33' = 300 × 0.567 = 170

The load kVAR = 200 tan 44° 46' = 200 × 0.992 = 198.4

Power diagram:

ABC is the power triangle for synchronous motor.

OAE is the power triangle of the factory load before connecting the synchronous motor. Thus, synchronous motor kVAR

= 198.4 – 170

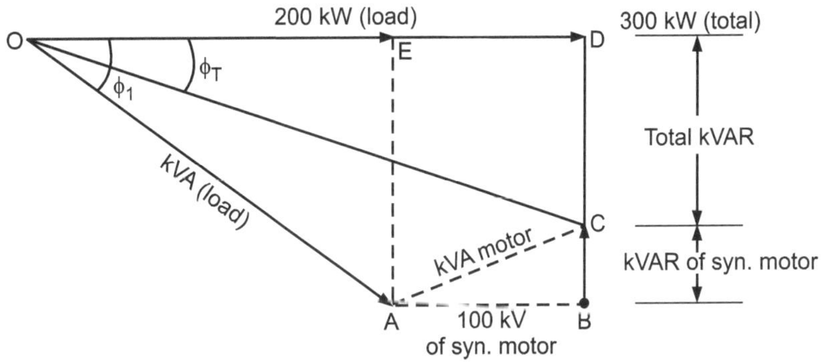

Fig. 6.7

$$\therefore \quad \text{Synchronous motor kVA} = \sqrt{(AB)^2 + (BC)^2} = \sqrt{(100)^2 + (28.4)^2} = \boxed{\textbf{104 kVA}}$$

Example 6.9: *A factory takes 300 kW at 110 volts from a 3-phase supply and power factor of 0.7 lagging. A synchronous motor is installed which takes an additional 150 kW. What must be the kVA rating of this motor to raise the power factor of the system to 0.85 lagging?*

Solution: Given: P_L = 300 kW, V = 110 V, $\cos \phi$ = 0.7 lagging,

Improved p.f. = 0.85 lag; P_m = 150 kW, $\tan \phi_L = 1$, $\tan \phi_{new} = 0.6197$.

$$\text{Reactive power, } Q_L = P_L \tan \phi_L = 300 \times 1 = 300 \text{ kVAR (lagging)}$$

Reactive power after connecting synchronous motor,

$$Q_{new} = (P_L + P_m) \tan \phi_{new} = (300 + 150) \times 0.6197$$

$$= 278.8849 \text{ kVAR (lagging)}$$

For p.f. improvement, reactive power taken by synchronous motor

$$= \phi_L - \phi_{new} = 300 - 278.8849 = \boxed{\textbf{21.115 kVAR (leading)}}$$

kVA rating of synchronous motor,

$$S_m = \sqrt{(P_m + \phi_m)^2} = \sqrt{(150)^2 + (21.115)^2} = 151.4788 \text{ kVA}$$

Power factor of synchronous motor,

$$\cos \phi_m = \frac{P_m}{S_m} = \frac{150}{151.4788} = \boxed{\textbf{0.9902 (leading)}}$$

Example 6.10: *A factory load takes a line current of 141.4 Amps at 0.707 lagging p.f. from 3 phase 11,000 V mains. A synchronous motor is used as synchronous condenser. The line current reduces to 115.5 Amp. Calculate the reactive kVA taken by the synchronous motor.*

Solution: A factory line current = 141.4 Amp. lagging V by 45°

(as $\cos \phi_1 = 0.707$, $\phi_1 = 45°$)

So active component of the current = $I \cos \phi_1 = 141.4 \cos 45 = 100$ Amp.

$\therefore$ Reactive component of the current = $I \sin \phi_1 = 141.4 \cos 45 =$ **160 Amp.**

Now line current to be reduced to 115 Amp. Its active component should be same but reactive component will be reduced. That component is equal to the reactive component of synchronous motor taking leading current.

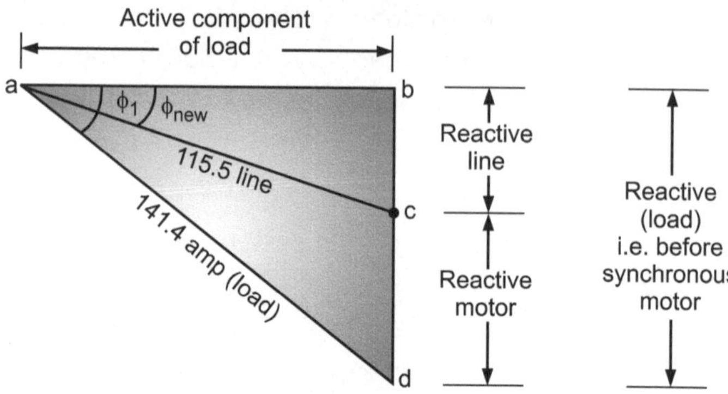

Fig. 6.8

From Δ abc,

$$bc = \sqrt{(ac)^2 - (ab)^2}$$

$$= \sqrt{(115.5)^2 - (100)^2}$$

$$cd = bd - bc$$

$$= 100 - 57.79 = 42.21$$

This is the reactive current of synchronous motor.

$$\text{kVAR of synchronous motor} = \frac{V \times I_{(reactive)}}{1000}$$

$$= \frac{11000 \times 42.21}{1000}$$

$$= \boxed{464.31 \text{ kVAR}}$$

Example 6.11: *A 400 V, 50 Hz, 3-phase line delivers 200 kW at 0.8 p.f. lagging. It is desired to raise the line power factor to unity by installing shunt capacitors. Calculate the capacitance of each unit if they are connected in (i) star and (ii) delta.*

Solution: Load P = 200 kW.

Power factor, $\cos \phi_1 = 0.8$ (lag);

$\phi = \cos^{-1} 0.8 = 36.87°$

and $\tan 36.87° = 0.75$

Power factor, $\cos \phi_2 = 1.0$;

$\phi = \cos^{-1} 1.0 = 0°$

and $\tan 0° = 0$

Leading kVAR supplied by static capacitors,

$$\text{Capacitive kVAR} = kVAR_1 - kVAR_2 = P(\tan \phi_1 - \tan \phi_2)$$

$$= 200\,(0.75 - 0) = 150$$

Since, reactive kVA output of capacitors when connected in star is given by $2\pi f C V_L^2 \times 10^{-3}$, where C is in μF.

$$\therefore \quad 2\pi f C V_L^2 \times 10^{-9} = 150$$

or

$$C = \frac{150 \times 10^9}{2\pi f V_L^2} = \frac{150 \times 10^9}{2\pi \times 50 \times (400)^2}$$

$$= \boxed{2984\ \mu F}$$

Since, reactive kVA output of capacitors when connected in delta is given as $6\pi f C V^2 \times 10^{-9}$ where C is in μF.

$$\therefore \quad C = \frac{150 \times 10^9}{6\pi f V_L^2} = \frac{150 \times 10^9}{6\pi \times 50 \times (400)^2} = \boxed{995\ \mu F}$$

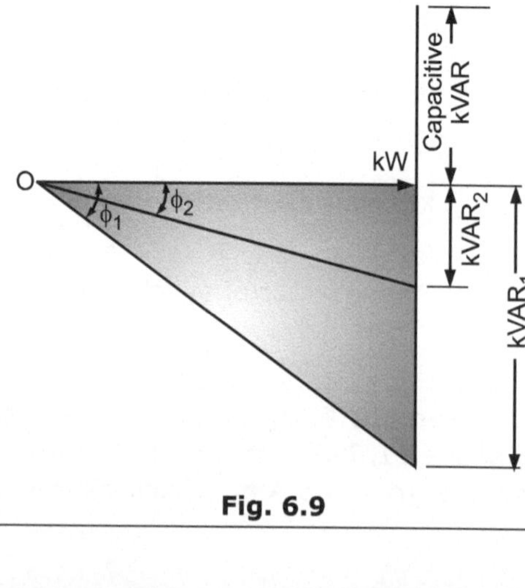

Fig. 6.9

Example 6.12: *A generator station is connected with the following loads:*

(i) A lighting load of 200 kW.

(ii) An induction motor of 200 hp (metric) p.f. 0.7 lagging and efficiency 80%.

(iii) A synchronous motor of 100 hp (metric) efficiency 90%.

Calculate the power factor of the synchronous motor so that the combined power factor becomes unity.

Solution: Input of induction motor $P_M = \dfrac{\text{Output in hp} \times 735.5}{\text{Motor efficiency}} = \dfrac{200 \times 735.5}{0.8}$

$= 1{,}83{,}875$ W or 183.875 kW

Total load = Lighting load + Induction motor load + Synchronous motor load

$$= 200 + 183.875 + \frac{100 \times 735.5}{0.9 \times 1{,}000}$$

$$= 200 + 183.875 + 81.\ 722 = 465.6 \text{ kW}$$

Leading kVAR drawn by synchronous motor

= Lagging kVAR drawn by induction motor

$= 183.875 \tan (\cos^{-1} 0.7) = 183.875 \times 1.02 = 187.6$

Since p.f. of lighting load is considered unity and p.f. after improvement is also unity,

$$\text{Power factor of synchronous motor} = \cos \tan^{-1} \frac{\text{Leading kVAR drawn by synchronous motor}}{\text{Power drawn by synchronous motor}}$$

$$= \cos \tan^{-1} \frac{187.6}{81.722} = \boxed{\mathbf{0.4\ (leading)}}$$

Example 6.13: *A 3-phase synchronous motor having a mechanical load (including losses) of 122 kW is connected in parallel with a load of 510 kW at 0.8 p.f. lagging. The excitation of the motor is adjusted so that the kVA input to the motor becomes 140 kVA. Determine the new power factor of the whole system.*

Solution: Load, $P_1 = 510$ kW

Load power factor, $\cos \phi_1 = 0.8$ lagging

kVAR (lagging) drawn by the load, $kVAR_1 = P_1 \tan \phi_1 = 510 \tan (\cos^{-1} 0.8) = 382.5$

kVA input to synchronous motor = 140

kW input to the synchronous motor, $P_2 = 122$ kW

kVAR (leading) drawn by the synchronous motor, $kVAR_2$

$= \sqrt{(kVA)^2 - (P_2)^2}$

$= 8.673$

Total kVAR drawn by the whole system $= kVAR_1 - kVAR_2 = 382.5 - 68.673 = 313.827$

Total load on the system, $P = P_1 + P_2 = 510 + 122 = 632$ kW

New power factor of the system, $\cos \phi = \cos \tan^{-1} \dfrac{313.827}{632}$

$= \boxed{\mathbf{0.8956\ (lagging)}}$

Example 6.14: *A 3-phase, 50 Hz, 400 V motor develops 100 HP, the power factor being 0.75 lagging and efficiency 93%. A bank of capacitors is connected in delta across the supply terminals and power factor raised to 0.95 lagging. Each of the capacitance unit is built of four similar 100 V capacitors. Determine the capacitance of each capacitor.*

Solution: Motor output = 100 HP = $100 \times 746 = 74600$ watt

$$\text{Motor input} = \frac{\text{Motor output}}{\text{Motor efficiency}} = \frac{74600}{0.93}$$

$$P = \text{Motor input} = 80215.053376 \text{ watt}$$

$$\text{Let power factor of motor} = \cos\phi_1 = 0.75 \text{ given}$$

$$\therefore \quad \phi_1 = 41.41^\circ$$

$$\tan\phi_1 = 0.8819$$

$$\cos\phi_2 = 0.95$$

$$\therefore \quad \phi_2 = 18.19$$

$$\tan\phi_2 = 0.3286$$

The leading kVAR is to be supplied by capacitors.

$$\text{Capacitive kVAR} = \text{kVAR}_1 - \text{kVAR}_2 = P(\tan\phi_1 - \tan\phi_2)$$

$$= 80215.05376\,(0.8819 - 0.3286) = 44382.9892$$

Since capacitance bank is in delta,

$$44382.9892 = 6\pi fc\, V_L^2$$

$$3\omega = 32\pi f$$

$$\therefore \quad 44382.9892 = 6\pi \times 50 \times c \times 400^2 \times 10^{-9}$$

$$\therefore \quad C = 29432^{-4} \times 10^{-4} \text{ Farad} = 294323.8387\ \mu F$$

Now this is the total capacitance of one branch which consists of 4 similar capacitors connected in series.

$$\therefore \quad \frac{1}{C_1} + \frac{1}{C_2} + \frac{1}{C_3} + \frac{1}{C_4} = 29432 \times 10^{-4} \text{ F}$$

Let $C = C_1 = C_2 = C_3 = C_4$

$$\therefore \quad \frac{4}{C} = 294323.8387 \text{ Farad}$$

$$\therefore \quad \boxed{C = 73580.95\ \mu F}$$

6.5 ENERGY CONSERVATION

6.5.1 Need and Importance of Energy Conservation

1. Describe the need of energy conservation in induction motor.
2. State the need of energy conservation in electrical motor.

6.5.2 (A) Energy Conservation in Electric Drives and Transformers

3. Explain the necessity of energy conservation in electrical motors.

- India is a growing economy and the economic growth of any country largely depends on industrialisation and energy consumption.
- The growing cost of energy due to limited stock of energy sources calls for power saving at each possible step in an industry.
- The three-phase induction motors consume about 70% of the electricity used in an industry.
- Hence the running cost of these induction motors is the real challenge.
- Earlier, induction motor design and fabrication was based more with the initial cost of the motor rather than how much energy it consumes.

- The materials such as aluminium or copper wires and steel laminations, were selected to meet the performance requirements of the motor.
- Hence induction motors have low power factors, more losses and low efficiency. After 90's the cost of energy starts increasing which directly affected the motor cost as the motors were less efficient.
- This leads the customers to search for other more energy efficiency options.
- Therefore the manufacturers started to figure out different methods to increase the efficiency of motors.
- To achieve these, manufacturers started use of good and thin steel lamination materials having low losses and more number of copper wires in windings to increase the efficiency of induction motor.
- The standard induction motors in a process plant consume large amount of energy due to low efficiency.
- To save energy consumption in a process plant, the use of **energy efficient motors** are chosen over standard induction motor.
- Energy efficient motors have better efficiency and power factor than standard induction motors.
- The use of energy efficient motors reduces energy consumption of the plant. The payback period for energy efficiency motors is very less as compared to standard motors.
- **"The motors having high efficiency, good power factor and fewer losses are called energy efficient motors".** Energy efficient motors have efficiency 2-3% more than standard motors.
- In the improvement of energy efficient motors, different technologies are made.
- The changes **in slot shape increase the efficiency** and thus reduce the internal **motor losses**.
- The use of energy efficient motors offers the utilities for possibility of achieving sustainable energy and demand reduction.
- **Need of energy conservation can be summarised as need of energy conservation is felt because:**
  1. Cost of electricity is increasing, so conservation methods to motors be provided.
  2. Motor must work at higher efficiency so provisions of reducing losses be provided.
  3. Inefficient motors need large power for the same work output.
  4. To save power, therefore energy efficient newly designed motors be provided (replacing old design motors).
  5. Good quality materials for core and windings be used by the manufacturers to improve efficiency by conserving energy.
  6. New design changed the shapes of slots to get more efficiency.
  7. Saving in cost of operation compels us to go for modern designed "energy efficient" motors working at lower losses, higher efficiency and hence conserving the energy.
- **Factors affecting the motor efficiency and minimising motor losses in operation:**

Power Supply Quality:

- Motor performance is affected considerably by the quality of input power, that is the actual volts and frequency available at motor terminals vis-a-vis rated values as well as voltage and frequency variations and voltage unbalance across the three phases.

Table 6.1: General Effects of Voltage and Frequency Variation on Induction Motor Characteristics

	Percent-tage	Starting and max. running torque	Syn-chronous speed %	% slip	Full-load speed %	Efficiency			Power factor			Full-load Current %	Starting current %	Max. Overload Capacity %
						100% load	75% load	50% load	100% load	75% load	50% load			
Voltage	120	↑ 44	None	↓ 30	↑ 1.5	Small incr.	↓ 5 to 2 pts	↓ 7 to 20 pts	↓ 5 to 15 pts	↓ 10 to 30 pts	↓ 15 to 40 pts	↓ 11	↑ 25	↑ 44
	110	↑ 21	None	↓ 17	↑ 1	↑ 0.5 to 1 pt.	Almost none	↓ 1 to 2 pts	↓ 3 pts	↓ 4 pts	↓ 5 to 6 pts	↓ 7	↑ 10 to 12	↑ 21
	90	↓ 19	None	↑ 23	↓ 1.5	↓ 2 pts.	Almost none	↑ 1 to 2 pts.	↑ 1 pt	↑ 2 to 3 pts	↑ 4 to 5 pts	↑ 11	↓ 10 to 12	↓ 19
	Function of	(volt)^2	Constant	(volt)^2	(Synch. speed slip)								(volt)	(volt)^2
Freq.	105	↓ 10	↑ 5	Almost none	↑ 5	...... slight increase each load						Slight	↓ 5 to 6	Slight ↓
	95	↑ 11	↓ 5	Almost none	↓ 5	...... slight decrease each load						Slight ↑	↑ 5 to 6	Slight ↑
	Function of	(freq.)^2	(freq.)		(Synch. speed slip)								(freq.)$^{-1}$	

The options available for an energy manager to ensure near to rated voltage at motor terminals include :

(i) Transformer tap changing as required in case of consistent and continuous low voltage situations.

(ii) Load end power factor improvement by providing matching.

(iii) Minimizing line/cable voltage drops from sub-station to motor terminals.

- Motors in India must comply with standards set by the Bureau of Indian Standard (BIS) for tolerance to variations in input power quality.
- The BIS standards specify that a motor should be capable of delivering its rated output with a voltage variation of ± 6% and frequency variation of ± 3%. Fluctuations much larger than these are quite common in utility-supplied electricity in India. Voltage fluctuations can have detrimental impacts on motor performance.
- The general effects of voltage and frequency variation on motor performance are presented in Table 6.1.
- Voltage unbalance, the condition where the voltages in the three phases are not equal, can be still more detrimental to motor performance and motor life.
- Unbalance typically occurs as a result of supplying single-phase loads disproportionately from one of the phases.
- It can also result from the use of different sizes of cables in the distribution system. An example of the effect of voltage unbalance on motor performance is shown in Table 6.2.

Table 6.2: Example of the effect of voltage unbalance on motor performance

Parameter	Percent unbalance in voltage*		
	0.30	**2.30**	**5.40**
Unbalance in current (%) …………	0.4	17.7	40.0
Increased temperature rise (°C) ………	0	30	40

* Percent unbalance in voltage is defined as 100 $(V_{max} - V_{avg})/V_{avg}$, where, V_{max} and V_{avg} are the largest and the average of three-phase voltages, respectively.

The options that can be exercised to minimize voltage unbalance include:

(i) Segregating any single phase load which disturb the load balance and feed them from a separate line/transformer.

(ii) Balancing any single phase loads equally among all the three phases.

Motor loading:

Measuring load: % loading of the motor can be estimated by the following relation:

$$\% \text{ loading} = \frac{\text{Input power drawn by the motor (kW) at existing load} \times 100}{\text{(Name plate full load kW rating/name plate full load motor efficiency)}}$$

$$\text{or} \quad \% \text{ loading} = \frac{\text{Input power drawn by the motor (kW) at existing load} \times 100}{\sqrt{3} \times \text{kV} \times \text{I} \cos \phi}$$

- Never assume power factor.
- Loading should not be estimated as the ratio of currents.

Reducing Under-loading:

- The most common practice contributing to sub-optimal motor efficiency is that of under-loading. Under-loading results in lower efficiency and power factor and higher than necessary first cost for the motor and related control equipment.
- Original equipment manufacturers tend to use a large safety factor in motors they select.
- Under-loading of the motor may also occur from under-utilization of the equipment. For example, machine tool equipment manufacturers provide for a motor rated for the full capacity loads of the equipment, for example, depth of cut in a lathe machine.
- The user may need this full capacity rarely, resulting in under-loading operation most of the time.
- Another common reason for under-loading is selection of a larger motor to enable the output to be maintained at the desired level even when input voltages are abnormally low.
- Finally, under-loading also results from selecting a large motor for an application requiring high starting torque where a special motor, designed for high torque, would have been suitable.

- A careful evaluation of the load would determine the capacity of the motor that should be selected. Another aspect to consider is the incremental gain in efficiency achievable by changing the motor.
- Larger motors have inherently higher rated efficiencies than smaller motors. Therefore, the replacement of motors operating at 60-70% of capacity or higher is generally not recommended.
- However, there are no rigid rules governing motor selection, the savings potential needs to be evaluated on a case-to-case basis.
- When downsizing, it may be preferable to select an energy efficient motor, the efficiency of which may be higher than that of a standard motor of higher capacity.
- For motors, which consistently operate at loads below 40% of rated capacity, an inexpensive and effective measure might be to operate in star mode. A change from the standard delta operation to star operation involves re-configuring the wiring of the three phases of power input at the terminal box.
- Operating in the star mode leads to a voltage reduction by a factor of $\sqrt{3}$. Motor is electrically downsized by star mode operation, but performance characteristics as a function of load remain unchanged. Thus, full-load operation in star mode gives higher efficiency and power factor than partial load operation in the delta model.
- However, motor operation in the star mode is possible only for applications where the torque-to-speed requirement is lower at reduced load.
- As speed of the motor, reduces in star mode this option may be avoided in case the motor is connected to a production facility whose output is related to the motor speed.
- For applications with high initial torque and low running torque needs, delta-star starters are also available in market, which help in load the following derating of electric motors after initial start-up.

Sizing to Variable Load:

- Industrial motors frequently operate under varying load conditions due to process requirements. A common practice in cars where such variable-loads are found is to select a motor based on the highest anticipated load.
- In many instances, an alternative approach is typically less costly, more efficient, and provides equally satisfactory operation.
- With this approach, the optimum rating for the motor is selected on the basis of the load duration curve for the particular application.
- Thus, rather than selecting motor of high rating that would operate at full capacity for only a short period, a motor would be selected with a rating slight lower than the peak anticipated load and would operate at overload for a short period of time.
- Since, operating within the thermal capacity of the motor insulation is of greatest concern in a motor operating at higher than its rated load, the motor rating is selected as that which would result in the same temperature rise under continuous full-load operation as the weighted average temperature rise over the actual operating cycle. Under extreme load changes, e.g. frequency starts/stops or high inertial loads, this method of calculating the motor rating is unsuitable since it would underestimate the healing that would occur.
- Where loads vary substantially with time, in addition to proper motor sizing, the control strategy employed can have a significant impact on motor electricity use.
- Traditionally, mechanical means (e.g. throttle valves in piping systems) have been used when lower output is required.
- More efficient speed control mechanisms include multi-speed motors, eddy-current couplings, fluid couplings, and solid-state electronic variable speed drives.

Power Factor Correction:

- As studied earlier, induction motors are characterized by power factors less than unity, leading to lower overall efficiency (and higher overall operating cost) associated with a plant's electrical system.
- Capacitors connected in parallel (shunted) with the motor are typically used to improve the power factor.
- The impacts of p.f. correction include reduced kVA demand (and hence reduced utility demand charges), reduced I_2R losses in cable upstream of the capacitor, reduced voltage drop in the cables and an increase in the overall efficiency of the plant electrical system.
- It should be noted that p.f. capacitor improves power factor from the point of installation back to the generating side. It means that, if a p.f. capacitor is installed at the starter terminals of the motor, it would not improve the operating p.f. of the motor, but the p.f. from starter terminals to the power generating side will improve that is the benefit of p.f. would be only on upstream side.
- The size of capacitor required for a particular motor depends upon the no-load reactive kVA (kVAR) drawn by the motor, which can be determined only from no-load testing of the motor. In general, the capacitor is then selected to not exceed 90% of the no-load kVAR of the motor.
- Alternatively, typical power factors of standard motors can provide the basis for conservative estimates of capacitor ratings to use for different size motors. The capacitor rating for power connection by direct connection to induction motors is shown in Table 6.3.

Table 6.3: Capacitor ratings for power factor correction by direct connection to induction motors

Motor rating (HP)	Capacitor rating (kVAR) for motor speed					
	3000	**1500**	**1000**	**750**	**600**	**500**
5	2	2	2	3	3	3
7.5	2	2	3	3	4	4
10	3	3	4	5	5	6
15	3	4	5	7	7	7
20	5	6	7	8	9	10
25	6	7	8	9	9	12
30	7	8	9	10	10	15
40	9	10	12	15	16	20
50	10	12	15	18	20	22
60	12	14	15	20	22	25
75	15	16	20	22	25	30
100	20	22	25	26	32	35
125	25	26	30	32	35	40
150	30	32	35	40	45	50
200	40	45	45	50	55	60
250	45	50	50	60	65	70

- From the above table, it may be noted that required capacitive kVAr increases with decrease in speed of the motor. As the magnetizing current requirement of a low speed motor is more in comparison to the high speed motor for the same HP of the motor.
- Since a reduction is line current, and associated energy efficiency gains, are reflected backwards from the point of application of the capacitor.
- The maximum improvement in overall system efficiency is achieved when the capacitor is connected across the motor terminals, as compared to somewhere further upstream in the plant's electrical system.
- However, economies of scale associated with the cost of capacitors and the labour required to install them will place an economic limit on the lowest desirable capacitor size.

Maintenance:

- Inadequate maintenance of motors can significantly increase losses and lead to unreliable operation. For example, improper lubrication can cause increased friction in both the motor and associated drive transmission equipment.
- Resistance losses in the motor, which rise with temperature, would increase.
- Providing adequate ventilation and keeping motor cooling ducts clean can help dissipate heat to reduce excessive losses.
- The life of the insulation in the motor would also be longer: for every 10°C increase in motor operating temperature over the recommended peak, the time before rewinding would be needed is estimated to be halved.
- A list of good maintenance practices to help ensure proper motor operation is as follows:
    1. Inspecting motors regularly for wear in bearings and housings (to reduce frictional losses) and for dirt/dust in motor ventilating ducts (to ensure proper heat dissipation).
    2. Checking load conditions to ensure that the motor is not over or under loaded. A change in motor load from the last test indicates a change in the driven load, the cause of which should be understood.
    3. Lubricating appropriately. Manufacturers generally give recommendations for how and when to lubricate their motors. Inadequate lubrication can cause problems, as noted above. Over-lubrication can also create problems, e.g. excess oil or grease from the motor bearings can enter the motor and saturate the motor insulation, causing premature failure or creating a fire risk.
    4. Checking periodically for proper alignment of the motor and the driven equipment. Improper alignment can cause shafts and bearings to wear quickly, resulting in damage to both the motor and the driven equipment.
    5. Ensuring that supply wiring and terminal box are properly sized and installed. Inspect regularly the connection at the motor and starter to be sure that they are clean and tight.

Age:

- In India, most motor cores are manufactured from silicon steel or de-carbonized cold-rolled steel, the electrical properties of which do not change measurably with age.
- However, poor maintenance, e.g. inadequate lubrication of bearings, insufficient cleaning of air cooling passages, etc. can cause a deterioration in motor efficiency over time
- Ambient conditions can also have a detrimental effect on motor performance.
- For example, excessively high temperatures, high dust loading, corrosive atmosphere, and humidity can impair insulation properties; mechanical stresses due to load cycling can lead to misalignment. However, with adequate care, motor performance can be maintained.

Selection of motor:

- The primary technical consideration defining the motor choice for any particular application is the torque required by the load, especially the relationship between the maximum torque generated by the motor and the torque requirements for start-up and during acceleration periods.
- The duty/load cycle determines the thermal loading on the motor.
- One consideration with Totally Enclosed Fan Cooled (TEFC) motors is that the cooling may be insufficient when the motor is operated at speeds below its rated value.
- Ambient operating conditions affect motor choice; special motor designs are available for corrosive or dusty atmospheres, high temperatures, restricted physical space etc.
- An estimate of the switching frequency (usually dictated by the process), whether automatic or manually controlled, can help in selecting the appropriate motor for the duty cycle.
- The demand of a motor will place on the balance of the plant electrical system is another consideration - if the load variations are large, for example, as a result of frequent starts and stops of large components like compressors, the resulting large voltage drops could be detrimental to other equipment.

- **Reliability:** In many cases, however, designers and process engineers seeking reliability will grossly oversize equipment, leading to sub-optimal energy performance.
- Good knowledge of process parameters and a better understanding of the plant power system can aid in reducing oversizing with no loss of reliability.
- **Inventory:** Many large industries use standard equipment, which can be easily serviced or replaced, thereby reducing the stock of spare parts that must be maintained and minimizing shut-down time.
- This practice affects the choice of motors that might provide better energy performance in specific applications. Shorter lead times for securing individual motors from suppliers would help to reduce the need for this practice.
- **Price:** Many users are first-cost sensitive, leading to the purchase of less expensive motors that may be more costly on a lifecycle basis because of lower efficiency.
- For example, energy efficient motors or other specially designed motors typically save within a few years an amount of money equal to several times the incremental cost for an energy efficient motor, over a standard-efficiency motor.

Some salient features are given below:

1. Reactive power drawn (kVAR) by the motor.
2. The Indian Standard IS 8789 addresses technical performance of Standard Motors while IS 12615 addresses the efficiency criteria of High Efficiency Motors. Both follow IEC 34-2 test methodology wherein, stray losses are assumed as 0.5% of input power. By the IEC test method, the losses are understated and if one goes by IEEE test methodology, the motor efficiency values would be further lowered.
3. Indian Standard 325 for standard motors allows 15% tolerance on efficiency for motors upto 50 kW rating and 10% for motors over 50 kW rating.
4. In the selection process, the power drawn at 75% of loading can be a meaningful indicator of energy efficiency.
5. The cost benefits can be worked out on the basis of premium required for high efficiency versus worth of annual savings.
6. The energy savings by motor replacement can be worked out by the simple relation:
$$\text{kW saving} = \text{kW output} \times [1/\eta_{old} - 1/\eta_{new}]$$
where, η_{old} and η_{new} are the existing and proposed motor efficiency values.
7. It would be prudent for buyers to procure motors based on test certificates rather than labelled values.

6.5.2 (B) Need of Energy Conservation in Transformer

1. State the need of energy conservation in transformer. Explain the use of Epoxy Resin Cast/ Encapsulated dry type transformer from energy conservation point of view.

- Transformer is a vital power equipment in a power system, which plays significant role in generation, transmission and distribution of electrical energy.
- Transformer operating efficiency directly affects the costs and benefits of the whole power system.
- In a typical power distribution grid, electric transformer power loss contributes to about 40-50% of the total transmission and distribution loss.
- In distribution system, the demand of distribution transformer is ever increasing.
- The power loss generated by transformers can be considerable due to its large transportation energy and long operating time.
- According to the statistics, total power loss generated by the transformer can be about 10% of total generation capacity. Therefore, study of transformer energy conservation is necessary.

6.5.2.1 Opportunities of Energy Conservation Technique in Transformer

- Attempts to reduce energy losses related with transformer operation in electrical power system be made.
- Distribution transformer's losses are much more 30 to 35% of energy loss.
- These losses are to be given attention and if so, it will considerably reduce green house emission.
- Most modern transformers are so designed, that they operate giving upto 99% efficiency.
- Transformers are required everywhere supplying power to variety of consumers - commercial, industrial, rural, domestic.
- Thus even a small positive change in transformer efficiency will lead to remarkable reduction in generation capacity requirement.

6.5.3 Energy Conservation Techniques in Induction Motor

1. Explain the energy conservation technique "by improving power quality" for induction motor.
2. Explain energy conservation method in induction motor by improving power quality.
3. Explain energy conservation technique in induction motor by improving mechanical power and transmission efficiency.
4. Explain the following energy conservation methods of electrical motor:
   (a) Rewinding of motor (b) Operating in star mode.

(a) Improving Power Quality:

1. List out the energy conservation techniques to be adopted to reduce losses in the induction motor.

- Motor performance can be increased by maintaining the voltage level within in BIS standards i.e. within the tolerance limit of ± 6% and keeping the frequency within the tolerance of ± 3%. This is achieved by avoiding voltage unbalance, maintaining voltage and frequency values and avoiding harmonic distortion.
- **Voltage unbalanace:** 3-phase motors are designed to operate on balanced supply. Unblanace voltage causes excessive heating and vibrations. This leads to increase I^2R losses. Remedy on this is to distribute single-phase loads on 3-phases.
- Harmonic filters are used to reduce harmonics in the system, maintaining frequency as constant.

(b) Motor Survey:

- Motor survey is important aspect in energy conservation. Conducting a motor survey is the best way to correctly size a replacement motor. A motor survey should begin by reviewing and cataloging the nameplate information on the existing motor to obtain the parameters such as rating of motor, rated speed, efficiency, full-load current, etc. The motor nameplate is the first step of a motor survey. It gives available information, such as speed and full-load current, which is helpful while selecting the right size motor. Motors operate most efficiently near full load, so determining load requirements accurately is important. The highest efficiency is achieved above 80 percent of full- load torque, below that efficiency starts to drop dramatically.

(c) Matching Motor with Loading:

- The characteristics of motors vary widely with their load and the type of duty they are expected to perform. For example, the applications like constant speed, constant torque, variable speed, steep/sudden start, continuous/ intermittent duty, frequent start/stops, etc. should be taken into consideration carefully when deciding for the type of a motor for that specific application. Proper selection of the range of the motor according to load requirements will reduce the power consumption. The efficiency of motors operating at loads below 40% is likely to be poor and energy savings are possible by replacing these with properly sized motors, new or interchanging with another load. An incorrect selection of motor always leads to various problems such as premature failure of the motor causing severe production loss.

(d) Minimizing the Idle and Redundant Running of Motor:

- By minimizing the idle and redundant running of motors, prolonged idle running of machine tools, conveyors, exhaust fan, lights etc. can be avoided. Idle running of auxiliaries like cooling towers, air compressors, pumps etc. during prolonged stoppage of production machines can be avoided.

(e) Operating in Star Mode:

- In loads operating at less than 30% of the full load, i.e., at light loads, operation of 'Delta' connected motor in 'Star' connection can save energy. If a motor is oversized and continuously loaded below 30% of its rated shaft load, energy can be saved by permanently connecting the motor in Star. In many cases, the load is below 30% most of the time, but sometimes the load exceeds 50% in this condition automatic Star-Delta changeover switches can be installed. This can save upto 5 to 15% of the existing power consumption.

(f) Rewinding of Motor:

- Rewinding of electric motors when they fail can be a cost-effective option in man ways.
- While the rewinding process is expected to be extensive and capable of bringing the motor back to a like-new condition, most rewinders do not follow every step or the precision needed to be followed in every step of rewinding.
- This has major impact on the efficiency of a rewound motor. Rewinding will decrease the efficiency by 2-5% in most of the motors.
- New technologies are available which help reduce the losses in a rewound motor; rare cases have shown that a motor's efficiency increased after being rewound.
- A common problem occurs when heat is applied to strip old windings: the insulation between laminations can be damaged, thereby increasing eddy current losses.
- A change in the air gap may affect power factor and output torque.
- Efficicncy an he improved by changing the winding design, though the power factor could be affected in the process.
- Using wires of large cross section would reduce stator losses thereby increasing efficiency.

(g) Energy Efficient Motor

- An EEM generates the same shaft output power, but uses less input power than a standard efficiency motor.
- Energy efficient motors apply less electricity and last longer than standard motors of the same size.
- Energy-efficient motors are defined as the motor in which design improvements are incorporated specifically to increase efficiency of motors.
- Design improvements focus on reducing intrinsic motor losses.
- Improvements include the use of lower-loss silicon steel, a longer core (to increase active material), thicker wires (to reduce resistance), thinner laminations, smaller air gap between stator and rotor, copper instead of aluminium bars in the rotor, superior bearings and a smaller fan, etc.
- Energy-efficient motors now available in India operate with efficiencies that are typically 3 to 4 percentage points higher than standard motors.

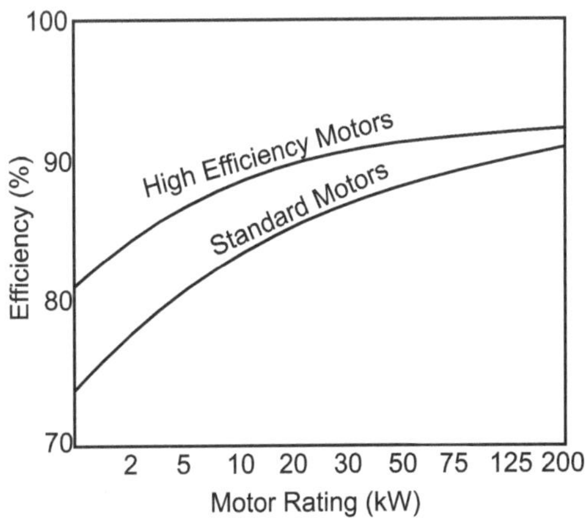

Fig. 6.10

(h) Periodic Maintenance:

- **Machine cleaning:** To ensure that ventilation and motor cooling is proper.
- **Machine set up and alignment:** To ensure that the belt drives and couplings are set up properly.

- **Bearing lubrication:** Verify that they are lubricated and sealed properly.
- **Condition assessment:** Vibration, unusual temperature rise etc. indicate problems.
- **Performance assessment:** Regularly measure supply voltage variations. Voltage imbalance leads to higher losses.
- **Maintenance of electrical connections in the starter and motor terminal box:** The loose connections are unsafe and source of heat losses.

  This will definitely save the energy in induction motor.

  Other factors are:

(i) By phase balancing.

(j) Matching the motor in star-mode.

(k) Use high quality bearings to reduce friction losses.

(*l*) Use small cooling fan to reduce fan load.

(m) Lower air gap reduces magnetic losses.

6.5.3.1 The Energy Conservation Techniques in Some Cases

1. Suggest the energy conservation techniques in following cases:
   (i) Motor is running with 70% loaded condition.
   (ii) Motor is continuously loaded at 50%.
   (iii) Motor runs with 30% loaded condition but sometimes rises to 50% loading condition.
   (iv) Motor runs continuously under no-load condition.

Sr. No.	Case	Technique
1.	Motor running with 30% loaded condition but sometime rises to 50% loading condition.	1. Convert Δ to Y connection for lightly loaded. 2. Use APFC or IPSC for better p.f. improvement. 3. Use cooling fans for motor. 4. if possible use low capacity motor.
2.	Motor running 70% in loaded condition.	1. Use APFC or IPSC for better p.f. improvement. 2. Use cooling fan for motor.
3.	Motor runs continuously under no-load condition.	1. Convert Δ → Y for highly loaded motor. 2. Switch off the motor. 3. Use low capacity motor. 4. Use cooling fans for motor. 5. Use APFC and IPFC for better p.f. improvement.
4.	Motor is continuously loaded at 50%.	1. Convert Δ → Y for highly loaded motor. 2. Use cooling fan. 3. Use APFC or IPFC for better p.f. improvement.

6.5.4 Energy Conservation Techniques in Transformer

1. State any two opportunities for energy conservation techniques in transformer.
2. State the opportunities for energy conservation techniques in transformer.
3. State the need of energy conservation in transformer.

(a) Load Sharing:

- The problems like overloading, variation in voltage and heating are very common in transformers.
- It takes a lot of time for its repairing and maintenance and also involves high cost.
- Load sharing protects the transformer under overload condition.
- Due to overload the efficiency drops and the secondary winding gets overheated or it may burn because of excess heat.
- So the transformer can be protected by reducing the extra load.

- This can be done by operating another transformer in parallel with the main transformer.
- When the loads on transformer are constant, the total active power loss and reactive power consumption of transformer will be changed with the variation of load distribution between the transformers.
- Therefore, the transformer total active power loss and reactive power consumption can be reduced to minimum values according to economic dispatch of loads between transformers, so as to achieve transformer energy conservation.

(b) Parallel Operation:

1. State how 'parallel operation of transformers' helps in energy conservation.
2. Explain parallel operation of transformer in context of energy conservation.

- The transformer is said to be in parallel operation when their primary windings are connected to a common voltage supply and the secondary windings are connected to a common load.
- The connection diagram of the parallel operation of a transformer is shown in Fig. 6.11.
- The transformer capacity is a most important parameter when selecting transformer.
- If the transformer capacity is selected over large, then the investment of transformer is increased, as well as the transformer no-load loss.

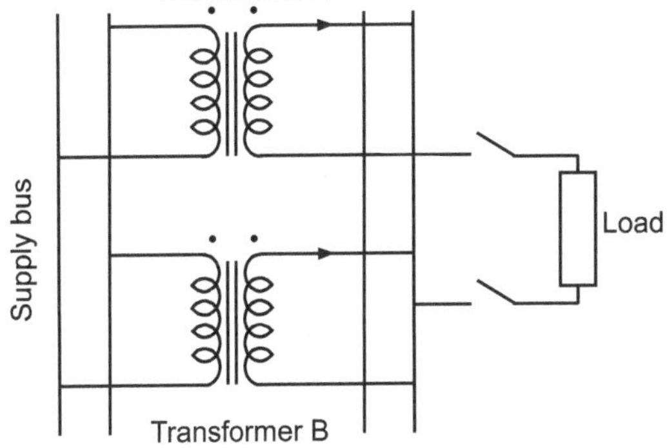

Fig. 6.11

- Mostly power transformer gives maximum efficiency at full load.
- If one runs number of transformers in parallel, one can switch on only those transformers which will give the total demand by running nearer to its full-load rating for that time.
- When load increases, one can switch on one by one other transformer connected in parallel to fulfill the total demand. In this way, one can run the system with maximum efficiency.

(c) Isolating Techniques:

- Isolation is defined as cutting of electrical supply to the system in order to ensure the safety of those working on the equipment by making dead those parts which are live in normal service.
- An isolator is a mechanical device which is operated manually and used to open or close a circuit off load. An isolator switch must be provided close to supply point, so that the transformer can be made safe for maintenance.
- Devices which are suitable for isolation are isolation switches, fuse links, circuit-breakers.
- On three phase L.T. supplies the triple pole switch with neutral link are provided at the time of isolation, the switch should break only the live conductors and the solid link in the neutral should not be removed before opening the switch.
- When transformers are operating in parallel to share a particular load, it is necessary to remove and isolate some transformers when load demand is less, so that other transformers will supply energy at their maximum efficiency.
- At the time of removing the transformer from circuit, first the secondary side load is removed by tripping the circuit breaker by opening the load switches (ICTP) having high rupturing capacity fuses.
- So that transformer will remain connected to bus, without sharing any load.
- Now the isolator switch on the primary side is opened as they operate on no load only, so that transformer is removed from bus-bar completely.
- The isolator in this condition is locked and earthed so that no one else than the operator can use it.
- When it is required to reconnect the transformer again in parallel, earthing of isolator switch is removed first.

- Then the isolator is closed so that transformer is connected to primary side, then either the circuit breaker or load switch is closed so that the transformer will start sharing load in proportion to its kVA capacity.

(d)Replacement by Energy Efficient Transformer:

- Energy efficient transformers are therefore an important means to reduce transmission and distribution loss.
- With the improvement of silicon steel properties, the losses of a transformer can be reduced.
- With new magnetic materials, it is possible to achieve even higher efficiency.
- The amorphous metal transformer is a modern example of energy efficient transformer.
- An energy-efficient transformer is very appealing giving its non-stop operation and 25 years service life.
- These savings translate into reductions in peak loading, lower electricity bills and greater reliable of supply.
- Payback periods vary with the equipment and electricity costs and can be as short as one year or as long as six years or more.
- A transformer can be made more energy efficient by improving the materials of construction (e.g. better-quality core steel or winding material) and by modifying the geometric configuration of the core and winding assemblies.
- Making a transformer more energy efficient (i.e. reducing electrical losses) is often a trade off between more expensive, lower-loss materials and designs, and the value a customer attaches to those losses.
- For a given efficiency level, the no-load and load losses are generally inversely related: reducing one usually increases the other. By using energy efficient transformers, efficiency improves from 95% to 97%. By using amorphous transformers, efficiency improves from 97% to 98.5%.
- By using epoxy resin cast/encapsulated dry type transformer, efficiency improves from 93% to 97%.

Advantages of energy efficiency transformers:

1. They have less operating losses.
2. Less heat is generated due to less losses.
3. They have longer life.
4. Better cost effective.
5. Problem of small thickness of core can be overcome to some extent with the development of amorphous metal strips.
6. Low field magnetization, reduces hysteresis loss.
7. High electrical resistivity of core material reduces eddy currents and eddy current loss.
8. Low magnetizing current and low iron losses, reduce no load current.

(e)Periodic Maintenance:

- Periodic maintenance is a vital part of loss control activities. Periodic maintenance of transformer monitors the deterioration and detects or predicts insulation, equipment and system flaws. Periodic maintenance reduces the frequency of losses. Being transformers, static machines without any moving and turning parts, they are very reliable machines and if maintained properly, can last for 40 years or more. Transformers are frequently overloaded and allowed to operate well beyond its capacity. Other causes of deterioration may be load changes, circuit changes, improperly selected protective devices and changing voltage conditions.
- Scheduling of a regular periodic maintenance program may minimize the risk of equipment failure and the resulting problems of that failure, the detection of latent faults and the first step for troubleshooting which will in turn achieves the energy efficiency.

Maintenance schedule of distribution transformer as per ISS 1186 – 1967:

- Before carrying out maintenance work on transformer, it should be disconnected from supply by disconnecting load from secondary, opening contacts of drop out fuses and opening, Air-break switch.
- The incoming supply line is then properly earthed. Then the maintenance work should commence.
- The maintenance work should be carried out by authorized persons only who should carry out proper line clear certificate before commencement of work.
- He should not smoke while working, and if there is conservator etc. fitted to the tank the oil should be taken out from conservator before opening tank.

Table 6.4: Maintenance schedule upto 1000 kVA transformers

Sr. No.	Frequency of inspection	Inspected	Inspection notes	Action required
1.	Hourly	Amperes, volts and temperature	Check against rated figures	Start fans if necessary
2.	Daily	Dehydrating breather	Check that air passage is clear, check colour of active agent.	If silica gel is pink, the old may be used by treatment
3.	Monthly	Oil level in transformer	Check transformer oil level	If low, top up with dry oil check any leaks
4.	Quarterly	Bushings	Examine for cracks and dirt deposits.	Clean and replace
5.	Half yearly	No conservator	Check for moisture under cover	Improve ventilation, check oil
6.	Yearly	Oil in transformer	Check for dielectric strength and water content, check for acidity and sludge	Take suitable action to restore quality of oil
7.	Yearly	Earth resistance	–	Take suitable actions if earth resistance is high
8.	Yearly	Relays, alarms, their circuits etc.	Examine relay and alarm contacts, their operation, fuses etc. check relay accuracy etc.	Clean the components and replace contacts and fuses if necessary. Change the setting if necessary.
9.	2 yearly	Non-conservator transformers	Internal inspection above core	Filter oil regard of condition
10.	5 yearly or after internal fault	–	Overall inspection, lifting of core and coils	Wash by housing-down with clean dry-oil

6.5.4.1 Energy Conservation Opportunities in Transformer

- Opportunities are:
  1. Use amorphous core containing ferromagnetic elements like Iron, Cobalt alloy of high resistivity than silicon steeel. So low core losses.
  2. Use encapsulated dry type transformer.
  3. Use autowound (saving in copper) or tapped transformer.
  4. Use energy efficient transformer.
  5. Use low resistant better quality copper conductors for winding to reduce copper losses.
  6. Use good quality insulation materials to improve overload capacity.
  7. Keep proper maintenance of transformer.
  8. Keep power quality by maintaining operating voltage, form factor and frequency at rated values. This results in low losses and increases efficiency of transformer.

6.5.4.2 Reasons for Energy Conservation Techniques to be Adopted in Transformers Even Though its Efficiency is 90%

1. Why energy conservation techniques should be adopted in transformers even though its efficiency is 90%?

- Transformer's performance depends on its efficiency.
- Loss of efficiency reduces its performance.
- Therefore users should try different methods to improve its efficiency.
- Transformer shall be on load at near about its full-load capacity. At less load the efficiency is lesser.
- Try the combination of transformers to run in parallel to find the stage that combination results in better efficiency.
- Duty of transformers in the line is all along the day of 24 hours. So see that how the losses be reduced.
- Use different methods and techniques to run the transformer even at 90% or more efficiency.

6.5.5 Energy Conservation Equipments

6.5.5.1 Comparison Between Soft Starter with D.O.L. Starter

1. State the comparison between soft starter and conventional DOL starter.
2. Compare soft starter with conventional starter (any four points).

Sr. No.	Parameter	Soft Starter	Direct On Line (DOL) Starter
1.	Starting	Smooth starting, no jurks.	Starts with jurks.
2.	Current peak	Low.	Very high (6 to 8 × motor current)
3.	Maintenance	Zero.	More.
4.	Voltage dip	Negligible.	Causes flickering of lights.
5.	Starting torque	Variable.	Uncontrolled, so huge mechanical stresses on equipments.
6.	Energy saving	Saves energy.	Affects M.D.

6.5.5.2 Soft Starters

1. State four advantages of soft starter.

- When starting, AC induction motor develops high torque than its required value at full speed.
- This mechanical stress is transferred to the mechanical transmission system resulting in excessive wear-tear and premature failure of chains, belts, gears, etc. Additionally, sudden rise in speed also has a large impact on input supply with high inrush currents drawings + 600% of the normal current.
- The use of Star Delta can only provide a partial solution to the problem.
- Soft starter provides a reliable and economical solution to these problems by delivering a controlled release of power to the motor, thereby providing smooth, steepless acceleration and deceleration.
- Motor life will be extended as damage to windings and bearings is reduced. Soft Start and Soft Stop is built into 3-phase units, providing controlled starting and stopping with a selection of ramp times and current limit settings to suit all applications.

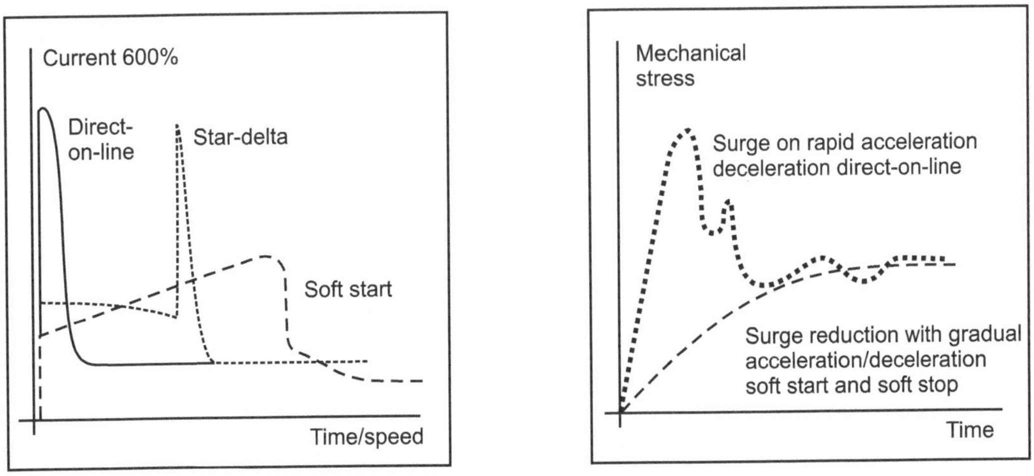

Fig. 6.12

- Soft starters are essentially stator voltage controllers; helps to overcome above problem. It helps to restrict starting current and also provide smooth start and stop operation.

Application	Number of working hours	Number of jobs	Energy consumed in kWh	Savings in kWh	Savings in %
Grinding machine:					
Without soft starter	7	55	168.0	–	–
With soft starter	7	51	136.8	31.2	18.6
Lathe machine (5.5 kW)					
Without soft starter	7	231	96.4	–	–
With soft starter	7	228	76.4	19.6	20.4

Advantages of soft starter:

1. System efficiency improves.
2. Energy losses in line reduces.
3. Small line voltage drop in operation of motor.
4. Power factor improves.
5. Low starting torque.
6. Motor is protected from mechanical stresses.
7. Smooth starting operation.
8. M.D. is controlled as current is limited.
9. Less mechanical maintenance.
10. Saving in operating cost.

Working of soft starter:

- This delivers a controlled power to the motor. This provides stepless/smooth acceleration/ deacceleration.
- **Construction:** It consists of thyristor in main circuit.
- The motor voltage is regulated with a printed circuit board.
- So, as the voltage is low at the time of starting, the current and torque is lower.
- At start it provides low voltage to motor.
- This enables to adjust the play between the gear wheels or stretching driving belts or chains etc.
- In other words, it eliminates unnecessary jurks during start.
- Gradually, the voltage and the torque increases so that the machine starts to accelerate.
- The line voltage drops, so the losses at start are very low.
- It provides a reliable and economical solution to overcome problems related with starting.

6.5.6 Variable Frequency Drives

- The VDF operates on a very simple principle.
- The rotational speed of an AC induction motor depends on the number of poles in that stator and the frequency of the applied AC power.

- Although number of poles in an induction motor cannot be altered easily, variable speed can be achieved through changes in frequency.
- The VFD changes standard 50 cycle AC line power to DC, then converts the DC to a variable frequency AC output. Motors connected to VFD provide variable speed mechanical output with high efficiency.
- These devices are capable of upto a 9:1 speed reduction ratio (11 percent of full speed) and a 3:1 speed increase (300 percent of full speed).
- In recent years, the use of digital microprocessor control, along with high switching frequency IGBTs (Insulated Gate Bipolar Transistors) power devices for variable frequency drives (VFD) has evolved largely.
- The two most significant benefits from the evolution in technology have been that of cost and reliability, in addition to the significant reduction in physical size.

6.5.6.1 Benefits of VFD

1. State two benefits and applications of variable frequency drives.
2. State two benefits and two applications of variable frequency drives.

- Following are the benefits of VFD:
  (1) Energy saving, (2) Less cost, (3) Smooth starting, (4) Better process control, (5) Less maintenance, (6) Improves p.f., (7) Bearings, motor's life is large, (8) Improved power quality.

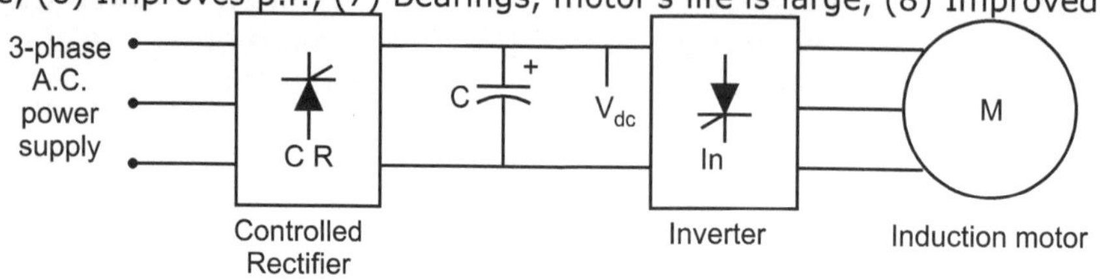

Fig. 6.13

6.5.6.2 Applications of VFD

- Following are the applications of VFD:
  1. For controlling motor driven, fans, pumps, blowers.
  2. Mining, boaring, oil drilling platform, machine starting/controlling.
  3. For machine tools, conveyors, production line equipments.
  4. Textile mills.

6.5.7 Automatic Star-Delta Starter

- Automatic Star Delta Starter is used to save energy during variable load conditions in three-phase induction motor. This is done by using a microcontroller. The three-phase winding inside the motor can be connected in star or delta. About 25-30% of energy gets saved by incorporating delta star conversion module. The starting current of any heavy electric motor can be 10 times more than the normal load current it draws, when it gains speed and has reached its normal running output and temperature. So we would require large circuit breakers, which can withstand the starting current to pass without immediately shutting it off. If the windings of a 3-phase motor are connected in star the voltage applied to each winding is reduced to only $(1/\sqrt{3})$ of the voltage applied to the winding when it is connected directly across two incoming power service line phases in DELTA. The current per winding is reduced to normal running current taken when it is connected in DELTA. This can save 25% of power when compared to delta mode. This arrangement leads to better power factor, size of conductor can be reduced and minimizing the starting current will reduce the stress in other equipment.

6.5.8 Automatic Power Factor Controller (APFC)

1. Explain the working principle of automatic power factor controller.

- It is used to improve the system power factor by automatic sensing of power factor correction and control.
- APFC results in decreased system current and hence reduced voltage drop and power loss automatically.

- Improvement in power factor to unity will not only result in savings in electricity consumption but also it helps to get the savings in the electricity bill too.
- Power factor correction is not an energy saving measure (real power remains the same).
- However, it is a cash saving measure. Given that one of the main purposes of energy efficiency is to save money, power factor correction is typically carried out along with energy efficiency retrofits.
- Standard practice is to connect power capacitors or automatic power factor controller in the power system at appropriate places to compensate the inductive nature of load.
- Bulk correction method of power factor and control.

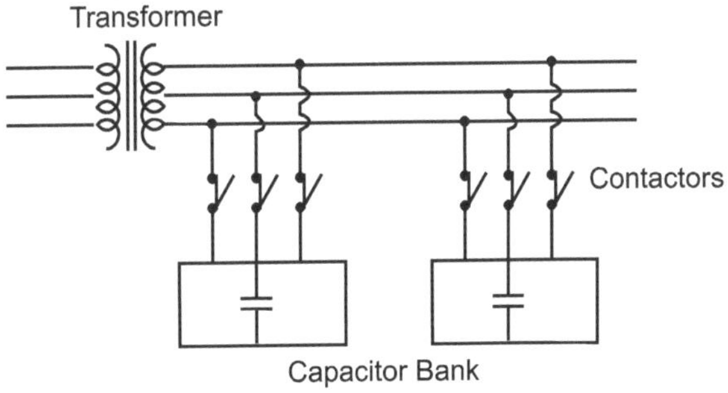

Fig. 6.14

- Individual starter controls the capacitors located at distribution panel.
- Centralized compensation is employed in large industrial plants where so many motors are in use at a time.
- Capacitor-racks are installed at the incoming distribution panel of the industry.
- These racks are sub-divided into steps.
- The current supplied to the distribution panel is monitored by the controller for power factor.
- Automatic control system switches sub-divided racks of capacitor units in steps to maintain power factor more than preset limit (0.95).

6.5.9 Intelligent Power Factor Controller (IPFC)

Effects of Low Power Factor:

(a) For same kW demand of load, at low power factor current flowing through armature winding of alternator or windings of transformer increases, hence kVA rating required increases.
(b) As current supplied is increases for obtaining proper voltage regulation, conductor size of transmission lines etc. increases.
(c) The cost of generation and transmission increases for the same power output.
(d) As due to low p.f., current flowing through transmission lines increases, the line loss will increase or line efficiency reduces and voltage drop in lines will increase.
(e) kW capacity of transformer will be reduced and voltage drop in it will increase.
(f) Conductors required for bus-bar will need more cross-sectional area and also contact surface of the switchgears must be increase as current flowing through them increases.
(g) kW capacity of generator reduces.
(h) Power required from exciters increases.
(i) As current flowing through armature winding increases, it will increase copper losses and cause more temperature rise, reducing efficiency of generator and transformer.
(j) Prime mover efficiency will be reduced.

Power Factor Improvement

- Power factor improvement is the fastest way of achieving energy conservation and improving the bottom line.
- Some of the techniques included for the purpose are as follows:
  (a) Study of various types of loads.
  (b) Determination of the percentage of inductive loads.
  (c) Determination of transformer capacity.
  (d) Determination of power factor under full load conditions of transformer by calculating impedance.
- From the data collected the kVAR required to improve the power factor to the desired value can be determined or calculated.
- After this, it would be to determine the capacitor-banking pattern in the electrical system, depending upon whether the loads are major or minor.

- Then, capacitors are installed at all major loads and finally to install capacitors for the fine turning of power factor, preferably with the minimum banking pattern.
- Effects of current voltage, harmonics and temperature are to be considered as they produce degradations in capacitors and contactors, but they remain unnoticed because the equipment operate silently in a corner, until a severe breakdown occurs.
- Power factor compensation gives the best return on investment.
- Due to power factor correction (PFC), the power distribution system operates at its maximum efficiency.
- All power suppliers use either active PFC or passive PFC.

Methods of Improving Power Factor

1. Using capacitors.
2. Using synchronous condensers (synchronous motor).
3. Using phase advancers.

Role of Intelligent Power Factor Controller (IPFC)

- For any industry with varying loads, automatic power factor compensation gives best return on investment.
- The kVAR investment required is smaller than with fixed capacitors needed to meet the entire load.
- Automatic power factor correction helps to avoid leading p.f. situations by switching off extra capacitors.

Advantages of Installing IPFC

(a) On equipment:

(i) It protects the equipment.
(ii) Field failures are reduced.
(iii) Wear and tear of contactors is minimized.
(iv) It switches off the system for excessive harmonics, voltage and current.
(v) It switches off banks of capacitors on main faults and also on micro-interruptions.
(vi) It safeguards and increases the life of the equipment.
(vii) Intelligent alarm functions are included in IPFC.

(b) At site:

(i) Simple inspection procedure is available.
(ii) Gives signals on likely p.f. penalties (irregular compensation) i.e. when p.f. is not compensated to desired level.
(iii) Manual control can be minimized.
(iv) It prevents improper manual switching.
(v) It clears alarm until discharge time has elapsed to prevent false alarms.
(vi) Past events are stored in memory.
(vii) Gives alarm on malfunctions.
(viii) Before proper discharged, connection is not possible.
(ix) Gives signals on system failures.
(x) Manual modes are also available for the equipment check-up.

Advantages of Intelligent Power Factor Controller for the Improvement in Energy Efficiency

(i) Increase in system voltage can be avoided with it.
(ii) Capacitor's life increases.
(iii) Plant efficiency increases.
(iv) Utilization of demand is optimized.
(v) Line losses are reduced.
(vi) Extra load can be connected without sanction of additional demand.
(vii) Demand penalty can be avoided.
(viii) Zero cost maintenance is possible.

6.5.10 Active Harmonic Filters (AHF)

- Harmonic currents are caused by non-linear loads connected to the distribution system. A load is said to be non-linear when current it draws does not have same wave shape as the supply voltage. The flow of harmonic current through the system impedances causes voltage distortion in the distribution system. Following are the major harmonic generators:
  1. Industrial Processes viz. Welding, Impact Loads, Induction Furnaces, Battery Charger, DC Power Supplies.
  2. Variable Speed Drives for AC and DC motors.
  3. Uninterrupted Power Supplies.
  4. Office Equipment (PCs, Printers, Servers, Displays etc.)
- Harmonic currents increase the RMS current in electrical systems and deteriorate the quality. They stress electrical network and potentially damage the equipment. They may disrupt normal operation of devices and increase operating costs.
- Office buildings, commercial establishments having predominant non-linear loads such as data servers, UPS systems and mix of singe phase loads leads to harmonic generation in the supply network which create several problems such as overheating of cables, tripping of breakers, blowing of fuses, additional heating in the transformers etc. High levels of triplex (3rd order and multiple) harmonics specifically overload the neutral conductor and often result in unacceptable neutral to earth potential. Therefore the load currents in such installations are highly polluted due to the presence of high level of harmonic current.
- Improvement upto the extent of 95% can be achieved in the harmonic current distortion (THDI) at certain feeders after installation of active harmonic filters. Excessive THDI at high load current is the main cause of higher heating losses, tripping of breakers and other problems. The active harmonic filters not only reduce the high harmonic distortion but also improve the power factor close to unity (0.99 lag). This has resulted in to 20% reduction in kVA consumption by the UPS system. Low power factor is characterized by high load current which in turn results in additional line losses, overheating and in some cases penalties causing direct financial losses. A leading power factor is not desirable and may require derating of UPS systems. The active harmonic filters correct the leading power factor to unity. The combined effect of harmonic filtering and power factor improvement brings additional benefits to the customer. Reduced losses lower the operating temperature of the equipment which indirectly increases the life of the electrical equipment.

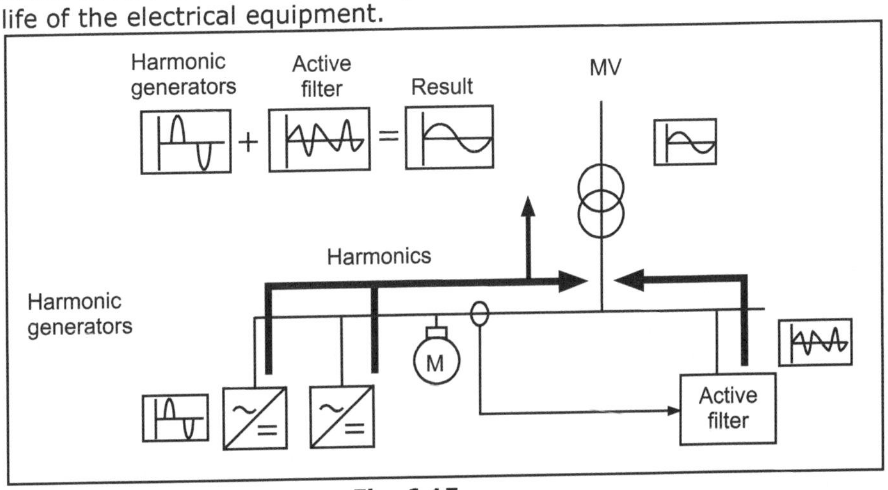

Fig. 6.15

- Active harmonic filters are designed to measure the load current, calculate variance from objectives set and inject the right amount of current to meet the required level of harmonics, displacement power factor or load balancing.
- When harmonic mitigation is required, the logic measures the load current and calculates the amplitude and phase angle for every harmonic to as high as 50th order. The control signal is generated and semiconductors are directed to duplicate the control signal as injected into the supply. This reduces the supply side harmonics.

6.5.11 Transformer Energy Conservation Method

- After selection of a transformer, adopt a transformer economic operation scheme to arrange operations, such as make the transformer operate in economic operation interval, economic switching of transformer between each operation mode and economic dispatch between transformer loads, etc.
- Adopting these schemes can improve transformer operating efficiency and reduce its operation loss, thus achieve transformer energy conservation. Adopt a reactive power compensation method, making the reactive power of transformer compensated from the loads, thus can reduce the reactive power loss and comprehensive power loss of transformer.
- Reactive power compensation methods include adoptions of conventional static reactive power compensator SVC, static reactive power generator SVG and active power filters (APF), etc.
- Also we can adopt new-type of distribution transformer integration SVC technique to realize comprehensive compensation and control reactive power and power quality in junction positions in high and low voltage levels.

6.5.12 Energy Efficient Motors

- As stated earlier, the energy efficient motor is that motor which operates giving efficiencies (2 to 4% more than other standard motors), operates at reasonably good power factor and loss of power is lesser.
- These are manufactured with the change in design of slot shapes.

6.5.12.1 Speciality of this Motor and its Constructional Improvement

- Energy Efficient Motors (EEM) are the ones in which, design improvements are incorporated specifically to increase operating efficiency over motors of standard design (Refer Fig. 6.16).
- Design improvements focus on reducing intrinsic motor losses.
- Improvements include the use of lower-loss silicon steel, a longer core, thicker wire, thinner laminations, smaller air gap between stator and rotor, copper instead of aluminium bars in the rotor, superior bearings and a smaller fan, etc.
- Energy efficient motors now available in India operate with efficiencies that are typically 3 to 4 percentage points higher than standard motors.
- In keeping with the stipulations of the BIS, energy-efficient motors are designed to operate without loss in efficiency at loads between 75% and 100% of rated capacity. This may result in major benefits in varying load applications.
- The power factor is about the same or may be higher than for standard motors.

6.5.12.2 Comparison of Characteristics of Energy Efficient Motor with Standard Motor

- Furthermore, energy efficient motors have lower operating temperature and noise levels, greater ability to accelerate higher-inertia loads and are less affected by supply voltage fluctuations.

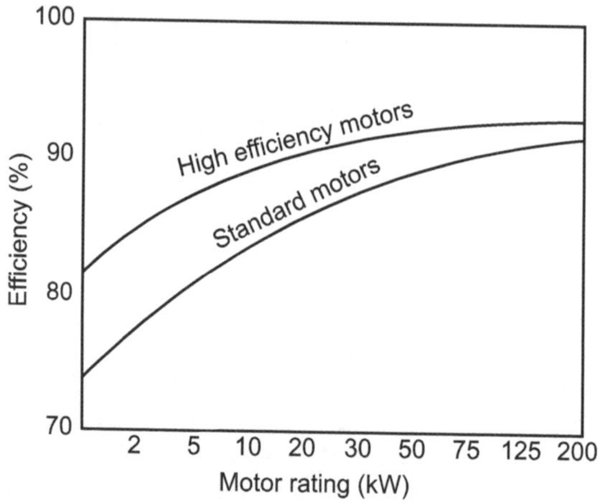

Fig. 6.16: Standard versus high efficiency motors

6.5.12.3 How Different Losses are Minimized in Energy Efficient Motor

- Measures adopted for energy efficiency address each loss specifically as under:

(A) Core Losses:

- Core losses are found in the stator-rotor magnetic steel and are due to hysterisis effect and eddy current effect during 50 Hz magnetization of the core material.
- These losses are independent of load and account for 20-25% of the total losses.
- The hysterisis losses which are a function of flux density are be reduced by utilizing low-loss grade of silicon steel laminations.
- The reduction of flux density is achieved by suitable increase in the core length of stator and rotor. Eddy current losses are generated by circulating current within the core steel laminations. Core losses are reduced by using thinner laminations.

(B) Stator and Rotor I^2R losses:

- These losses are major losses and typically account for 55% to 60% of the total losses. I^2R losses are heating losses resulting from current passing through stator and rotor conductors. I^2R losses are the function of conductor resistance and the square of current.
- Resistance of conductor is a function of conductor material, length and cross-sectional area. The suitable selection of copper conductor size will reduce the resistance.
- Reducing the motor current is most readily accomplished by decreasing the magnetizing component of current. This involves lowering the operating flux density and possible shortening of air gap. Rotor I^2R losses are a function of the rotor conductors (usually aluminium) and the rotor slip. Utilization of copper conductors will reduce the winding resistance.
- Motor operation closer to synchronous speed will also reduce rotor I^2R losses.

(C) Friction and Windage Losses:

- Friction and windage losses results from bearing friction, windage and circulating air through the motor and account for 8-12% of total losses.
- These losses are independent of load. The reduction in heat generated by stator and rotor losses permit the use of smaller fan. The windage losses also reduce with the diameter of fan leading to reduction in windage losses.

(D) Stray Load Losses:

- These losses vary according to square of the load current and are caused by leakage flux induced by load currents in the laminations and account for 4 to 5% of total losses. These losses are reduced by careful selection of slot numbers, tooth/slot geometry and air gap.
- Energy efficient motors cover a wide range of ratings and the full load efficiencies are higher by 3 to 7%. The mounting dimensions are also maintained as per IS 1231 to enable easy replacement.
- As a result of the modification to improve performance, the costs of energy-efficient motors are higher than those of standard motors. The higher cost will often be paid back rapidly in saved operating costs, particularly in new applications or end-of life motor replacements.
- In cases, where existing motors have not reached the end of their useful life, the economics will be less clearly positive.
- Because the favourable economics of energy-efficient motors are based on savings in operating costs, there may be certain cases which are generally economically ill-suited to energy efficient motors.
- These include highly intermittent duty or special torque applications such as hoists and cranes, traction drives, punch presses, machine tools and centrifuges. In addition, energy efficient designs of multi-speed motors are generally not available.
- Furthermore, energy efficient motors are not yet available for many special applications, e.g. for flame-proof operation in oil-field or fire pumps or for very low speed applications.

- Also most energy efficient motors produced today are designed only for continuous duty cycle operation.
- Given the tendency of over sizing on the one hand and ground realities like; voltage, frequency variations, efficacy of rewinding in case of a burnout, on the other hand, benefits of EEMs can be achieved only by careful selection, implementation, operation and maintenance efforts of energy managers.

6.5.13 Measures for Energy Conservation in Electric Heating

- Measures for energy conservation in Electric Heating.
- Taking the case of heat in Boilers, to there are various types of Boilers like:

Boilers
- (i) Fire Tube
- (ii) Water Tube

- Areas of energy conservation are found where there is energy losses and these are to be minimized.
- Air is required for combustion of fuels in the boiler.
- Too much air results in excessive heat losses as the surplus air does not take part in combustion but carries away heat to the atmosphere from the boiler furnace.
- Also surplus air lowers the furnace temperature and thereby the heat transfer rate, and results in lower combustion efficiency.
- So appropriate air shall be provided to reduce losses.

Other areas of energy conservations:

- The waste heat from the flue gases can be recovered by methods - feed water preheating by these gases, in this 1% fuel is saved.
- Combustion air can be preheated by heat carried away by flue gases.
- Returning steam condensate to the boiler can also reduce level consumption. Condensate carries sensible heat which can be as high as 20 to 25% of the total heat generated by boiler.

6.5.14 Measures for Energy Conservation in Refrigeration/Air-conditioning

- Refrigeration is the process of removing heat at a low temperature level and rejecting it at a relatively higher temperature level.

Systems used
- (1) Vapour compression system
- (2) Vapour absorption system
- (3) Steam jet refrigeration cycle

Energy conservation measures:

(i) By increasing the chilled water temperature set point, the aim is to avoid unnecessary super cooling without affecting the production, quality and safety.
Increasing energy costs have forced many industries to experiment and stabilize operation at higher temperatures.

(ii) In recent developments, the fan motors in the refrigerators uses an electronic speed controller unit. This derive is called as A.C. VSD i.e. A.C. variable speed drive. This electronic circuitary changes the speed by varying frequency. Since, the power required for driving a centrifugal fan or pump is proportional to the cube of the fan/pump speed, large reductions in power consumption are obtained at reduced speed, so this VSD drive saves the power consumption.

6.5.15 Energy Conservation in Traction

- Energy saving in traction is achieved by electric drives explained in above Sub-Section (i)

6.5.16 Energy Conservation in Illumination

1. State the criteria for lamp replacement with energy deficient lamps and use of light control gear as a energy conservation technique.
2. With example explain how the replacement of lamps in lighting system contributes for energy conservation?
3. Discuss the role of replacement of old lamps by new more energy efficient lamps in the conservation of energy.

(a) By replacing lamp sources:

- The lamp sources used for lighting are:
  - (i) Filament lamps.
  - (ii) Fluorescent lamp. While following types of lamps are mostly used for street lighting, factories etc.
  - (iii) Mercury vapour lamp.
  - (iv) Sodium vapour lamp.
- Halogen lamps can be used for public halls, factories, studios, play grounds, car parking purposes etc.
- When we consider general lighting system, then mostly filament lamps or fluorescent lamp i.e. tube lights are mostly used.
- Filament lamps are more sensitive to voltage variation. A small reduction in voltage reduces its luminous output greatly, while a small increase in voltage increases its output but at the same time reduces its life.
- The same variations in case of fluorescent lamps are comparatively less.
- The luminous output of filament lamps goes on reducing with usage, which is due to slow vapourization of the lamp filament and partly due to black deposit formed inside the lamp.
- In case of fluorescent lamps a change of ± 1% of the supply voltage produces a change of about ± 1% in luminous output.
- The increase or decrease in voltage reduces lamp life, with increase voltage heating effect of electrodes is increases and they loose electron emissive material by evaporation.
- With reduced voltage, the current reduces causing sputtering at the electrodes reducing their life.
- The luminous output of incandescent lamps of 40 watt is nearly 10 lumens/watt while that of 40 watt fluorescent tube or lamp is nearly 40-50 lumens watt.
- Hence, replace incandescent lamps by fluorescent lamps of proper size.
- Compact fluorescent lamps (CFL) of smaller wattage will provide same illumination.
- Life of CFL lamps is nearly 10 to 15 times more than ordinary incandescent lamps.
- Energy consumption can still further be reduced by using electronic ballasts instead of electromagnetic ballast.

	Lamp		Replaced by
1.	Incandescent	→	Compact fluorescent lamp (CFL).
2.	Conventional fluorescent lamp	→	Energy efficient fluorescent lamp.
3.	Mercury vapour or sodium vapour lamp	→	Halide lamp.
4.	HPMV lamp	→	HPSV lamp.
5.	Filament lamps on panels	→	LED.

(b) Electric chokes/starters/ballast:

1. State advantages of electronic ballast as compared to electric ballast.

- Saving can be achieved by selecting a ballast base for the CFL lamps separate from the bulb, which will help to change the bulb only at its life end.

- In case of compact fluorescent lamps, their colour rendering index should be seen, we should select lamps having CRI nearest to 100, which is best.
- Linear fluorescent tubes are suitable for kitchen, bathroom, utility area select suitable styles and lengths of these instead of incandescent lamps.
- If incandescent lamps of large size are used for light focus, halogen lamps with a special reflector may be used, replacing incandescent lamps, as they consume 40 to 50% less energy giving output similar to incandescent lamps. Hence, they should be used for lighting of gardens and pathways.
- Efficiency of lamps, whether incandescent or fluorescent goes on reducing upto life end, hence we should not wait till the lamps blows out, it must be replaced by new one when it is observed that it is giving less luminous output.

Advantages of high frequency ballast:

1. Instant light.
2. Light in weight.
3. Operates at low voltage.
4. Power saving ≈ 35%.
5. Power factor improvement.
6. Heat output is negligible, that reduces load on air-conditioning.
7. Enhances life of lamp.

Advantages of electronic ballast over electric ballast:

1. Instant light.
2. Lighter in weight.
3. 35% power saving.
4. Power factor improvement.
5. Heat output is less.
6. Operates at low voltage.
7. Enhances life of lamp.

(c) Using energy efficient luminaires:

- As explained previously, the incandescent lamps use only 5 to 10% of their energy to produce light and the rest is waisted in developing heat, hence their luminous efficiency is low as 10 to 12 lumens/watt.
- While the CFL or fluorescent lamps have a luminous efficiency of 40 to 50 lumens/watt, they are available in different sizes and shapes and can suitably replace the incandescent lamps, in position where they are required.
- Incandescent lamps of large capacity used for gardens, parks, roads, play grounds, flood-lighting etc. be replaced by halogen lamps, sodium vapour lamps or mercury vapour lamps having more luminous efficiency.

  Criteria for selecting energy efficient lamps (replacement option):

  1. While replacing the lamps by higher energy efficiency, we must ensure that the required current rendering (CRI) is maintained, otherwise it has an adverse effect on the quality and rate of work output. Also the cost involved must also be considered.
  2. By any lighting method, electrical and mechanical accidents should be less.
  3. The glare to human eyes should be less by any lighting method.
  4. By proper selection of lighting method, surrounding conditions should be pleasant to every body.

     Energy efficient lamps → CFL, LEDs.

(d) Using light controlled gears:

> 1. Describe the following energy conservation methods of lighting system:
>   (i) By replacing lamp sources. (ii) Using light control gears.

- Generally, the outside lighting is made on and off at particular time of the day irrespective of the requirement.
- However, the wastage of such energy when it is not required can be saved by using light sensers i.e. a device that will use photoelectric controls, which will make the circuit on or off.
- When the illumination level goes below a particular level or when it goes above a particular level. So that wastage of energy can be saved.

- Use of controlled switches may be made in the wiring system, so that when necessary high capacity bulbs may be made on and when low illumination is required, less capacity bulbs can be used.
- Only a particular bulb can be kept on and other can be made off when not required giving much more saving in energy.
- Using light control gear:
  1. Install a microprocessor/infra-red controlled dimming or switching.
  2. Movement detectors/lighting sensors used feed signals to the controller.
  3. Optimum use of day lighting in combination with electric lighting.
  4. Installation of exclusive transformer for lighting (this reduces voltage related problems).
  5. This increases lighting efficiency.
  6. Install high frequency (HF) electronic ballast in place of conventional ballasts, this saves 35% energy.

(e) By installation of separate transformer/servo stabilizer for lighting:

> 1. Explain the energy conservation techniques in lighting system by installation of separate transformer/servo stabilizer for lighting.

- The luminous efficiency of lamps depends upon the voltage applied across its terminals.
- It gives its best output at rated voltage. Small reduction in applied voltage greatly reduces its luminous output.
- When it is necessary that the voltage applied across them is properly maintained constant. This can be achieved by installing a transformer sub-station near the load so that variation in voltage will be negligible.
- However to such sub-stations different categories of loads are connected, which may act in such a way that there may be sudden voltage drop taken place and when such loads are made off, it may result in sudden rise of voltage, which will certainly affect on the luminous efficiency of the lamps.
- To obtain constant luminous output in particular installation it is necessary to stabilize the voltage.
- Hence for such purposes servo stabilizers are connected in the system of such premises, which will maintain the voltage constant in that installation for giving best luminous efficiency of lamps.
- To obtain constant luminous output in particular installation, it is necessary to stabilize the voltage. Hence for such purposes, servo-stabilizers are connected in the system of such premises which will maintain the voltage constant in that installation for giving best luminous efficiency of lamps.

(f) Periodic survey and adequate maintenance programs:

- Depreciation in lumen output of the lamp occurs with time due to blackening of lamp, evaporation of element material or reduction of electron emitting materials from fluorescent tube etc.
- The actual amount of light to be provided by the source is greater by 50 to 100% on account of dust depositing on the reflector surfaces.
- Hence, a periodic survey is necessary for a particular installation, in which we have to clean all shades of lamps, surface of lamps and remove dirt, dust etc. accumulated on the surface inter and outer of lamp shades and outer surfaces of lamps etc.
- See that blackening of lamps at the electrodes of fluorescent lamps has taken place or not if it is observed replaced them by new ones. If the lamps are in use for long time do not wait for blowing out, replace them for better luminous efficiency.

6.5.17 Energy Conservation Techniques in Fans and Electronic Regulators

> 1. List out the energy conservation techniques in fans.
> 2. Explain any two energy conservation techniques in fan.

- While purchasing and installing fans, precaution should be taken to purchase energy efficient fans. They must be of proper size of a room, means a fan of 48 cm size is suitable for a room of 10 feet × 10 feet. More than this size will cause unnecessary wastage of energy.

- When installed, we should see that the blades are properly balanced and produce no noise and vibration, which will increase power input.
- See that the bearing is good, it do not cause unnecessary friction and causing heating effect and loss of additional energy.
- The regulators used for the fan should not be electrical type, which use barrate type resistance or chokes, in which power loss is more, hence provide electronic regulator which will need nearly no electrical energy.
- Preferably, use electronic controlled regulators so that one can easily control the speed from where he is seating in the room.
- Or conveniently make it off thus saving the energy required. Speed of fans is also the factor that will affect on the energy consumption.
- Keep proper speed of fans in the room.
- Greater speed not necessary which will cause wastage of energy. Proper height from the working plane is also the factor that affects on energy consumption.
- If the height of fans is more, we will have to keep speed high and it needs more electrical energy, hence use proper height of fans from the working plane to save electrical energy.
  1. Use electronic regulator in place of conventional regulator.
  2. Use fans at low speed instead of at high speed.
  3. Adjust direction of ceiling fan so as to air blow downwards.
  4. Turn off fans when not required.
  5. Maintain, repair, use properly in good condition.
  6. Use the fan which has energy efficient motor.
  7. Use properly designed blade fans.
  8. Use windows to allow natural air circulations.

Practice Questions

1. Give the classification of costs and explain each in brief. [Sec. 6.2.1]
2. Give the meaning of the term "tariff". [Sec. 6.3.1]
3. Give objectives/characteristics of tariff. [Sec. 6.3.2]
4. Explain the principle factors involved in fixing tariff. [Sec. 6.3.3]
5. Give the types of tariffs. [Sec. 6.3.4]
6. Explain any four types of tariffs. [Sec. 6.3.4]
7. Compare block rate tariff with flat rate tariff. [Sec. 6.3.4.1]
8. Explain Time of Day (TOD) tariff. [Sec. 6.3.5]
9. Compare two-part and three-part tariffs. [Sec. 6.3.5]
10. What are the different methods of improving power factor? [Sec. 6.4]
11. Explain use of static capacitors for power factor improvement. [Sec. 6.4]
12. Explain the advantages of power factor improvement. [Sec. 6.4.2]
13. Derive the formula for most economical power factor. [Sec. 6.4.3]
14. Explain the need of energy conservation. [Sec. 6.5.1]
15. Explain the need of the energy conservation in transformer. [Sec. 6.5.2 (A)]
16. Explain the need of energy conservation techniques in induction motor. [Sec. 6.5.3]
17. Explain the need of energy conservation techniques in transformer. [Sec. 6.5.4]
18. Give comparison between soft starter and D.O.L. starter. [Sec. 6.5.5.1]
19. Write on variable frequency drives giving benefits and applications. [Sec. 6.5.6.1]
20. Explain with figure automatic power factor control. [Sec. 6.5.8]
21. Write a note on intelligent power factor controller (IPFC). [Sec. 6.5.8]
22. Write a note on active harmonic filters (AHF). [Sec. 6.5.10]
23. State the specialities of energy efficient motors. [Sec. 6.5.12]

■■■